Simone Marinai and Hiromichi Fujisawa (Eds.)

Machine Learning in Document Analysis and Recognition

Studies in Computational Intelligence, Volume 90

Editor-in-chief
Prof. Janusz Kacprzyk
Systems Research Institute
Polish Academy of Sciences
ul. Newelska 6
01-447 Warsaw
Poland
E-mail: kacprzyk@ibspan.waw.pl

Further volumes of this series can be found on our homepage: springer.com

Vol. 66. Lakhmi C. Jain, Vasile Palade and Dipti Srinivasan (Eds.)
Advances in Evolutionary Computing for System Design, 2007
ISBN 978-3-540-72376-9

Vol. 67. Vassilis G. Kaburlasos and Gerhard X. Ritter (Eds.)
Computational Intelligence Based on Lattice Theory, 2007
ISBN 978-3-540-72686-9

Vol. 68. Cipriano Galindo, Juan-Antonio Fernández-Madrigal and Javier Gonzalez
A Multi-Hierarchical Symbolic Model of the Environment for Improving Mobile Robot Operation, 2007
ISBN 978-3-540-72688-3

Vol. 69. Falko Dressler and Iacopo Carreras (Eds.)
Advances in Biologically Inspired Information Systems: Models, Methods, and Tools, 2007
ISBN 978-3-540-72692-0

Vol. 70. Javaan Singh Chahl, Lakhmi C. Jain, Akiko Mizutani and Mika Sato-Ilic (Eds.)
Innovations in Intelligent Machines-1, 2007
ISBN 978-3-540-72695-1

Vol. 71. Norio Baba, Lakhmi C. Jain and Hisashi Handa (Eds.)
Advanced Intelligent Paradigms in Computer Games, 2007
ISBN 978-3-540-72704-0

Vol. 72. Raymond S.T. Lee and Vincenzo Loia (Eds.)
Computation Intelligence for Agent-based Systems, 2007
ISBN 978-3-540-73175-7

Vol. 73. Petra Perner (Ed.)
Case-Based Reasoning on Images and Signals, 2008
ISBN 978-3-540-73178-8

Vol. 74. Robert Schaefer
Foundation of Global Genetic Optimization, 2007
ISBN 978-3-540-73191-7

Vol. 75. Crina Grosan, Ajith Abraham and Hisao Ishibuchi (Eds.)
Hybrid Evolutionary Algorithms, 2007
ISBN 978-3-540-73296-9

Vol. 76. Subhas Chandra Mukhopadhyay and Gourab Sen Gupta (Eds.)
Autonomous Robots and Agents, 2007
ISBN 978-3-540-73423-9

Vol. 77. Barbara Hammer and Pascal Hitzler (Eds.)
Perspectives of Neural-Symbolic Integration, 2007
ISBN 978-3-540-73953-1

Vol. 78. Costin Badica and Marcin Paprzycki (Eds.)
Intelligent and Distributed Computing, 2008
ISBN 978-3-540-74929-5

Vol. 79. Xing Cai and T.-C. Jim Yeh (Eds.)
Quantitative Information Fusion for Hydrological Sciences, 2008
ISBN 978-3-540-75383-4

Vol. 80. Joachim Diederich
Rule Extraction from Support Vector Machines, 2008
ISBN 978-3-540-75389-6

Vol. 81. K. Sridharan
Robotic Exploration and Landmark Determination, 2008
ISBN 978-3-540-75393-3

Vol. 82. Ajith Abraham, Crina Grosan and Witold Pedrycz (Eds.)
Engineering Evolutionary Intelligent Systems, 2008
ISBN 978-3-540-75395-7

Vol. 83. Bhanu Prasad and S.R.M. Prasanna (Eds.)
Speech, Audio, Image and Biomedical Signal Processing using Neural Networks, 2008
ISBN 978-3-540-75397-1

Vol. 84. Marek R. Ogiela and Ryszard Tadeusiewicz
Modern Computational Intelligence Methods for the Interpretation of Medical Images, 2008
ISBN 978-3-540-75399-5

Vol. 85. Arpad Kelemen, Ajith Abraham and Yulan Liang (Eds.)
Computational Intelligence in Medical Informatics, 2008
ISBN 978-3-540-75766-5

Vol. 86. Zbigniew Les and Mogdalena Les
Shape Understanding Systems, 2008
ISBN 978-3-540-75768-9

Vol. 87. Yuri Avramenko and Andrzej Kraslawski
Case Based Design, 2008
ISBN 978-3-540-75705-4

Vol. 88. Tina Yu, David Davis, Cem Baydar and Rajkumar Roy (Eds.)
Evolutionary Computation in Practice, 2008
ISBN 978-3-540-75770-2

Vol. 89. Ito Takayuki, Hattori Hiromitsu, Zhang Minjie and Matsuo Tokuro (Eds.)
Rational, Robust, Secure, 2008
ISBN 978-3-540-76281-2

Vol. 90. Simone Marinai and Hiromichi Fujisawa (Eds.)
Machine Learning in Document Analysis and Recognition, 2008
ISBN 978-3-540-76279-9

Simone Marinai
Hiromichi Fujisawa (Eds.)

Machine Learning in Document Analysis and Recognition

With 142 Figures and 41 Tables

Springer

Simone Marinai
Dipartimento di Sistemi e Informatica
University of Florence
Via S. Marta, 3
50139 Firenze
Italy
marinai@dsi.unifi.it

Hiromichi Fujisawa
Hitachi Central Research Laboratory
1-280, Higashi-Koigakubo
Kokubunji-shi, Tokyo 185-8601
Japan
fujisawa@crl.hitachi.co.jp

ISBN 978-3-540-76279-9 e-ISBN 978-3-540-76280-5

Studies in Computational Intelligence ISSN 1860-949X

Library of Congress Control Number: 2007939058

Cover design: Deblik, Berlin, Germany

Printed on acid-free paper

9 8 7 6 5 4 3 2 1

springer.com

Preface

The objective of Document Analysis and Recognition (DAR) is to recognize the text and graphical components of a document and to extract information. With first papers dating back to the 1960's, DAR is a mature but still growing research field with consolidated and known techniques. Optical Character Recognition (OCR) engines are some of the most widely recognized products of the research in this field, while broader DAR techniques are nowadays studied and applied to other industrial and office automation systems.

In the machine learning community, one of the most widely known research problems addressed in DAR is recognition of unconstrained handwritten characters which has been frequently used in the past as a benchmark for evaluating machine learning algorithms, especially supervised classifiers.

However, developing a DAR system is a complex engineering task that involves the integration of multiple techniques into an organic framework. A reader may feel that the use of machine learning algorithms is not appropriate for other DAR tasks than character recognition. On the contrary, such algorithms have been massively used for nearly all the tasks in DAR. With large emphasis being devoted to character recognition and word recognition, other tasks such as pre-processing, layout analysis, character segmentation, and signature verification have also benefited much from machine learning algorithms.

This book is a collection of research papers and state-of-the-art reviews by leading researchers all over the world including pointers to challenges and opportunities for future research directions. The main goals of the book are identification of good practices for the use of learning strategies in DAR, identification of DAR tasks more appropriate for these techniques, and highlighting new learning algorithms that may be successfully applied to DAR.

Depending on reader's interests, there are several paths that can be followed when reading the chapters of the book. We therefore avoided grouping the chapters into sections; instead we provide a deep introduction to the field and to the book's contents in the first chapter.

It is our hope that this book will help readers identify the current status of the use of machine learning techniques in DAR. Moreover, we expect that it can contribute to stimulate new ideas, new collaborations and new research activities in this research arena.

We wish to express our warmest thanks to the Authors, without whose interesting work this book would not have materialized.

Simone Marinai, Hiromichi Fujisawa

July 2007

Contents

Contributors

Teresa M.A. Basile
Università di Bari
Dipartimento di Informatica
Via Orabona 4
70126 Bari - Italy
basile@di.uniba.it

Matthew J. Beal
Department of Computer Science and Engineering,
University at Buffalo,
The State University of New York,
Buffalo NY, USA
mbeal@cse.buffalo.edu

Abdel Belaid
Université Nancy 2
LORIA UMR 7503
France
Abdel.Belaid@loria.fr

Margherita Berardi
Università di Bari
Dipartimento di Informatica
Via Orabona 4
70126 Bari - Italy
berardi@di.uniba.it

Dorothea Blostein
School of Computing
Queen's University
Kingston, Ontario,
Canada, K7L 3N6
blostein@cs.queensu.ca

Michael Blumenstein
School of Information and Communication Technology
Gold Coast campus
Griffith University QLD 4222
Australia
m.blumenstein@griffith.edu.au

Horst Bunke
University of Bern
Institute of Computer Science and Applied Mathematics (IAM)
Neubrückstrasse 10
CH-3012 Bern, Switzerland
bunke@iam.unibe.ch

Michelangelo Ceci
Università di Bari
Dipartimento di Informatica
Via Orabona 4
70126 Bari - Italy
ceci@di.uniba.it

Siyuan Chen
Center of Excellence for
Document Analysis and Recognition
(CEDAR),
Buffalo NY, USA
schen7@cedar.buffalo.edu

James R. Cordy
School of Computing
Queen's University
Kingston, Ontario,
Canada, K7L 3N6
cordy@cs.queensu.ca

Nicola Di Mauro
Università di Bari
Dipartimento di Informatica
Via Orabona 4
70126 Bari - Italy
ndm@di.uniba.it

David Doermann
Laboratory for Language and Media Processing
Institute for Advanced Computer Studies
3451 AV Williams Building
University of Maryland
College Park, Maryland 20742
doermann@umiacs.umd.edu

Floriana Esposito
Università di Bari
Dipartimento di Informatica
Via Orabona 4
70126 Bari - Italy
esposito@di.uniba.it

Stefano Ferilli
Università di Bari
Dipartimento di Informatica
Via Orabona 4
70126 Bari - Italy
ferilli@di.uniba.it

Hiromichi Fujisawa
Central Research Laboratory,
Hitachi, Ltd.
1-280 Higashi-koigakubo
Kokubunji, Tokyo 185-8601
Japan
hiromichi.fujisawa.sb@hitachi.com

Venu Govindaraju
University at Buffalo
Dept. of Computer Science and Engineering
520 Lee Entrance, Suite 202, UB Commons
Amherst, NY 14228-2567
venu@cubs.buffalo.edu

Tatsuhiko Kagehiro
Central Research Laboratory,
Hitachi, Ltd.
1-280 Higashi-koigakubo
Kokubunji, Tokyo 185-8601
Japan
tatsuhiko.kagehiro.tx@hitachi.com

Stefan Jaeger
Institute for Advanced Computer Studies,
University of Maryland, College Park,
MD 20742, USA
jaeger@umiacs.umd.edu

Huanfeng Ma
Institute for Advanced Computer Studies,
University of Maryland, College Park,
MD 20742, USA
hfma@umiacs.umd.edu

Cheng-Lin Liu
Institute of Automation,
Chinese Academy of Sciences
Beijing 100080, P.R. China
liucl@nlpr.ia.ac.cn

Donato Malerba
Università di Bari
Dipartimento di Informatica
Via Orabona 4
70126 Bari - Italy
malerba@di.uniba.it

Simone Marinai
University of Florence
Dipartimento di Sistemi e Informatica
Via S. Marta, 3
50139 Firenze, Italy
marinai@dsi.unifi.it

Emanuele Marino
University of Florence
Dipartimento di Sistemi e Informatica
Via S. Marta, 3
50139 Firenze, Italy
marino@dsi.unifi.it

George Nagy
RPI ECSE DocLab,
Troy, NY
12180 USA,
nagy@ecse.rpi.edu

Yves Rangoni
Université Nancy 2
LORIA UMR 7503
France
Abdel.Belaid@loria.fr

Andreas Schlapbach
University of Bern
Institute of Computer Science and Applied Mathematics (IAM)
Neubrückstrasse 10
CH-3012 Bern, Switzerland
schlpbch@iam.unibe.ch

Giovanni Soda
University of Florence
Dipartimento di Sistemi e Informatica
Via S. Marta, 3
50139 Firenze, Italy
soda@dsi.unifi.it

Sargur N. Srihari
Center of Excellence for Document Analysis and Recognition (CEDAR),
Buffalo NY, USA
srihari@cedar.buffalo.edu

Harish Srinivasan
Center of Excellence for Document Analysis and Recognition (CEDAR),
Buffalo NY, USA
hs32@cedar.buffalo.edu

Sergey Tulyakov
University at Buffalo
Dept. of Computer Science and Engineering
520 Lee Entrance, Suite 202, UB Commons
Amherst, NY 14228-2567
tulyakov@cedar.buffalo.edu

Tamás Varga
University of Bern
Institute of Computer Science and Applied Mathematics (IAM)
Neubrückstrasse 10
CH-3012 Bern, Switzerland
varga@iam.unibe.ch

Sriharsha Veeramachaneni
SRA Division,
ITC-IRST, Trento, 38050,
Italy
sriharsha@itc.it

Richard Zanibbi
Department of Computer Science
Rochester Institute of Technology
102 Lomb Memorial Drive
Rochester, New York, USA,
14623-5608
rlaz@cs.rit.edu

Introduction to Document Analysis and Recognition

Simone Marinai

University of Florence
Dipartimento di Sistemi e Informatica (DSI)
Via S. Marta, 3, I-50139, Firenze, Italy
marinai@dsi.unifi.it

Document Analysis and Recognition (DAR) aims at the automatic extraction of information presented on paper and initially addressed to human comprehension. The desired output of DAR systems is usually in a suitable symbolic representation that can subsequently be processed by computers.

Over the centuries, paper documents have been the principal instrument to make permanent the progress of the humankind. Nowadays, most information is still recorded, stored, and distributed in paper format. The widespread use of computers for document editing, with the introduction of PCs and word-processors in the late 1980's, had the effect of increasing, instead of reducing, the amount of information held on paper. Even if current technological trends seem to move towards a paperless world, some studies demonstrated that the use of paper as a media for information exchange is still increasing [1]. Moreover, there are still application domains where the paper persists to be the preferred media [2].

The most widely known applications of DAR are related to the processing of office documents (such as invoices, bank documents, business letters, and checks) and to the automatic mail sorting. With the current availability of inexpensive high-resolution scanning devices, combined with powerful computers, state-of-the-art OCR packages can solve simple recognition tasks for most users. Recent research directions are widening the use of the DAR techniques, significant examples are the processing of ancient/historical documents in digital libraries, the information extraction from "digital born" documents, such as PDF and HTML, and the analysis of natural images (acquired with mobile phones and digital cameras) containing textual information.

The development of a DAR system requires the integration of several competences in computer science, among the others: image processing, pattern recognition, natural language processing, artificial intelligence, and database systems. DAR applications are particularly suitable for the incorporation of

S. Marinai: *Introduction to Document Analysis and Recognition*, Studies in Computational Intelligence (SCI) **90**, 1–20 (2008)
www.springerlink.com

machine learning techniques for two factors: first, classification algorithms are used at several processing levels, from image pre-processing to character classification; second, large collections of manually annotated document images are available and can be used for automatic training of classifiers. As a matter of fact, in the last decades isolated handwritten character recognition has been frequently used as a standard benchmark for evaluating and comparing machine learning algorithms. However, besides isolated character recognition there are several other sub-tasks that have been tackled with machine learning techniques.

In spite of these considerations, there are several systems described in the DAR literature that address relevant sub-tasks with manually tuned algorithms without resorting to machine learning techniques. The aim of this book is to link together the DAR research with the machine learning one. The chapters in the book cover the state of the art in several DAR sub-tasks and include also inspiring pointers to future research directions. The DAR literature is large and covered by several books and survey papers [3, 4, 5, 6, 7, 8]. In this introductory chapter, we provide a brief guide to the DAR field providing pointers to the relevant literature and to benchmark databases. We propose also an overview of the contents of this book: the remaining fifteen chapters cover many sub-tasks in DAR applications providing some insights in the current research trends.

1 DAR Applications

There are many examples of the use of DAR techniques into both commercial and research-driven systems. Some systems have been in use (with continuous improvements) for decades and are now widely identified as successful DAR applications. We can split the DAR applications into two broad categories: business-oriented and user-centered ones.

Office documents reach a total of more than 85% of the amount of new original information stored on paper in the world [1]. It is therefore not surprising that business-oriented applications received a great interest. In this category we can include the automatic check processing (including both amount reading and signature verification), the information extraction from forms and invoices, the automatic document organization that involves page classification.

One well known application of handwriting recognition is the automatic postal mail sorting based on address recognition on envelopes. Two main problems are addressed in these systems. First, the layout of the envelope image is processed so as to identify the address position. Second, the address is recognized, avoiding confusion between sender and recipient.

Among the user-centered applications we include the software tools, such as OCR software for general purpose PCs, that can be used to process personal information originated in paper form. Other applications that received

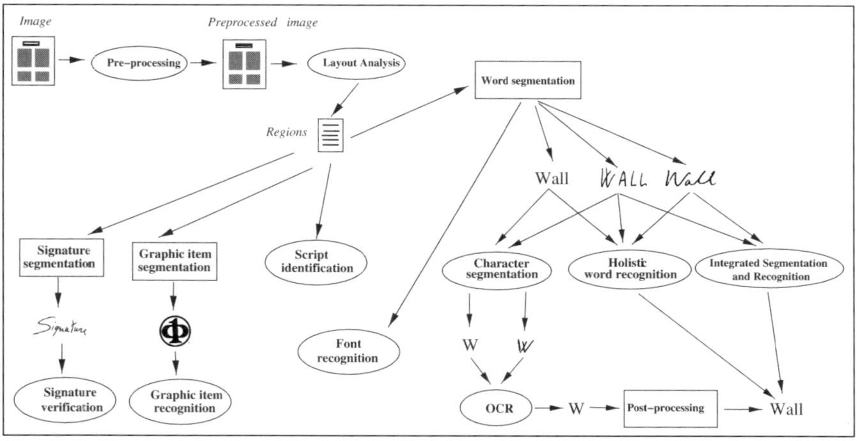

Fig. 1. General view of the document processing data-flow. Oval boxes correspond to tasks approached with significant results using machine learning-based methods. In the right part, printed, hand-printed, and handwritten instances of the word "Wall" are taken as examples to show different processing flows related to different writing sources

attention in recent years are related to the development of tools aimed at improving the access to the objects in digital libraries and processing of historical documents. In this context, large collections of digitized documents are now available in Internet, to both scholars and the general public. The information extraction from these documents (mostly book and journals) allows users to retrieve information related to the cultural heritage, as well as to identify novels and essays of interest. Further applications in this category are the recognition of printed music score and the analysis of drawings such as maps and cadastral maps. Recently, portable devices, such as mobile phones and PDAs, has been considered to provide input on-the-fly for camera-based document processing systems[1].

In most DAR applications the document content is conceptually described by means of the *physical* and the *logical* structures. The *physical* structure describes the visual aspect of the document by representing the basic objects and their mutual positions. The *logical* structure assigns to each object a suitable meaning.

2 Processing Steps in DAR

A complex DAR system is organized similarly to most Pattern Recognition (PR) systems including four principal components: pre-processing, object

[1] Some recent references can be found in the *Camera Based Document Analysis and Recognition* (CBDAR) workshop proceedings: http://www.m.cs.osakafu-u.ac.jp/cbdar2005

segmentation, object recognition, and post-processing. The *pre-processing* aims at improving the quality of the images. The *object segmentation* allows to identify the basic objects in the document. In DAR this task takes different names, depending on the application level considered. When dealing with regions in the page the segmentation is referred to as layout analysis (see Section 4). At a lover processing level we deal with various objects such as signature, word, character, and so on. The *object recognition*, or classification, deals with the objects identified in the previous step. We analyze in Section 6 the various levels at which object classification can occur in DAR applications. Lastly, the *post-processing* checks the results of the classification on the basis of contextual information.

To describe with more details the peculiarities of DAR systems we depict in Fig. 1 the relationships between the main document processing tasks. *Pre-processing* operations in DAR are used in order to improve the input image for subsequent analysis. Common tasks are de-skew, image enhancement (noise reduction) and character thinning. *Layout analysis* methods are aimed at extracting the physical and/or logical structure of the document image. When dealing with the textual parts of a document, the words are usually located by *word segmentation* with low-level methods such as morphological processing and connected components clustering. It is important to remark that in some languages, like Japanese and Chinese, there is no word separation by spaces as in western languages such as English. Word segmentation is required for the recognition of printed, hand-printed, and cursive text. The subsequent word recognition can be based either on the segmentation and subsequent recognition of the individual characters, or on the recognition of the whole word image as a single unit. In the former approach, based on a "divide-et-impera" scheme, a preliminary step of *character segmentation* is required and the isolated characters are subsequently recognized. In *holistic word recognition* the whole word is considered as a single object to be recognized. In so doing errors due to wrong character segmentations can be avoided. However, this approach can be effectively used only in the presence of a limited dictionary for instance in postal applications and check reading. An alternative approach is based on *integrated segmentation and recognition*, where the two operations take place at the same time. *Graphical items* are recognized with methods similar to those applied to character recognition, whereas specific approaches are considered for dealing with *signature verification*.

Important DAR tasks are also *script identification* that helps processing multi-lingual languages, and *font recognition* that can contribute to improve the performance of text reading. In addition to text-intensive documents, a large group of DAR applications deals with graphical documents such as technical drawings and maps, where different recognition methods are adopted.

Similarly to other PR applications there are two principal approaches to document analysis: *top-down* and *botttom-up*. If the broad structure of the document layout is known in advance, then a model-driven (top-down) approach can be used. When the physical structure is not known in advance,

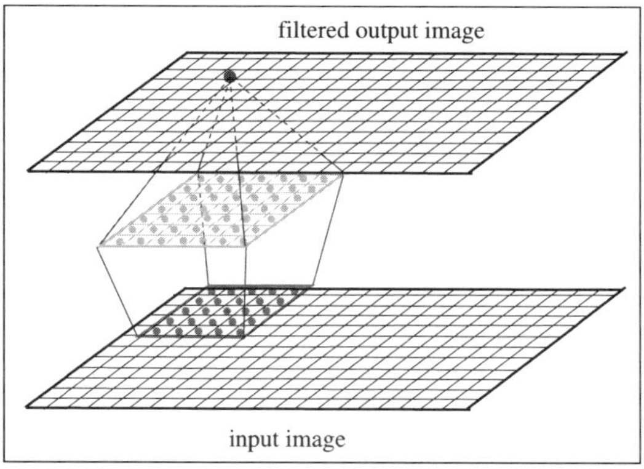

Fig. 2. An MLP acting as a filter. The input is transformed onto the output by moving the input window in raster order

then a data-driven (bottom-up) approach should be considered. In the latter case all the objects in the image must be located and recognized in order to identify the desired information. The peculiarities of top-down and bottom-up approaches in layout analysis are discussed in particular in the chapter by Belaïd and Rangoni.

In the following sections we provide a deeper analysis of the most important tasks in DAR with pointers to appropriate chapters in this book.

3 Pre-processing

The document acquisition is the process of obtaining an electronic image of a paper-based document. In most cases a flat-bed scanner is used, however in digital libraries also book scanners can be considered, whereas in recent years portable devices such as digital cameras and mobile phones are used as well.

Pre-processing operations in document image analysis transform the input image into an enhanced image more suitable for further analysis. Image-to-image transformations in DAR belong to four main classes [9]: filtering, geometrical transformations (e.g. skew detection), object boundary detection, and thinning.

The filtering transforms the input into an image whose value in a generic position (i,j) is usually a function of the input values in a neighborhood of the point (i,j). The three main classes of filtering operations in DAR ([4], page 3) are *binarization*, *noise reduction*, and *signal enhancement*. An example of machine-learning applications for filtering is based on the well-known property of universal approximation of neural networks, and in particular of the Multi-Layer Perceptron (MLP) [10]. To this purpose, a trained MLP (or other supervised classifiers) is fed with the pixels of a fixed-size sliding window providing a suitable output image (Fig. 2).

Thinning algorithms are needed to compute features based on the symbol skeleton. The simplest approaches are based on a recursive erosion of the object contour (for instance by using morphological operations). Other approaches rely on clustering-based skeletonization algorithm (CBSA) that is also discussed in the chapter by Marinai, Marino, and Soda.

4 Layout Analysis

Printed pages contain various types of information (e.g. text, equations, graphics, images) that should be processed with suitable techniques. Document layout analysis is performed to segment the document image into regions having an homogeneous content and to assign a meaning to the regions (Fig. 3). The segmentation step is called *physical layout analysis*, and is used to identify the geometric page structure. The *logical layout analysis*, or functional labeling, assigns a logical meaning to each region generating the logical structure of the document. Further remarks on these topics can be found in the chapter by Malerba, Ceci, and Berardi. A layout analysis sub-task that is particularly difficult to tackle is table recognition that is addressed in this book in the chapter by Zanibbi, Blostein and Cordy.

4.1 Physical Layout Analysis

The physical layout analysis aims at extracting regions with uniform features from the document. Segmentation algorithms in image processing can be grouped into three main categories: pixel classification, edge-based segmentation, and region-based segmentation. More specifically, document segmentation approaches pertain to pixel classification and region-based segmentation.

In pixel classification a label is attached to each pixel considering its color and those of the neighboring pixels. It was initially applied to the binarization of document images (e.g. [11]) and then extended to deal with additional classes (e.g. text, graphics, and line drawing). The regions are subsequently extracted by removing small noisy elements, merging similar neighboring regions, and locating connected components in the resulting image.

Most segmentation methods in DAR belong to the region-based family, comprising bottom-up and top-down methods. Bottom-up approaches can be regarded as merging methods and are mostly based on the location of connected components, with subsequent aggregation in higher level structures. In the top-down analysis a page is segmented from larger components to smaller subcomponents (e.g. [12, 13]). For instance, a page can be split into columns, in paragraphs, text lines, words and characters. Several methods (e.g. the XY tree) are based on the computation of the projection profile by computing the number of black pixels accumulated in horizontal and vertical directions to identify gaps between regions in the page. Top-down methods are usually faster but work well only for documents having a regular layout.

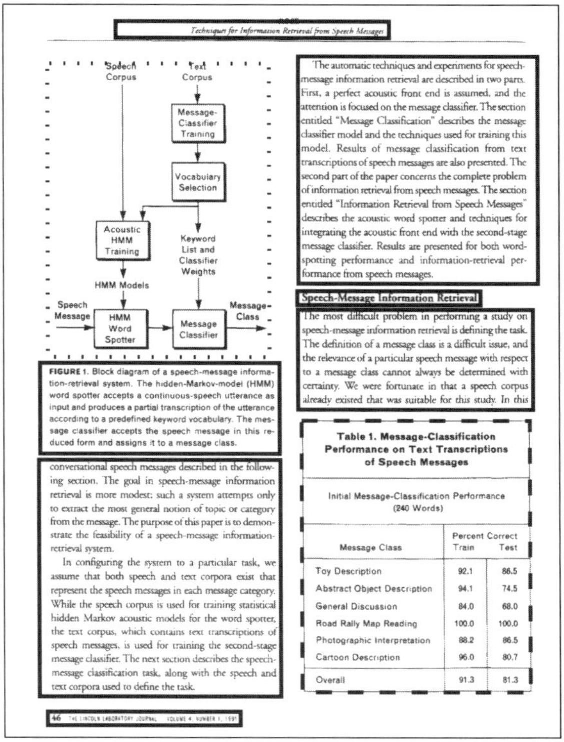

Techniques for Information Retrieval from Speech Messages

FIGURE 1. Block diagram of a speech-message information-retrieval system. The hidden-Markov-model (HMM) word spotter accepts a continuous-speech utterance as input and produces a partial transcription of the utterance according to a predefined keyword vocabulary. The message classifier accepts the speech message in this reduced form and assigns it to a message class.

conversational speech messages described in the following section. The goal in speech-message information retrieval is more modest: such a system attempts only to extract the most general notion of topic or category from the message. The purpose of this paper is to demonstrate the feasibility of a speech-message information-retrieval system.

In configuring the system to a particular task, we assume that both speech and text corpora exist that represent the speech messages in each message category. While the speech corpus is used for training statistical hidden Markov acoustic models for the word spotter, the text corpus, which contains text transcriptions of speech messages, is used for training the second-stage message classifier. The next section describes the speech-message classification task, along with the speech and text corpora used to define the task.

The automatic techniques and experiments for speech-message information retrieval are described in two parts. First, a perfect acoustic front end is assumed, and the attention is focused on the message classifier. The section entitled "Message Classification" describes the message classifier model and the techniques used for training this model. Results of message classification from text transcriptions of speech messages are also presented. The second part of the paper concerns the complete problem of information retrieval from speech messages. The section entitled "Information Retrieval from Speech Messages" describes the acoustic word spotter and techniques for integrating the acoustic front end with the second-stage message classifier. Results are presented for both word-spotting performance and information-retrieval performance from speech messages.

Speech-Message Information Retrieval

The most difficult problem in performing a study on speech-message information retrieval is defining the task. The definition of a message class is a difficult issue, and the relevance of a particular speech message with respect to a message class cannot always be determined with certainty. We were fortunate in that a speech corpus already existed that was suitable for this study. In this

Table 1. Message-Classification Performance on Text Transcriptions of Speech Messages

Initial Message-Classification Performance (240 Words)

Message Class	Percent Correct Train	Percent Correct Test
Toy Description	92.1	86.5
Abstract Object Description	94.1	74.5
General Discussion	84.0	68.0
Road Rally Map Reading	100.0	100.0
Photographic Interpretation	88.2	86.5
Cartoon Description	96.0	80.7
Overall	91.3	81.3

46 THE LINCOLN LABORATORY JOURNAL VOLUME 4, NUMBER 1, 1991

Fig. 3. Example of layout segmentation. Text blocks are identified with continuous lines. A dotted line and a dashed one identify one table region and one picture region, respectively

Bottom-up methods are more expensive from a computational point of view, but can deal with more complex layouts.

4.2 Logical Layout Analysis

The logical layout analysis assigns a meaning to the regions identified by the physical layout analysis. Examples of features considered are the size of the blocks, their mutual position, and some information on the textual parts such as the predominant font, character size and spacing. In this book the use of the logical classification of textual regions in a digital library context is described in the chapter by Esposito, Ferilli, Basile, and Di Mauro.

The physical and logical analysis can be performed together [14] so as to assign a meaning to blocks during their segmentation. In most cases this is not feasible, since the class can be defined only after analyzing the region position with respect to other parts of the page (or even after the reading of its content).

5 Text Recognition

The text recognition is executed to convert an image containing printed or handwritten text into a format that can be understood by a computer (e.g. ASCII or unicode). State-of -the-art OCR packages can read printed text with very high recognition rates. However, the research on printed text recognition is still in progress in order to deal with noisy documents and non-standard fonts (see e.g. the chapter by Marinai, Marino, and Soda), as well as large character sets (see e.g. the chapter by Jaeger, Ma, and Doermann and the survey chapter by Liu and Fujisawa).

Text reading techniques can be split into two main categories (*on-line* and *off-line*) on the basis of the input device used. The on-line recognition uses tablet PCs as input and the text is acquired and processed in real time when written by the user, whereas scanners are often used in off-line systems. One significant difference between on-line and off-line systems is the nature of the signal considered as input to the recognizer. In the on-line recognition the signal is dynamic and represents the text drawn by the user with one sequence of values that describes the pen position and pressure information. In off-line processing the input signal is a raster image. The position of the text in the page is usually unknown and should be identified by layout analysis algorithms.

In on-line systems there are some advantages with respect to off-line processing since the temporal information of the writing is known. This information is helpful to segment individual characters and to identify character's strokes. A more subtle advantage of on-line systems is the interactivity that is established between the user and the system. This interaction helps the automatic training of reading systems, but is mostly important for user adaptation to the system (related issues of user interaction are addressed in the chapter by Nagy and Veeranachaneni). The combination of on-line and off-line recognition into a multiple classifier framework is proposed in the chapter by Jaeger, Ma, and Doermann.

Some salient features of off-line reading systems are discussed in the rest of this section.

5.1 Character Recognition

In the *divide-et-impera* paradigm, that is typical of several pattern recognition applications, the text recognition systems frequently first split the words into characters and then assign one class to each isolated object. When dealing with printed text this approach is generally referred to as Optical Character Recognition (OCR) that is exploited by several software tools available on the market.

Summarizing, the character recognition process can be divided into three main steps: 1) *segmentation and preprocessing*, are used to identify the characters and improve the image quality; 2) *classification* is typically based on a sequential application of a feature extraction module and a supervised classifier; 3) *contextual processing* is adopted to check the recognition results on the basis of contextual information such as domain-specific dictionaries.

The first patent of a system for the automatic recognition of printed characters was registered in 1929 [3]. The system used electro-optical techniques to recognize characters. There is a large literature on character recognition and OCR with many technical papers dating back to the 1960's. We do not have the ambition of discussing in detail this literature. Instead, we point out a few survey papers [3, 6, 8] and the chapter by Liu and Fujisawa in this book that reviews the learning-based classification methods that have been applied to character recognition.

5.2 Word Recognition

The segmentation-based word recognition cannot be exploited when the location of segmentation points is impossible or unreliable, as in cursive handwriting. In this case one alternative is to use a holistic recognition where words are recognized as single units. Holistic recognition is effective when dealing with a reduced lexicon. For instance, the number of basic words required to fill a check is limited to a few dozen (32 words are needed for writing legal amounts of English checks [15], 30 words can be used for French checks [16]).

When the problem at hand requires a larger lexicon, then segmentation-based reading is appropriate, nevertheless segmentation requires some feedback from recognition. Integrated segmentation and recognition (ISR) techniques are related to the contextual development of segmentation and recognition modules. Two well known ISR techniques are Heuristic Over Segmentation (HOS) and Time-Delay Neural Networks (TDNN).

In HOS a segmentation algorithm is applied to a word image to locate a large number of candidate cutting points. Subsequently, a recognizer is employed to score the alternative segmentations generated and to find the best character sequence. The basic idea behind this approach is to over-segment the word in the hope of including all the correct segmentation points among those extracted. The use of MLP for labeling valid segmentation points is described in the chapter by Blumenstein, whereas a related application in the field of postal address reading is discussed in the chapter by Kagehiro and Fujisawa. In the chapter by Tulyakov and Govindaraju the combination of handwritten word recognizers based on the HOS approach is analyzed.

Time-Delay Neural Networks (TDNN) have been initially proposed to deal with temporal sequences. The output of a TDNN depends on its current and previous inputs, which are delayed by one or more time units. If the input signal is a vector of m values and we take a delay of n time units into account, then a TDNN can be implemented with an MLP having $n \cdot m$ input units ([17] page 256). TDNNs have been used for on-line and off-line word recognition. In on-line recognition the meaning of "time" is quite straightforward, whereas in the off-line case the horizontal axis in the window containing the word is considered as a temporal scale. In this book the chapter by Belaïd and Rangoni contains a deep analysis of TDNN.

6 Classification in DAR

Classifiers trained by supervised learning are key components of many pattern recognition systems and DAR systems are not exceptions. In this section we summarize the various DAR sub-tasks that have been addressed with supervised classifiers. Another important family of methods is based on unsupervised classifiers whose applications in DAR are discussed in the chapters by Nagy and Veeramachaneni and by Marinai, Marino, and Soda. Since the book is at the cross-road of document analysis and machine learning, it is not surprising that many chapters deal with the topics summarized below.

6.1 Pixel and Region Classification

Pixel classification (Section 4.1) can be used for both pre-processing and layout analysis. Document image binarization is a simple pre-processing task that can be performed with pixel classification. The assumption behind most pixel classification approaches is that textual and graphical regions have different textures, and thus the membership of each pixel can be estimated by analyzing a small region around it (e.g. [18]).

Region classification is used in most cases in conjunction with global features describing the whole region to be labeled. The classification was initially carried out using linear classifiers operating on these features according to user-defined parameters [19]. The basic hypothesis that is exploited in region classification is the assumption that the content of the region is homogeneous, and consequently some general features of the whole region can be used as inputs to a trainable classifier (e.g. [20]). The output of the classifier is a logic role that can be associated to each region.

6.2 Reading Order Detection

The task of text reading cannot be limited to the simple recognition of the individual characters or words. Instead, the identification of the correct reading order allows human readers to correctly understand the document content. The identification of the correct word sequence can be extremely complex when dealing with multi-column documents with footnotes and when figure and tables are intermixed in the text. The chapter by Malerba, Ceci, and Berardi formulates the reading order problem representing reading order chains in first order logic formalism. In the proposed system the learned rules state that one block "follows in reading" another block on the basis of the block position in the page and other textual features.

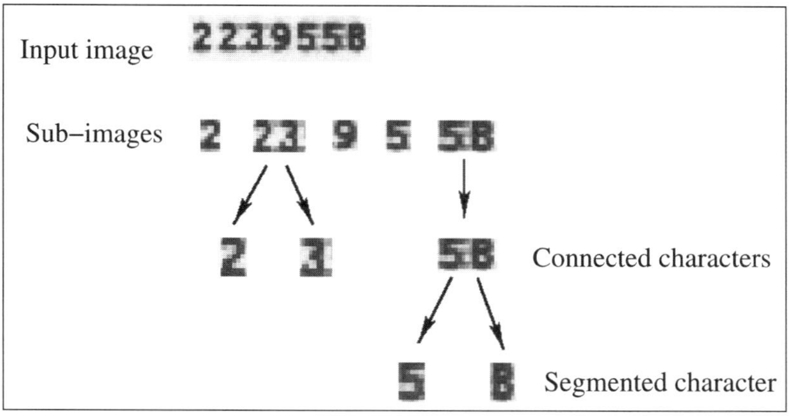

Fig. 4. Segmentation of connected strings. Supervised classifiers can be used for both identifying and segmenting the connected characters (as in '5B')

6.3 Text Recognition

In Section 5 we summarized the main features of text recognition systems. Various chapters in this book address the problem of word and character recognition. Liu and Fujisawa provide an updated survey of the use of learning methods in handwritten character recognition. Varga and Bunke propose the use of synthetic training data to improve the performance of an off-line cursive handwriting recognition system. One important research topic is the study of suitable techniques for combining classifiers so as to improve the overall classification performance. The chapter by Tulyakov, Jaeger, Govindaraju, and Doermann provides an overview of the classifier combination methods that have been addressed in several DAR tasks and in particular in character and word recognition.

Strictly related with word recognition is word retrieval that aims at efficiently retrieving occurrences of user-defined query words. The application of unsupervised learning to the task of word retrieval is discussed in the chapter by Marinai, Marino, and Soda.

6.4 Character segmentation

Supervised classifiers can be used in character segmentation with two main goals: the identification of touching characters and the location of cutting points (Fig. 4). In the former case the classifier is trained to distinguish isolated characters from pairs of touching characters (e.g. [21]) or more generally to estimate the number of characters in a string of connected digits (e.g. [22]). In the latter case the classifier is used for locating cutting points, i.e. the position in the word image that divides touching characters. Similarly to neural filters a classifier can be horizontally "moved" across the input image with the aim of labeling the corresponding position.

When dealing with handwritten strings the segmentation points can be located by analyzing the primitives (horizontal strokes) instead of the raw image. This approach is considered in [23] where an MLP is used in order to identify strokes that can correspond to cutting points. In this book the chapter by Blumenstein proposes a neural-based technique for classifying segmentation points in a cursive character segmentation algorithm.

6.5 Script Identification

Multi-lingual documents are important in particular in countries, like India, where there are multiple scripts that can be mixed together in the same text [24]. The script identification is a preliminary stage, with respect to the language recognition, that aims at recognizing the script used in a fragment of text. In some applications the identification is addressed at the page level. In such cases texture-based algorithms can be considered. In other applications there is more interest in the identification of the script of single words. In the chapter by Jaeger, Ma, and Doermann the combination of classifiers with informational confidence is tested, among other tasks, to the script identification in bilingual (e.g., Arabic-English) dictionaries.

6.6 Signature Verification

The task of signature verification systems is to evaluate whether unknow signatures are genuines or forgeries. At first one can imagine to solve the problem with a classifier trained to distinguish among two classes: genuine and forgery. However, several problems arise when trying to implement actual systems. For instance, we should deal both with random forgeries (affixed without knowing the actual signature) and with skilled ones (where the right signature is known to the forger). These issues are discussed in the chapter by Srihari, Srinivasan, Chen, and Beal that addresses the use of machine learning techniques for signature verification.

6.7 Writer Identification

Together with signature verification, writer identification, that is aimed at identifying the author of an handwritten text, is one of the oldest biometrics used for forensic purposes. In addition to these traditional applications, writer identification can be suitably incorporated in biometrics technology that is being adopted in various security applications. Writer identification systems can be either text-independent or text dependent. The former approach is the most difficult, since any text can be used to establish the identity of a writer. In this book the chapter by Schlapbach and Bunke presents an off-line, text-independent system for writer identification and verification.

6.8 Page Classification

In page classification the input to the classifier is the whole page. This task is useful when processing large collections of documents, for instance to perform an automatic document indexing. Page classification methods can be used in many application domains, such as form processing systems and digital libraries. Earlier applications concerned form classification methods that are aimed at selecting an appropriate reading strategy for each form to be processed. These methods often take the presence of ruling lines in the preprinted form layout into account [25]. Other typical classes are business letters and technical papers [26]. In the last few years the classification of journal and book pages received greater attention for digital library applications [27, 28].

Page classification in the domain of scientific documents is discussed in the chapter by Esposito, Ferilli, Basile, and Di Mauro where multi-page documents are indexed by matching their first page against some automatically learned models of document classes. Page classification for invoice processing is addressed in the chapter by Marinai, Marino, and Soda.

6.9 Document Categorization

Page classification has some similarities with document categorization, where documents are classified on the basis of the topic they address. Document categorization can be obtained by taking into account the document textual content (possibly recognized with an OCR software). The application of document categorization techniques to PDF documents is discussed in the chapter by Esposito, Ferilli, Basile, and Di Mauro. In some application domains it is possible to attempt the document categorization relying on the word image representation [29]. The latter strategy is addressed in the chapter by Marinai, Marino, and Soda.

7 Training Data

The tuning and comparison of machine learning algorithms heavily relies on the use of large collections of annotated data to be used to train the algorithms. In particular, the recognition of isolated handwritten digits has been considered for a long time as a benchmark for the comparison of trainable classifiers. For example the dataset "Optical Recognition of Handwritten Digits" is hosted since 1998 in the well known *UCI Machine Learning Repository* [2]. We survey in this section some of the most widely used public domain databases in the DAR field. We attempted to mention the most important datasets (including those collected to perform competitions at main conferences), without aiming to be exhaustive. Digital libraries are another important source of data

[2] http://mlearn.ics.uci.edu/MLSummary.html

for research in DAR. Some researchers built databases starting from widely available data such as free DLs or web sites of main publishers (e.g. IEEE). Examples of these datasets are adopted in the chapters by Esposito et al., and Marinai et al.

7.1 Public Databases

The databases listed in this section have been used by different research groups to compare the performance of DAR systems addressing several tasks. In most cases the data are available on the Internet and can be freely downloaded for research purposes.

UW databases

The UW databases are three document collections that have been gathered and manually annotated by the Intelligent Systems Laboratory, at the University of Washington (WA) in the late 1990's under the supervision of Prof. Robert Haralick. The databases were distributed as CD-ROMs containing document images and software for research and development.

The UW-III is the third in the series (it was published in 1996) and contains pages of chemical and mathematical equations, pages of line drawings and engineering drawings. There are also 33 pages containing English text with the corresponding character level groundtruth, 979 pages from UW-I, and 623 pages from UW-II corrected for skew, and the word bounding boxes for each word on these pages. These CD-ROMs, distributed for research purposes under the payment of a small fee, have been a reference for many years for the research on printed text.

NIST databases

In the 1990's the National Institute of Standards and Technologies (NIST) produced several CD-ROMs aimed at supporting the research on OCR software and information retrieval systems. A fully-automated process developed at NIST was used to derive the groundtruth information for the document images. The method involves matching the OCR results from a page with typesetting files for an entire book. The databases included scanned images, SGML-tagged groundtruth text, commercial OCR results, and image quality assessment results.

NIST's SD-3 (Special Database 3) and SD-1 contained binary images of handwritten digits. NIST originally designated SD-3 as training set and SD-1 as test set. However, SD-3 is much cleaner and easier to recognize than SD-1 and this fact is a limit for comparing different algorithms.

MNIST dataset

To overcome the previously mentioned limits of the NIST datasets Yann LeCun [30] designed the MNIST database of handwritten digits with a training set of 60,000 examples, and a test set of 10,000 examples[3]. MNIST is a subset of a larger collection of data available from NIST. The digits have been size-normalized and centered in a fixed-size image. The database has been widely used in the last few years for comparing various classifiers on real-world data [30, 31].

MediaTeam database

The MediaTeam Oulu Document Database[4] is a collection of 500 scanned document images with related groundtruth information about the physical and logical structure of the documents. The images cover a broad range of document types including journal papers, maps, newsletters, form, music, dictionaries and can be used for comparing various tasks in DAR.

Infty project

The InftyProject is a voluntary R&D organization consisting of researchers from different universities and research institutes in Japan with the shared objective of investigating and developing new systems to process scientific information by computer. Starting from 2005 three datasets have been distributed[5]. InftyCDB-1 [32] consists of mathematical articles in English containing 688,580 objects (characters and mathematical symbols) from 476 pages. The image of each object is recorded together with appropriate groundtruth information. InftyCDB-2 has the same structure of InftyCDB-1 and contains some documents in German and French, as well as many in English. InftyCDB-3 is a database of single alphanumeric characters and mathematical symbols. Unlike InftyCDB-1 and InftyCDB-2, word and mathematical expression structure is not included. The images are of individual characters only for a total of 259,389 symbols.

IFN/ENIT database

The IFN/ENIT-database[6] is based on more than 2,200 Arabic handwritten forms (filled by Tunisian people) containing about 26,000 word images together with groundtruth information [33]. This database has been used as a basis for a competition on Arabic handwriting recognition that is described below.

---

[3] http://yann.lecun.com/exdb/mnist
[4] http://www.mediateam.oulu.fi/downloads/MTDB
[5] http://www.inftyproject.org/en/database.html
[6] http://www.ifnenit.com

MARG database

MARG[7] is a freely-available repository of document page images and their associated groundtruth information on the textual and layout content. The pages are representative of articles drawn from the corpus of important biomedical journals. The database is suitable for the development of algorithms to locate and extract text from the bibliographic fields typical of articles within such journals. These fields include the article title, author names, institutional affiliations, abstracts and possibly others. Only the first page of each article is available, or the second page if the abstract runs over [34].

IAM database

The IAM-Database[8] includes over 1,500 scanned forms of handwritten text from more than 600 different writers [35]. The groundtruth information is provided at the word, line, and page levels allowing several types of experimentations. Overall, the database contains 115,320 labeled words that have been extracted using an automatic segmentation scheme and have been subsequently manually verified.

7.2 Competitions

Besides "static" databases, that are built with significant efforts and aim at a long term research, there are also several competitions that are run at conferences and workshops in the DAR field. The data collected by the organizers are usually available on the Internet. The format of the competitions is generally based on the distribution of training and test data to participants in the months before the conference. The data are freely available for research purposes, but a registration is usually required. The evaluation of the results for system comparison can be based on three main approaches. In some cases the participants send to the organizers the results obtained by their systems on a pre-defined test set. In other cases the participants send the executable programs to the organizers that compare the systems "in house". In the last approach the participants run their systems during the conference on new data previously unknown.

These competitions are helpful for the DAR research not only for the test data that are collected, but also for the development of performance evaluation methods as well as approaches for the automatic generation of synthetic data that are required to run these events. We list in the following some of the most recent competitions that have been organized at DAR conferences.

---

[7] http://marg.nlm.nih.gov/index2.asp
[8] http://www.iam.unibe.ch/~ fki/iamDB

Page segmentation

Page segmentation is one important step in layout analysis and is particularly difficult when dealing with complex layouts. The page segmentation competition has been organized in the last ICDAR conferences (from 2001 onwards)[9]. The main objective of the competition is to compare the performance of layout analysis algorithms using digitized documents from common publications [36].

Arabic handwritten

Starting from the experience made with the IFN/ENIT-database a series of competitions to establish the state of the art of recognizing Arabic handwritten words has been organized at the ICDAR conference to give the opportunity to further develop methods and discuss results on Arabic recognition systems [37].

Symbol recognition

The GREC workshop is organized every two years by the IAPR Technical Committee on Graphics Recognition (TC 10). Graphics recognition is a subfield of DAR that deals with graphical entities in engineering drawings, maps, architectural plans, musical scores, mathematical notation, tables, and diagrams. In the last editions of the workshop some contests have been organized including an arc segmentation contest and a symbol recognition one[10]. The main goal of the latter contest is the comparison of various methods for recognizing linear graphic symbols, i.e. symbols made of lines, arcs and simple geometric primitives, which can be found in most graphic drawings.

Document image dewarping

In recent years the use of mobile devices for digitizing document images is becoming more and more important. One challenging task in the processing of camera-captured documents is the presence of page curl and perspective distortions. The goal of page dewarping is to flatten a document image in order to improve the recognition rate that can be achieved by state-of-the-art OCR systems. During the CBDAR 2007 workshop a page dewarping contest has been organized by the Image Understanding and Pattern Recognition group at the DFKI in Kaiserslautern (Germany). A database of camera captured documents with groundtruthed text-lines, text-zone, and ASCII text has been provided to participants[11].

---

[9] http://www.cse.salford.ac.uk/prima/ICDAR2007_competition

[10] http://symbcontestgrec05.loria.fr

[11] http://www.iupr.org/doku.php?id=didcontest

8 Concluding Remarks

In this introductory chapter we briefly introduced readers to the principal themes in the DAR research. We also attempted to present in a uniform way the main topics discussed in the chapters in the book that make, as a whole, an accurate picture of the current state of the art.

We hope that this book will contribute in two ways to the research in Document Analysis and Recognition. First, researchers in machine learning can contribute to the DAR research by starting from the problems addressed in this book and attempting to adopt innovative techniques previously studied in other domains. Second, we hope that researchers already active in the DAR field could employ machine learning techniques on several tasks modifying algorithms that are too frequently defined "by hand" by software developers.

References

1. Lyman, P., Varian, H.R.: How much information. Technical Report Retrieved from http://www.sims.berkeley.edu/how-much-info-2003 on 6-1-2007 (2003)
2. Sellen, A.J., Harper, R.: The myth of the paperless office. MIT press (2001)
3. Mori, S., Suen, C., Yamamoto, K.: Historical review of OCR research and development. Proc. IEEE **80** (1992) 1029–1058
4. O'Gorman, L., Kasturi, R.: Document Image Analysis. IEEE Computer Society Press, Los Alamitos, California (1995)
5. Casey, R.G., Lecolinet, E.: A survey of methods and strategies in character segmentation. IEEE Transaction on PAMI **18**(7) (1996) 690–706
6. Nagy, G.: Twenty years of document image analysis in PAMI. IEEE Transaction on PAMI **22**(1) (2000) 38–62
7. Plamondon, R., Srihari, S.N.: On-line and off-line handwriting recognition: a comprehensive survey. IEEE Transaction on PAMI **22**(1) (2000) 63–84
8. Marinai, S., Gori, M., Soda, G.: Artificial neural networks for document analysis and recognition. IEEE Transactions on PAMI **27**(1) (2005) 23–35
9. Ha, T., Bunke, H.: Image processing methods for document image analysis. In: Handbook of character recognition and document image analysis. World Scientific (1997) 1–47
10. Hornik, K., Stinchcombe, M., White, H.: Multilayer feedforward networks are universal approximators. Neural Networks **2** (1989) 359–366
11. Taxt, T., Flynn, P.J., Jain, A.K.: Segmentation of document images. IEEE Transaction on PAMI **11**(12) (1989) 1322–1329
12. Nagy, G., Seth, S.: Hierarchical representation of optically scanned documents. In: Int'l Conference on Pattern Recognition. (1984) 347–349
13. Watanabe, T., Luo, Q., Sugie, N.: Structure recognition methods for various types of documents. MVA **6**(6) (1993) 163–176
14. Nagy, G., Seth, S., Viswanathan, M.: A prototype document image analysis system for technical journals. IEEE Computer **25**(7) (1992) 10–22
15. Kim, J.H., Kim, K.K., Suen, C.Y.: An HMM-MLP hybrid model for cursive script recognition. PAA **3**(4) (2000) 314–324

16. Gilloux, M., Lemarié, B., Leroux, M.: A hybrid radial basis function network/ hidden Markov model handwritten word recognition system. In: Int'l Conference on Document Analysis and Recognition. (1995) 394–397
17. Fu, L.M.: Neural networks in computer intelligence. McGraw-Hill, New York, NY (1994)
18. Jain, A.K., Zhong, Y.: Page segmentation using texture analysis. Pattern Recognition **29**(5) (1996) 743–770
19. Shih, F.Y., Chen, S.S.: Adaptive document block segmentation and classification. IEEE Trans. SMC **26**(5) (1996) 797–802
20. Strouthopoulos, C., Papamarkos, N.: Text identification for document image analysis using a neural network. Image and Vision Computing **16**(12/13) (1998) 879–896
21. Wang, J., Jean, J.: Segmentation of merged characters by neural networks and shortest path. Pattern Recognition **27**(5) (1994) 649–658
22. Lu, Z.K., Chi, Z., Siu, W.C.: Length estimation of digits strings using a neural network with structure based features. SPIE/IS&T Journal of Electronic Imaging **7**(1) (1998) 79–85
23. You, D., Kim, G.: An approach for locating segmentation points of handwritten digit strings using a neural network. In: Int'l Conference on Document Analysis and Recognition. (2003) 142–146
24. Pal, U., Sinha, S., Chaudhuri, B.: Multi-script line identification from indian documents. In: Int'l Conference on Document Analysis and Recognition. (2003) 880–884
25. Ishitani, Y.: Flexible and robust model matching based on association graph for form image understanding. Pattern Analysis and Applications **3**(2) (2000) 104–119
26. Dengel, A., Dubiel, F.: Clustering and classification of document structure -a machine learning approach-. In: Int'l Conference on Document Analysis and Recognition. (1995) 587–591
27. Cesarini, F., Lastri, M., Marinai, S., Soda, G.: Encoding of modified X-Y trees for document classification. In: Int'l Conference on Document Analysis and Recognition. (2001) 1131–1136
28. van Beusekom, J., Keysers, D., Shafait, F., Breuel, T.M.: Distance measures for layout-based document image retrieval. In: Proc. Second Int'l Workshop on Document Image Analysis for Libraries. (2006) 232–242
29. Kaski, S., Honkela, T., Lagus, K., Kohonen, T.: Websom self-organizing maps of document collections. Neurocomputing **21**(1–3) (1998) 101–118
30. LeCun, Y., Bottou, L., Bengio, Y., Haffner, P.: Gradient-based learning applied to document recognition. Proceedings of the IEEE **86**(11) (1998) 2278–2324
31. Simard, P.Y., Steinkraus, D., Platt, J.C.: Best practices for convolutional neural networks applied to visual document analysis. In: Int'l Conference on Document Analysis and Recognition. (2003) 958–963
32. Suzuki, M., Uchida, S., Nomura, A.: A ground-truthed mathematical character and symbol image database. In: Int'l Conference on Document Analysis and Recognition. (2005) 675–679
33. Pechwitz, M., Maddouri, S.S., Magner, V., Ellouze, N., Amiri, H.: IFN/ENIT-database of handwritten arabic words. In: 7th Colloque International Francophone sur l'Ecrit et le Document. (2002)

34. Ford, G., Thoma, G.: Ground truth data for document image analysis. In: Symposium on Document Image Understanding and Technology. (2003) 199–205
35. Marti, U., Bunke, H.: A full english sentence database for off-line handwriting recognition. In: Int'l Conference on Document Analysis and Recognition. (1999) 705–708
36. Antonacopoulos, A., Gatos, B., Bridson, D.: ICDAR 2005 page segmentation competition. In: Int'l Conference on Document Analysis and Recognition. (2005) 75–79
37. Margner, V., Pechwitz, M., Abed, H.E.: ICDAR 2005 arabic handwriting recognition competition. In: Int'l Conference on Document Analysis and Recognition. (2005) 70–74

Structure Extraction in Printed Documents Using Neural Approaches

Abdel Belaïd and Yves Rangoni

University Nancy 2 - LORIA, Campus Scientifique, 615 rue du Jardin Botanique, 54600 Villers-Lès-Nancy, France
{abelaid,rangoni}@loria.fr

Summary. This chapter addresses the problem of layout and logical structure extraction from image documents. Two classes of approaches are first studied and discussed in general terms: data-driven and model-driven. In the latter, some specific approaches like rule-based or formal grammar are usually studied on very stereotyped documents providing honest results, while in the former artificial neural networks are often considered for small patterns with good results. Our understanding of these techniques let us to believe that a hybrid model is a more appropriate solution for structure extraction. Based on this standpoint, we proposed a Perceptive Neural Network based approach using a static topology that possesses the characteristics of a dynamic neural network. Thanks to its transparency, it allows a better representation of the model elements and the relationships between the logical and the physical components. Furthermore, it possesses perceptive cycles providing some capacities in data refinement and correction. Tested on several kinds of documents, the results are better than those of a static Multilayer Perceptron.

1 Introduction

Automatic structure extraction remains a very challenging problem due to the inherent complexity of documents. For raster images of documents, the gap between physical and logical structure is huge. It is difficult to model the intermediate steps and the relationships between the original image blocks and recognized layout structures, and to maintain consistency between the processing steps in the recognition process. It is also difficult to handle image noise, layout variations and artifacts produced during processing.

In spite of the numerous researches done in this way, the investigation made in this area is prudent:

- recognition has been limited to few structures (less than 10, let say 5 in average), essentially in editorial documents (i.e. books, articles, reports, etc.), often accompanied by a DTD (Document Type Definition), making the recognition more stereotyped;

A. Belaïd and Y. Rangoni: *Structure Extraction in Printed Documents Using Neural Approaches*, Studies in Computational Intelligence (SCI) **90**, 21–43 (2008)
www.springerlink.com

- recognition methodology has been limited to translating DTD knowledge and its application on the document. The methods were mainly oriented toward context-free grammars and tree or graph comparisons, and often considered as limited in their ability to handle complex situations.

Certainly, the literature provides many approaches to structural recognition, but their application to document analysis is not straightforward and their advantages often equal their drawbacks.

There are two main approaches to document layout analysis: those based primarily on information manipulation, and those based primarily of perceiving features in data. Considering the information manipulation aspect, two sub-categories exist:

- model-driven (e.g. systems using rules or grammars). They use and formalize knowledge well, and are precise and fast but are dependent on an expert to guide their actions. Unfortunately, they do not generalize well, and have been found sensitive to variation and noise;
- data-driven, starting from low-level data. Their classes should represent very well the structure elements, but data description is not easy and the convergence is not assured. However, contrary model-driven methods, they remain very general and flexible as their adaptation to new documents is easier.

Considering the perception aspect, here also two points of view can be distinguished:

- global to local which is often assimilated to top-down approach. The process is based on a segmentation refinement: here the progress seems to be made continuously and safely but if an error is introduced in the beginning, it remains during all the process.
- local to global or bottom-up approaches. These labeling-based methods start from fine to coarse building progressively the context. In this case, a lot of unused features have to be extracted and managed.

As indicated above, all the methods investigated in the literature present some limitations. Hence, the solution that seems to be appropriate for document structure analysis is a hybrid approach in the sense where it mixes both aspects: data consistency and perceptual approaches for the processing methodology.

This chapter is organized as follows: section 2 discusses the use of Artificial Neural Network (ANN) approaches in Document Analysis and Recognition (DAR) area, specifically for recognition tasks involving the physical structure of a document. Section 3 focuses on the ANN based solution for logical structure extraction. Finally, section 4 gives some perspectives about the use of ANN in logical structure analysis.

2 Neural Networks in Document Analysis and Recognition

2.1 Physical or Geometrical Layout Analysis

In Document Analysis and Recognition (DAR), Artificial Neural Networks (ANN) have been devoted mainly to preprocessing tasks or recognition of small patterns as isolated characters. As detailed by Marinai et al. in [1], such kind of use include binarization, noise reduction, skew detection, and character thinning. The MultiLayer Perceptron (MLP) is used for example in [2] to binarize images for character segmentation. After a segmentation phase based on gray level histogram analysis, the authors feed a MLP with pixel values within a 5×5 windows. In [3], a Self Organizing Map (SOM) and a MLP are applied on the image to classify the pixels according to their gray levels or color values. Another use of ANN is for noise elimination such as in [4] by applying Kalman filtering.

For images representing characters, various ANN models dealing with printed or handwritten scripts have been experimented. Main of them proceeds directly on the images as the inputs are often composed of the image pixel values. Le Cun et al. [5] have provided a very interesting survey on various ANN models related to handwritten words. Similar architectures (convolutional) were used by [6] for handwritten digits recognition, complemented by a SOM to correct the rejections. For each rejected character, a SOM is trained and associated to the MLP to make possible the correction. Garris et al. [7] used an enhanced MLP for the same problem, where the enhancements are focused on neuron activation functions, regularization and Boltzmann pruning.

Hence, these examples show clearly that ANN are able to deal with local variations in a document image during recognition.

2.2 Logical Structure Analysis

There are few works on logical structure recognition using ANN. Indeed most of the approaches are model-driven. The model contains the description of the physical elements of the document and their associated logical labels. The recognition procedure tries to identify these associations.

Usually, these models are either trees or grammar rules. In both cases, a syntactical analysis procedure is employed to perform the structure labeling [8]. For example, Brugger et al. [9] use a generalized n-gram (with n = 3) to represent geometrical relationship between the text blocks, then an optimization method to match the current input with a global model or a sub-tree of this model. Hu et al. [10] use dynamic parsing and fuzzy logic to be more flexible when analyzing the logical structure. A rule-based system is employed by Niyogi et al. [11] with a top-down backward-chaining strategy. Their system "DeloS" handles about 160 rules in three levels for classification, reading order and logical structure analysis.

Although this methodology seems natural as it transcribes a known structure hierarchy of the document and works very well for simple documents, its application on more complex document becomes difficult and causes many errors. In fact, the use of deterministic models fails because of the rigidity in the application of the rules. Furthermore, these kinds of models are often created manually, leading the operator to select and tune himself a lot of parameters. This can explain the limits of such models when applied on real images where the structure is complex and does not fit exactly the general model. The inherent noise of the input image can sometimes introduce errors in the interpretation of the elements.

To face this problem, a data-driven method seems more appropriate. ANN can provide a good solution because they learn from examples, are robust, insensitive to noise and have a generalization capacity. Furthermore, the ANN based solution will avoid the drawbacks of model-based method provided that knowledge must be integrated. Indeed as mentioned in [1], the classical use of MLP is not sufficient to tackle the problem. Existing methods are focused on the tuning of the MLP to resolve the problem and not really on new architectures. The idea is to use a model which is not only based on MLP but which can integrate the structural aspect of the problem.

Two types of ANN can be considered:

- static ANN (with MLP configurations) can adapt to structured patterns by cleverly integrating the structure in the topology as made by [12];
- dynamic ANN by transforming the temporal chain in structured version as in [13, 14].

These two architectures will be described in the following.

3 Neural Networks for Structured Patterns

Neural networks are suitable to handle classification problems with static information. For several applications including logical structure analysis, the patterns to deal with are in a structured domain. ANN are designed to classify unstructured patterns and cannot deal directly with tree or graph structures. However, we can find models which can take into account the structured patterns either in a dynamic or a static version.

3.1 Static Networks

The best know type of static network is the MLP because it is the easiest to implement, its training algorithm is well known and it has been applied successfully to different kinds of data.

As mentioned in section 2, its use is generally devoted to physical element recognition where there are no or few structures to interpret. All research done on this kind of network does not focus on the model topology but more on the way MLP is applied to the specific task.

3.2 Dynamic Networks

In order to take into account the temporal dimension of some real-world problems, a dynamic network can be applied rather than a static one.

The Time Delay NN (TDNN) is a straightforward solution that unfolds the time sequence onto several static models through a Tapped Delay Line (TDL) [15]. The same approach can be done with the RBFN (Radial Basis Function Network) to take into account the temporal dimension [16].

Feedback dynamic methods as recurrent networks integrate feedback contrary to feed forward systems. The learning is recursive and consequently more complex to undertake. The output Feedback based systems use the network outputs in a second TDL besides the classical one's as in TDNN [17].

State feedback methods feedback connections between neurons are introduced: each neuron contributes to all components of the state vector.

Time Hopfield Networks (THN) [18] are mono-layer networks in which all the possible interconnections are used. The Continuous THN (CTHN) is well known as it can handle oscillations or even chaotic phenomenon. The Discrete THN (DTHN) is similar to the previous one's but here the activation function is hard limiter and not a sigmoid.

Continuous Time Recurrent Neural Network (TRNN) [19] is quite similar to CTHN: there is one layer of fully connected neurons, the difference is in the differential equation managing the dynamic process. The same analogy is done for the Discrete Time Recurrent Neural Networks (DTRNN) with its hard-limiter function [20]. The DTRNN can simulate deterministic finite automate. In such ANN, the training stage is more complicated and two main solutions can be seen in the literature. The first totally converts the network into a feed forward version by unfolding the network over time. The second method consists in the use of the recursive version of the gradient descent.

3.3 Dynamic Networks for Structured Patterns

The previous dynamic networks have been developed to process sequences of patterns but adaptations to structured patterns can be found in the literature.

Küchler and Goller [13] propose an approach to classify structured patterns. The patterns considered are those represented by a Direct Acyclic Graph (DAG) or by a Rooted LDAG (i.e. a graph with only one root node, i.e. one node with in degree zero). The network topology, in a static view, corresponds to the folding of the DAG in a feed forward MLP. The first layers compute the folding part (i.e. inputs through DAG representation) and the following layers constitute the transformation part (Fig. 1).

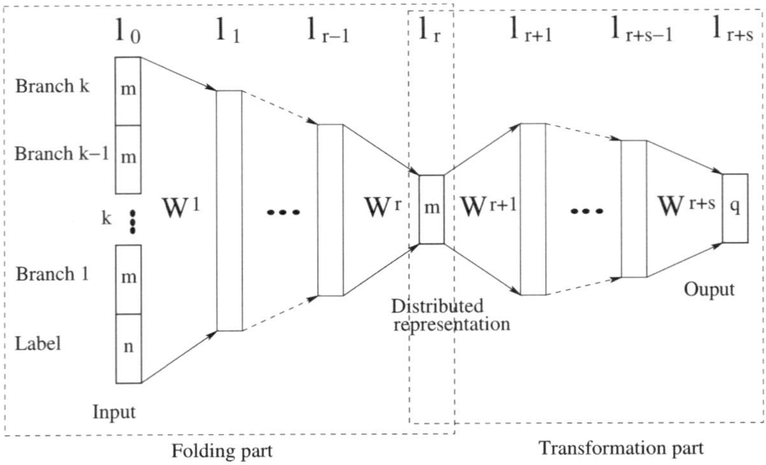

Fig. 1. Küchler et al. generic folding architecture

The input contains the vertex labels for distributed representation of DAG. The last layer corresponds to the task specific output. The network dynamics are defined as follows:

$$o_j^{(l+1)}(t) = f\left(\sum_i o_i^{(l)}(t) w_{ij}^{(l+1)} + \theta_j^{(l+1)}\right) \tag{1}$$

where $o_i^{(l)}(t)$ is the output of neuron i in the layer l at recursion stage t, $\theta_i^{(l)}$ is the bias associated with neuron i at layer l, $w_{ij}^{(l+1)}$ the weight of the connection between neuron i in layer l and neuron j in layer $l+1$ and f the sigmoid function.

The authors use a modified version of the Back-Propagation Through Time (BPTT) algorithm where the structure of a labeled DAG is incorporated in the error measurement

$$E = \sum_{i=1}^{p} \sum_{j=0}^{q-1} \frac{1}{2}\left([t_i]_j - o_j^{(r+s)}(root(s_i))\right)^2 \tag{2}$$

where *root* denotes the function mapping structures to their root nodes, s_i are in the general symbolic domain and t_i define by $\Xi(s_i) = t_i$ with Ξ being the function to be approximated.

Thanks to a special gradient descent technique called Back-Propagation Through Structure (BPTS), the network can be trained. Experimentation has been done on 2-classes classification problems on logical terms. The results are very promising: 99% for the training and 98% for the test.

Sperduti et al. [14] propose another dynamic NN extended to structural patterns. The main idea is to generalize a recurrent neuron in a "Generalized

Recursive Neuron" (GNR). The approach is different from the standard one which focuses on the tree-structure encoding in a fixed input vector. The GRN considers the outputs of the unit for all the vertices which are pointed by the current input vertex.

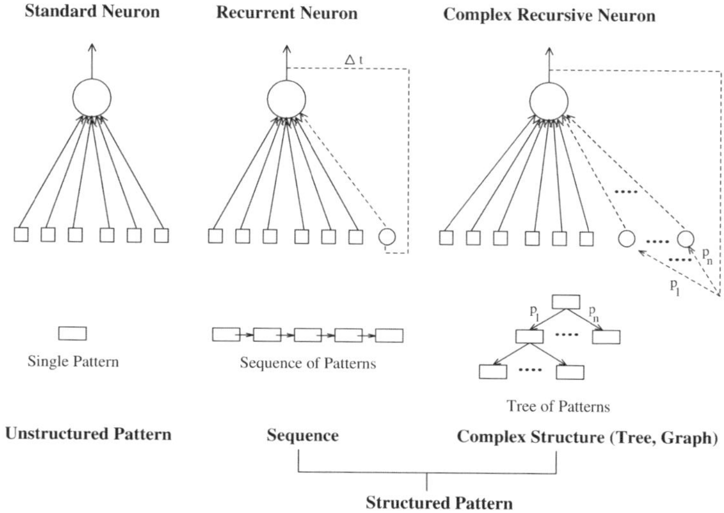

Fig. 2. Neuron models for different input domains

Figure 2 shows the standard models for structured and unstructured patterns, on the right side, the presented configuration can represent any graph or tree structure thanks to the proposed GRN. Usually, in a standard neuron the output is given by:

$$o^{(s)} = f\left(\sum_i w_i I_i\right) \tag{3}$$

where f is non-linear function such as the sigmoid, I the input vector and w the weight vector. In the recurrent version, the output depends on time:

$$o^{(r)}(t) = f\left(\sum_i w_i I_i(t) + w_s o^{(r)}(t-1)\right) \tag{4}$$

where $o^{(r)}(t-1)$ is the previous output at time $t-1$ that is weighted by w_s and added to the activation formulae. In the GRN the output $o^{(g)}(x)$ depends on a vertex in the graph and computed recursively on the output performed for all the vertices pointed by it. The output is given by:

$$o^{(g)}(x) = f\left(\sum_i^{N_L} w_i l_i + \sum_{j=1}^{\text{out\_degree}_X(x)} \hat{w}_j o^{(g)}(\text{out}_X(x,j))\right) \tag{5}$$

where x is a vertex of a graph X, N_L the unit number encoding the label l attached to the current input x, $\hat{w}_j$ the weights on the recursive connections and $out_X(x,j)$ the out nodes of the graph X attached to the node x. The graph is encoded to fit with the GRN representation (Fig. 3).

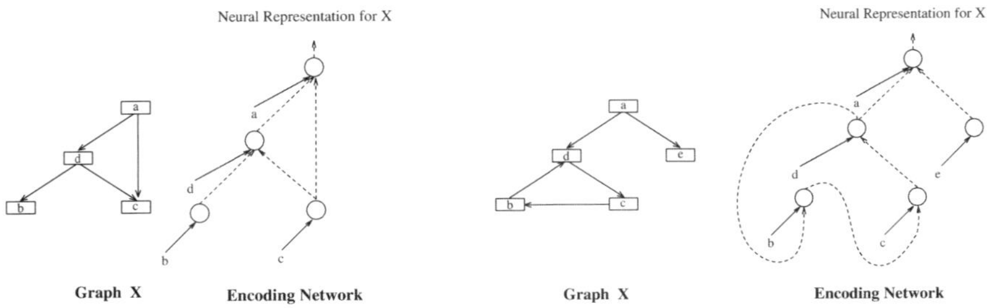

Fig. 3. On the left side, the network encoding for an acyclic graph is shown. On the right side, the encoding network for a cyclic graph is shown

The authors have extended five supervised algorithms for ANN to handle the GRN: back propagation through structure, real-time recurrent learning, LRAAM-based networks and simple recurrent networks, cascade-correlation for structures, and neural trees.

For example the Back propagation Through Structure (BPTS) is simply as in [13] an expression of the back propagation through time. The trick consists in unfolding through time the recurrent network in an equivalent and fully feed forward network. As a consequence, the transformed network can be trained using the back propagation algorithm. For the GRN, the network is decomposed into two parts: an encoding function Ψ and a classification function Φ such as

$$o(X) = \Phi(\Psi(X)) \tag{6}$$

Using standard back propagation, the weights are modified using (7) and (8):

$$\Delta \mathrm{W}_{\Phi} = -\eta \frac{\partial \mathrm{Error}(\Phi(y))}{\partial \mathrm{W}\Phi} \tag{7}$$

$$\Delta \mathrm{W}_{\Psi} = -\eta \frac{\partial \mathrm{Error}(\Psi(y))}{\partial \mathrm{y}} \frac{\partial y}{\partial \mathrm{W}_{\Psi}} \tag{8}$$

Two cases must be treated separately in the case of a DAG and graphs with cycles. With DAG, Küchler et al. [13] algorithm can be used. The training is computed by back propagation of the error from the feed forward network through the encoding network of each structure. For cyclic graphs, recurrent back propagation must be considered.

Real-Time Recurrent Learning can also be extended. For DAG the extension does not present particular problems, the cyclic graphs are more difficult

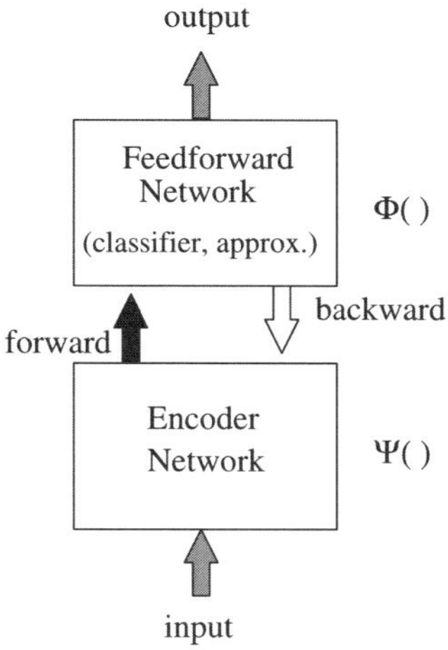

Fig. 4. The encoding part of the NN passes encoded structures to the classifier, the classifier returns the deltas used by the encoder to adapt its weights

to extend and require different situation according to global cycle presence. Thanks to the "Strongly Connected Component" and "Component Graph" notion, the cyclic graphs can be considered as many acyclic graphs and solved more easily.

Labeling Recursive Auto Associative Memory (LRAAM) [21], another model to represent labeled structures, is trained by a combination of a supervised method and an unsupervised one. For structured pattern recognition, Sperduti uses this LRAAM to produce a compressed representation of the structure, then he uses an MLP to carry out the classification.

GRN can be also extended to the cascade-correlation algorithm developed by Fahlmane and Lebiere [22]. This model generates a standard ANN by using an incremental approach for classification of unstructured patterns. The starting network $\mathcal{N}_0$ is a network with no hidden nodes trained using LMS. If $\mathcal{N}_0$ cannot resolve the problem, a hidden unit u_1 is added so that the correlation between the output of the unit and the residual error of the network $\mathcal{N}_0$ is maximized. The weights of u_1 are frozen and the remaining weights are retained. If the retained network $\mathcal{N}_1$ cannot solve the problem, the network is further grown by new hidden units which are connected (with frozen weights) with all the inputs and previous hidden units. The resulting network is a cascade of nodes. Sperduti et al. extend the output of the kth to GRN using (9) where $w_{(v,j)}$ is the weight of the kth hidden unit associated with the output of the vth hidden unit computed on the jth component pointed by x. $\bar{w}_q^{(k)}$ is the weight of the connection from q^{th} hidden unit and the kth hidden unit. Learning is performed as in standard cascade-correlation with the difference that the equations are recurrent on the structures.

$$
\begin{aligned}
o^{(k)}(x) &= f(\alpha + \beta + \gamma) \\
\alpha &= \sum_{i}^{N_L} w_i^{(k)} l_i \\
\beta &= \sum_{v=1}^{k} \sum_{j=1}^{\text{out\_degree}_X(x)} \hat{w}_{(v,j)}^{(k)} o^{(v)}(\text{out}_X(x,j)) \\
\gamma &= \sum_{q=1}^{k-1} \bar{w}_q^{(k)} o^{(q)}(x)
\end{aligned}
\tag{9}
$$

GRN can also be adapted to neural trees. The advantage of this kind of model is to build the structure on the fly and not be restricted to a static structure as in a feed forward network. New classes are learnt incrementally with supervised or unsupervised training. The extension of this network to a structured version is done by analogy: each discriminator associated with each node of the tree is replaced by a generalized recursive discriminator.

Experiments on GRN have been carried out on several classification tasks. The data are randomly generated. For small size structures (tree depth between 3 and 6) the results obtained on classification problems are nearly perfects for the training (near 100%) and very good for the testing (average of 95% and sometimes 100% with a good choice of hidden units and learning parameters).

There is more and more work about dynamic networks, and although they are not oriented directly towards logical structure extraction, it seems that the previous contributions can be easily extended to this kind of application. However, these recurrent techniques present some drawbacks compared to static ANN:

- they are time and memory consuming;
- the convergence is more difficult to reach as there are more local minima;
- the convergence is slower, decreasing the training step make the training more and more slow;
- there are more numerical errors that create serious repercussion on network's convergence;
- the gradient explosion occurs quickly on long sequences. The more the sequence is long and the more the global error is large.

On top of that, the presented dynamic neuronal methods can deal with logical structure recognition but are not sufficient. In addition to the inherent limitations, the structures to be recognized need to be known and fixed throughout the training and recognition. This it is not necessarily true in real world applications.

3.4 Transparent Neural Network for Handwriting Recognition

As seen in the previous section, dynamic ANN can be extended to deal with structured patterns. The well known static ANN such as MLP can be also improved to handle structured patterns.

In [23], Côté et al. propose a perceptual model, Perceptro, for handwritten word recognition. The proposed method is based on McClelland and Rumelhart's reading mode [24]. Two questions are explored: what kinds of features are detected and how the information concerning the meaning of a word is accessed. The key is to integrate a knowledge representation in the network. Indeed, trying to use a standard network with distributed representation, such as the MLP, cannot deal correctly with handwritten recognition. That is why in [23] a network with local representation is chosen for the kernel of their approach. The Interactive Activation Model of [24] is a neural network with local knowledge representation, parallel processing of information, and gradual propagation of activation between adjacent levels of neurons. The original activation is given by

$$\begin{aligned} A_i(t+\delta t) &= A_i(t) - \theta_i(A_i(t) - r_i) + E_i(t) \\ E_i(t) &= \begin{cases} n_i(t)(M - A_i(t)) & \text{if } n_i(t) > 0 \\ n_i(t)(A_i(t) - m) & \text{if } n_i(t) < 0 \end{cases} \\ n_i(t) &= \sum_j (\alpha_{ij} - \beta_{ij}) a_j(t) \end{aligned} \tag{10}$$

where θ_i is a decreasing constant, r_i the activation threshold, $E_i(t)$ the neighborhood contribution, M and m superior and lower activation bounds, α_{ij} and β_{ij} the positive and negative stimulation from j to i, and $a_j(t)$ the activation of node j.

Recognition is performed trough several bottom-up and top-down processes. The physical features extracted from the image are specific to the problem: primary (e.g. ascender, descender) secondary (e.g. loop, bar) and face-up/face-down valley (e.g. connected components of the background between the lower and upper contours of the word). The architecture of the system is general enough to handle hierarchical organized interpretation. The authors have chosen three levels of neurons: feature, letter, and word (Fig. 5).

The connections between adjacent levels are excitatory and bi-directional. The connections are only bottom-up between the feature letter and the letter level. The weights are determined according to a priori knowledge. Thanks to an active and passive neuron system, the network can reach the solution after several cycles (until saturation) of bottom-up and top-down processes called perceptual cycles. The system generates hypothesis, validates them and may insert letter candidates in the right place using already validated letters (Fig. 6).

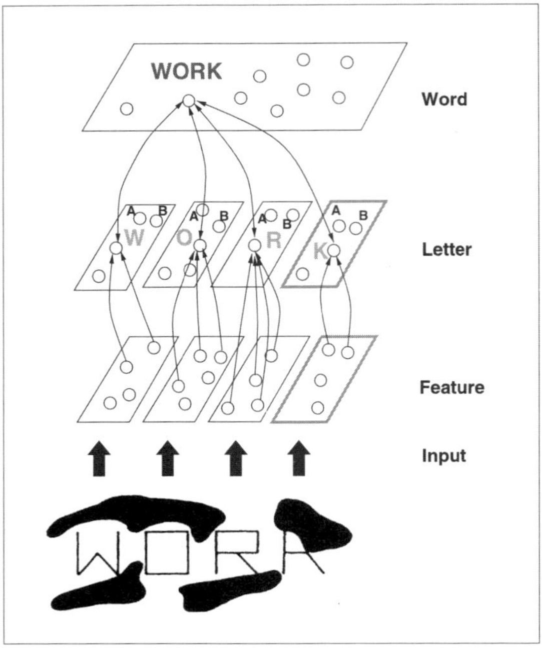

Fig. 5. Côté et al. hierarchically organized ANN model

Experiments have been made on CENPARMI database (French and English handwritten cheques), using 184 pattern for training and 2929 for testing achieve from 85.3% for word length 3 up to 100% with word length 9.

In [25], Maddouri et al. propose an extension of the Perceptro model [23]. They use a geometrical correction method to improve the performances of the Arabic handwritten word recognition system developed. The recognition proceeds in cycles of global and local observations. The global observations try to detect apparent features of the words. They create hypotheses on the word label. To carry out the recognition from different kinds of information, a normalization stage is done on the word edges to improve the local observations. Indeed, contrary to printed words or characters, the handwritten text needs a powerful normalization stage to handle the variability in position, size, rotation, slant, and distortion. The authors have chosen a Fourier based solution to eliminate this variability. The whole recognition process is summarized in Figure 7.

The top-down and bottom-up cycles are carried out thanks to the TNN model (right part of the schema), the local observation comes from the normalization of features such as ascenders, descenders, diacritics, and loops (left part of Fig. 7).

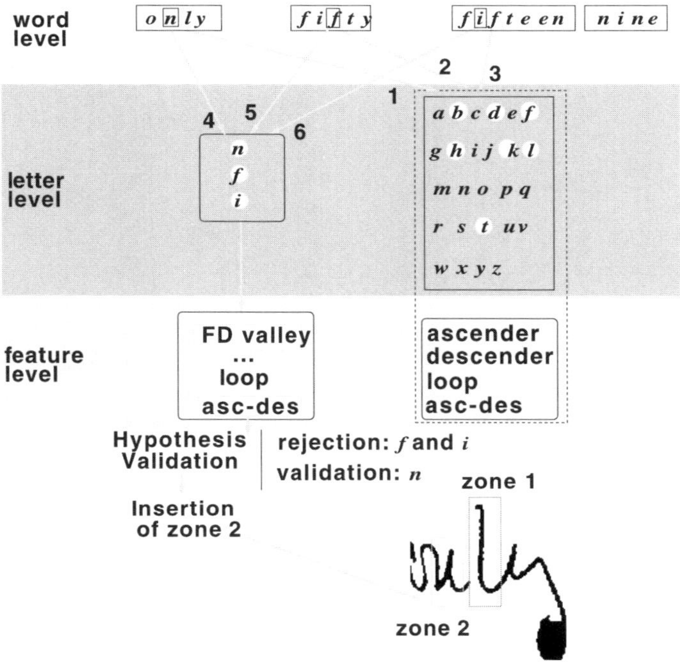

Fig. 6. Top-down process: feedback and insertion

The normalization is performed on the boundary of the word: a detection of the contour is done first, then a Freeman chain code is generated, the next step consists in computation of Fourier coefficient of the chain-encoded contour and finally, the coefficients are normalized to cope with variability. To obtain the final normalized character, a reverse Fourier transformation is applied to the latest normalized coefficient. The reader can refer to [25] to see how the boundary normalization is carried out. When the word is normalized, a metric distance is used to evaluate the difference between the current word and printed references.

3.5 Perceptive Structured Neural Network for Logical Structure Analysis

In [12], Rangoni et al. propose a quite similar TNN for logical structure recognition in document images. The hierarchically organized interpretation is kept and transposed to handle editorial documents. Each neuron corresponds to an interpretable concept and is attached to an element of the logical structure. Excluding the first layer composed of input physical features, the following

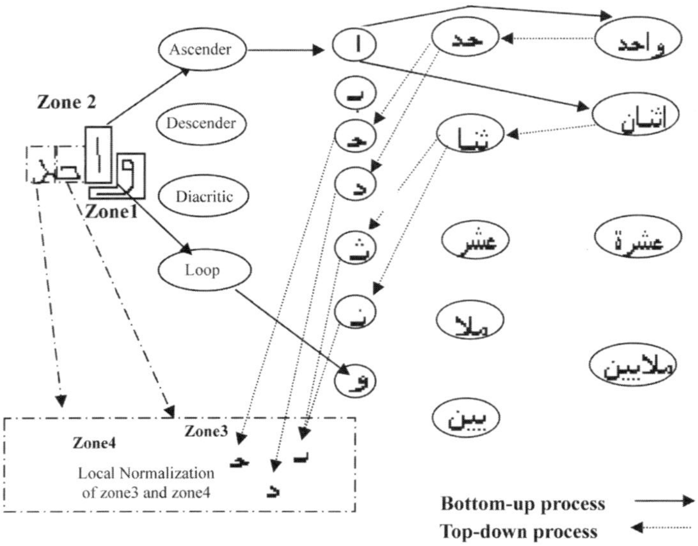

Fig. 7. Local normalization and global recognition

layers unfold the interpretation by introducing fine concepts in the first layers and general concepts in the latest layers (Fig. 8).

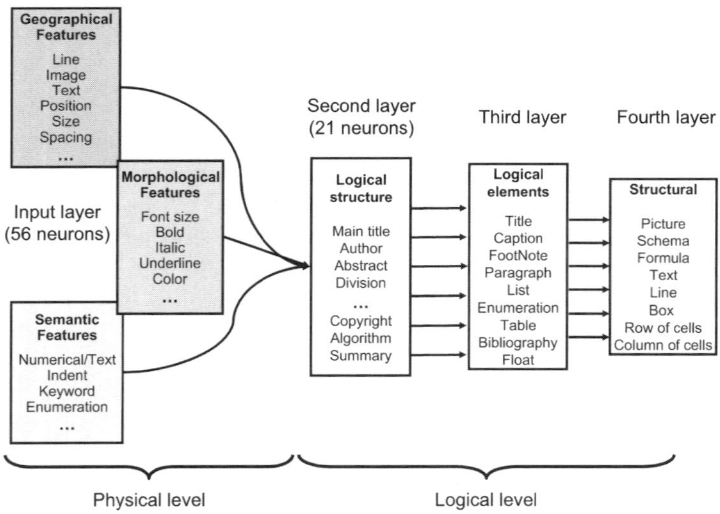

Fig. 8. Topology for scientific articles

If a DTD is present, it can be helpful to set the neurons meaning: the hierarchy included in the DTD can be unfolded to form the layers and the neurons. Contrary to other models [23, 25] the network is fully connected and the neurons can be inhibitors.

Training and recognition

As the relations between the layers are not straightforward, a training phase similar to MLP is proceeded to set all the weights.

In the back propagation algorithm, the error $E_p(w)$ between the desired output d_q and the computed output $o_{L,q}$ is minimized for each pattern p

$$\begin{aligned} E_p(w) &= \frac{1}{2}\sum_{q=1}^{N_L}(o_{L,q}(x_p) - d_q(x_p))^2 \\ o_{l,j} &= f\left(\sum_{i=0}^{N_{l-1}} w_{l,j,i}o_{l-1,i}\right) \end{aligned} \tag{11}$$

As a consequence, the weight between the unit i in layer l and unit j in layer $l+1$ is modified as follows

$$w_{l,i,j} \rightarrow w_{l,i,j} - \mu \sum_{p=1}^{P} \frac{\partial E_p(w)}{\partial o_{l,j}} f'\left(\sum_{m=0}^{N_{l-1}} w_{l,j,m}o_{l-1,m}\right) o_{l-1,i} \tag{12}$$

In case of [12], all the neurons carry interpretable concepts and the desired output is known for all the units. So, the partial term is given by:

$$\forall l, \frac{\partial E_p(w)}{\partial o_{l,j}} = o_{l,j}(x_p) - d_j(x_p) \tag{13}$$

and the network can be trained as a cascade of mono-layer perceptrons.

The model is on the one hand data-driven thanks to the training stage and on the other hand model-driven due to the integration of knowledge inside the topology. This kind of ANN is called Transparent Neural Network (TNN) in contrast to the "blackbox" aspect of MLP. For document logical layout analysis, we have named this the Perceptive Structured NN (PSNN).

The aim of the final layers is to bring context during the perceptive cycles as the previous authors used these to simulate the "word superiority effect" on letters. As the network is feed forward, the learning of the network is the same as an MLP but here the training is done separately between each consecutive pair of layers because all the desired outputs are known.

During the recognition step, the network is used as an MLP but after each propagation, the outputs are analyzed. If the output vector is close to a basis vector (14 & 15) the pattern is considered classified, otherwise the following layers are taking into account to bring context. $M(O)$ gives a vector with at least one component with high value, $\Gamma(O)$ give a vector where one component has a value very high compared to other components.

$$M(O) = \|O\|_\infty > \varepsilon \quad \text{with} \quad 0 \ll \varepsilon < 1 \tag{14}$$

$$\Gamma(O) = \frac{n((\sum O_i)^2 - \sum O_i^2)}{(n-1)(\sum O_i)^2} < \eta \quad \text{with} \quad 0 < \eta \ll 1 \tag{15}$$

As these layers contain more global information, they are more robust and accurate. They are used to generate hypothesis on the pattern. The context manages the correction of the input features. Once a label is supposed to be the good one, the input vector is modified according to this hypothesis and according to the knowledge extracted from the training database. Indeed, several representative samples are extracted from database and are matched with the current input. The input is modified to be close to a representative sample and another perceptive cycle is completed and so on until no ambiguities persist (Fig. 9).

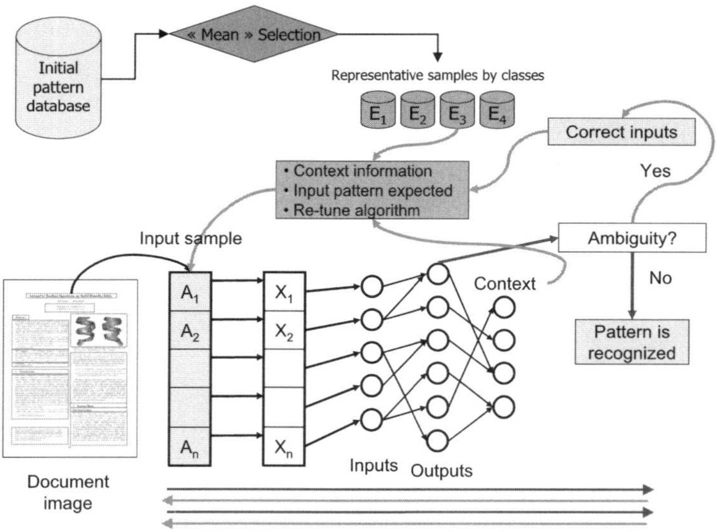

Fig. 9. Recognition in Perceptive Structured Neural Network

There are several methods to determine these representative samples; unfortunately there is no exact solution. Some approaches have been investigated. Methods using optimization produce mathematically perfect sample but they do not correspond to real-world interpretable solutions. Methods that are more straightforward can produce appropriate samples: mean sample for only one representative or a k-NN for select several samples per class. Others methods can be performed during the training stage: In [26], a new learning method is presented which can produce from a very small subset of the global database an MLP almost as efficient as trained on the whole database. The subset arisen from the algorithm will provide the representative samples.

Input feature clusterization

The perceptive cycles in this PSNN allow bottom-up and top-down resolution and refine the recognition. However, if too many recognition cycles have to be done, the task could be very time consuming because a lot of physical extraction must be completed. On top of that, some of the inputs are high-level (given by OCR) and slow down the logical structure recognition. In order to face this problem, a manual selection has been used to trim down the extractions. To simulate global and local vision, the input features are partitioned into clusters using a data categorization. Instead of feeding the network with the whole features for each cycle, the features are given progressively during the recognition and only if the pattern is too ambiguous (Fig. 10).

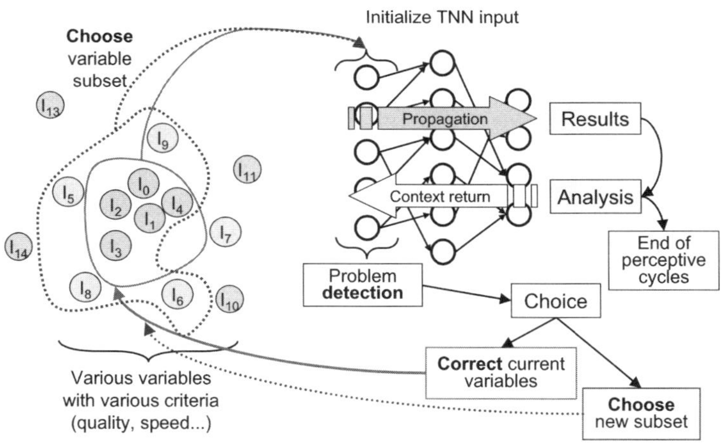

Fig. 10. Perceptive cycles: propagation, analyze, context return, correction, input feature selection

Subsets of feature are computed according to their extraction time and their predictive capacities. The first criterion is trivial as the extraction can be timed by experiments or by analyzing the algorithm complexity. Evaluating the predictive power of a set of features is more complicated as there is no optimal solution to do this. The literature proposes two main approaches: filter-based methods and wrapper methods [27]. The filter methods only use the sample database to score the feature, they are fast to evaluate the feature separately but do not produce good groups. On the other side, wrapper methods consider variables but they need the classifier to produce the groups. The method presented in [12] is based on a filter approach but can compute groups at the same time with ordered predictive power and less redundancy inside each group (Fig. 11).

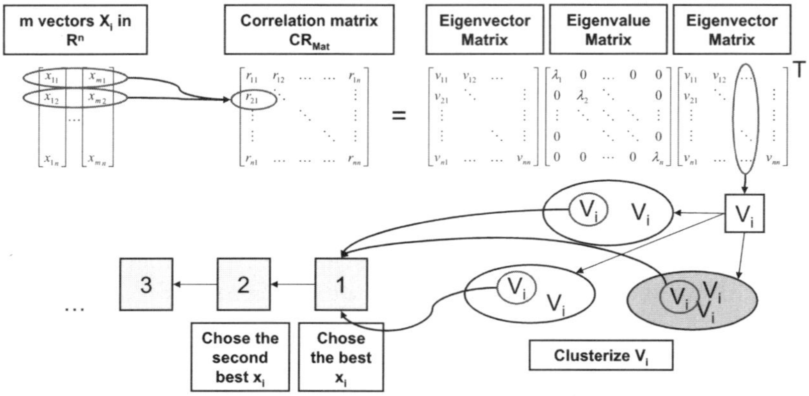

Fig. 11. Input feature clustering

By combining the input feature correction and selection, the PSNN is able to adapt the computation amount according to pattern complexity without adding too much processing time.

Experimentations

The system has been tested on scientific articles (Fig. 13). Physical inputs are mainly extracted by commercial OCR, others are computed by using existing ones. There are also 21 logical lables that cover any document image (Fig. 12).

After four perceptive cycles, the recognition rate increase to 91.7% which is 10 points better than a classical MLP (Tab. 1).

Logical	Physical		
	Geometrical	Morphological	Semantic
Title	Text	Bold	IsNumeric
Author	Image	Italic	KeyWords
Email	Table	Underlined	%KnownWords
Locality	Other	Strikethrough	%Punctuation
Abstract	x position	UpperCase	Bullet
Key words	y position	Small Capitals	Enum
CR Categories	Width	Subscript	Language
Introduction	Height	Superscript	Baseline
Paragraph	NumPage	Font Name	
Section	UpSpace	Font Size	
SubSection	BottomSpace	Scaling	
SubSubSection	LeftSpace	Spacing	
List	RightSpace	Alignment	
Enumeration		LeftIndent	
Float		RightIndent	
Conclusion		FirstIndent	
Bibliography		NumLines	
Algorithms		Boxed	
Copyright		Red/Green/Blue	
Acknowledgments			
Page number			

Fig. 12. Input feature clustering

Classes	MLP	PSNN			
		C_1	C_2	C_3	C_4
Whole	81.6%	45.2	78.9	90.2	91.7%
Best	86.9%	66.7	85.3	85.3	99.3%
Worst	0.0%	0.0	0.0	4.0	28.6%
Time	1	0.7	1.45	1.85	2.40

Table 1. Logical structure classification for MLP and for PSNN

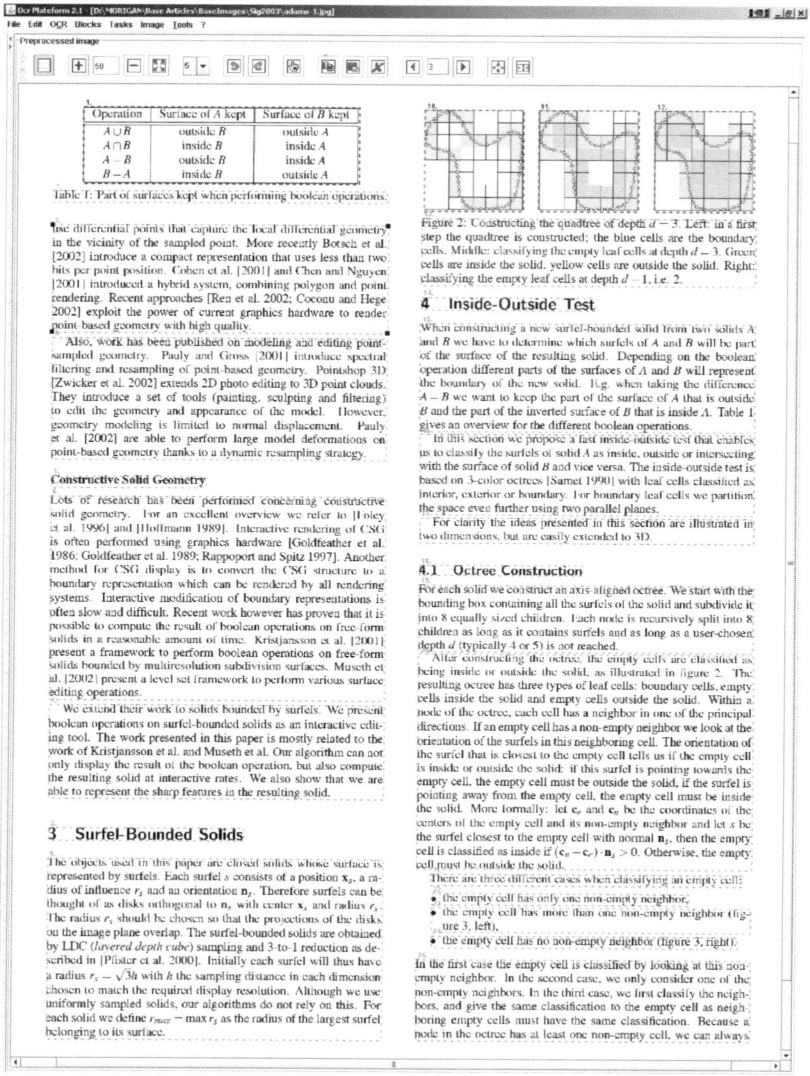

Operation	Surface of A kept	Surface of B kept
$A \cup B$	outside B	outside A
$A \cap B$	inside B	inside A
$A - B$	outside B	inside A
$B - A$	inside B	outside A

Table 1: Part of surfaces kept when performing boolean operations.

use differential points that capture the local differential geometry in the vicinity of the sampled point. More recently Botsch et al. [2002] introduce a compact representation that uses less than two bits per point position. Cohen et al. [2001] and Chen and Nguyen [2001] introduced a hybrid system, combining polygon and point rendering. Recent approaches [Ren et al. 2002; Coconu and Hege 2002] exploit the power of current graphics hardware to render point-based geometry with high quality.

Also, work has been published on modeling and editing point-sampled geometry. Pauly and Gross [2001] introduce spectral filtering and resampling of point-based geometry. Pointshop 3D [Zwicker et al. 2002] extends 2D photo editing to 3D point clouds. They introduce a set of tools (painting, sculpting and filtering) to edit the geometry and appearance of the model. However, geometry modeling is limited to normal displacement. Pauly et al. [2002] are able to perform large model deformations on point-based geometry thanks to a dynamic resampling strategy.

Constructive Solid Geometry

Lots of research has been performed concerning constructive solid geometry. For an excellent overview we refer to [Foley et al. 1996] and [Hoffmann 1989]. Interactive rendering of CSG is often performed using graphics hardware [Goldfeather et al. 1986; Goldfeather et al. 1989; Rappoport and Spitz 1997]. Another method for CSG display is to convert the CSG structure to a boundary representation which can be rendered by all rendering systems. Interactive modification of boundary representations is often slow and difficult. Recent work however has proven that it is possible to compute the result of boolean operations on free-form solids in a reasonable amount of time. Kristjansson et al. [2001] present a framework to perform boolean operations on free-form solids bounded by multiresolution subdivision surfaces. Museth et al. [2002] present a level set framework to perform various surface editing operations.

We extend their work to solids bounded by surfels. We present boolean operations on surfel-bounded solids as an interactive editing tool. The work presented in this paper is mostly related to the work of Kristjansson et al. and Museth et al. Our algorithm can not only display the result of the boolean operation, but also compute the resulting solid at interactive rates. We also show that we are able to represent the sharp features in the resulting solid.

3 Surfel-Bounded Solids

The objects used in this paper are closed solids whose surface is represented by surfels. Each surfel s consists of a position $\mathbf{x}_s$, a radius of influence r_s and an orientation $\mathbf{n}_s$. Therefore surfels can be thought of as disks orthogonal to $\mathbf{n}_s$ with center $\mathbf{x}_s$ and radius r_s. The radius r_s should be chosen so that the projections of the disks on the image plane overlap. The surfel-bounded solids are obtained by LDC (*layered depth cube*) sampling and 3-to-1 reduction as described in [Pfister et al. 2000]. Initially each surfel will thus have a radius $r_s = \sqrt{3}h$ with h the sampling distance in each dimension chosen to match the required display resolution. Although we use uniformly sampled solids, our algorithms do not rely on this. For each solid we define $r_{max} = \max r_s$ as the radius of the largest surfel belonging to its surface.

Figure 2: Constructing the quadtree of depth $d = 3$. Left: in a first step the quadtree is constructed; the blue cells are the boundary cells. Middle: classifying the empty leaf cells at depth $d = 3$. Green cells are inside the solid, yellow cells are outside the solid. Right: classifying the empty leaf cells at depth $d - 1$, i.e. 2.

4 Inside-Outside Test

When constructing a new surfel-bounded solid from two solids A and B we have to determine which surfels of A and B will be part of the surface of the resulting solid. Depending on the boolean operation different parts of the surfaces of A and B will represent the boundary of the new solid. E.g. when taking the difference $A - B$ we want to keep the part of the surface of A that is outside B and the part of the inverted surface of B that is inside A. Table 1 gives an overview for the different boolean operations.

In this section we propose a fast inside-outside test that enables us to classify the surfels of solid A as inside, outside or intersecting with the surface of solid B and vice versa. The inside-outside test is based on 3-color octrees [Samet 1990] with leaf cells classified as interior, exterior or boundary. For boundary leaf cells we partition the space even further using two parallel planes.

For clarity the ideas presented in this section are illustrated in two dimensions, but are easily extended to 3D.

4.1 Octree Construction

For each solid we construct an axis-aligned octree. We start with the bounding box containing all the surfels of the solid and subdivide it into 8 equally sized children. Each node is recursively split into 8 children as long as it contains surfels and as long as a user-chosen depth d (typically 4 or 5) is not reached.

After constructing the octree, the empty cells are classified as being inside or outside the solid, as illustrated in figure 2. The resulting octree has three types of leaf cells: boundary cells, empty cells inside the solid and empty cells outside the solid. Within a node of the octree, each cell has a neighbor in one of the principal directions. If an empty cell has a non-empty neighbor we look at the orientation of the surfels in this neighboring cell. The orientation of the surfel that is closest to the empty cell tells us if the empty cell is inside or outside the solid: if this surfel is pointing towards the empty cell, the empty cell must be outside the solid, if the surfel is pointing away from the empty cell, the empty cell must be inside the solid. More formally: let $\mathbf{c}_e$ and $\mathbf{c}_n$ be the coordinates of the centers of the empty cell and its non-empty neighbor and let s be the surfel closest to the empty cell with normal $\mathbf{n}_s$, then the empty cell is classified as inside if $(\mathbf{c}_n - \mathbf{c}_e) \cdot \mathbf{n}_s > 0$. Otherwise, the empty cell must be outside the solid.

There are three different cases when classifying an empty cell:

- the empty cell has only one non-empty neighbor,
- the empty cell has more than one non-empty neighbor (figure 3, left),
- the empty cell has no non-empty neighbor (figure 3, right).

In the first case the empty cell is classified by looking at this non-empty neighbor. In the second case, we only consider one of the non-empty neighbors. In the third case, we first classify the neighbors, and give the same classification to the empty cell as neighboring empty cells must have the same classification. Because a node in the octree has at least one non-empty cell, we can always

Fig. 13. Document sample

4 Conclusion and Perspectives

We outlined in this chapter several categories of approaches used for the extraction of document structures from raster images. After a description of their properties, advantages and drawbacks, we focused on neural approaches for their capacities in noise absorption and generalization capacities. Their application in document analysis was a real challenge for us as they were not considered for this kind of structured data. The idea of McClelland to propose a perceptive model with different cycles allowing a dynamic and progressive approximation of the problem was the basis of our investigations. After the study of others dynamic models, we proposed a specific one called Perceptive Structured Neural Network which can be applied for logical structure recognition. This model allows us to process several categories of structures based only on physical data. After few cycles, the behavior of the system is better than an MLP's. Besides, it gives us the possibility to refine the outputs by correcting the inputs accordingly.

Although dynamic ANN are able to deal with structured patterns, they are not still used for document logical layout analysis. Besides, static networks have been used far less than pure model-driven approaches. All the works presented in the section 3 show how to extend classical models to deal with such a problem. The neuronal approach is accessible and can be as competitive as grammar or rule based systems. It is obvious that, as mentioned in Nagy et al. [28], domain specific knowledge appears essential for document interpretation.

The proposed PSNN can be improved in a different way: the data-driven methods may be improved by introducing hidden layers between each layer of interpretable concepts. The "transparency" property will be lost but the system will be more accurate and have better generalization capacities.

Another approach could integrate transparency in a dynamic network or adding dynamic properties to PSNN. A simply output feedback-based PSNN will have more feedback information when using the context. On top of that, the context will be taking into account not only during the recognition but also during the training stage.

References

1. Marinai, S., Gori, M., Soda, G.: Artificial neural networks for document analysis and recognition. Pattern Analysis And Machine Intelligence **27**(1) (2005) 23–35
2. Chi, Z., Wong, K.: A two-stage binarization approach for document images. International Symposium on Intelligent Multimadia, Video and Speech Processing (2001) 275–278
3. Hamza, H., Smigiel, E., Belaïd, A.: Neural based binarization techniques. International Conference on Document Analysis and Recognition **4**(8) (2005) 317–321
4. Whichello, A.P., Yan, H.: Linking broken character borders with variable sized mask to improve recognition. Pattern Recognition **29**(8) (1995) 1429–1435

5. Le Cun, Y., Bottou, L., Bengio, Y., Haffner, P.: Gradient-based learning applied to document recognition. Proceedings of the IEEE **86**(11) (1998) 2278–2324
6. Cecotti, H., Belaïd, A.: Rejection strategy for convolutional neural network by adaptive topology applied to handwritten digits recognition. International Conference on Document Analysis and Recognition (8) (2005) 765–769
7. Garris, M.D., Wilson, C.L., Blue, J.L.: Neural network-based systems for handprint ocr applications. IEEE Transactions on Image Processing **7**(8) (1998) 1097–1112
8. Mao, S., Rosenfeld, A., Kanungo, T.: Document structure analysis algorithms: A literature survey. SPIE Electronic Imaging **50**(10) (2003) 197–207
9. Brugger, R., Zramdini, A., Ingold, R.: Modeling documents for structure recognition using generalized n-grams. International Conference on Document Analysis and Recognition **1**(4) (1997) 56–60
10. Hu, T., Ingold, R.: A mixed approach toward an efficient logical structure recognition from document images. Electronic Publishing: Origination, Dissemination, and Design **6**(4) (1993) 457–468
11. Niyogi, D., Srihari, S.N.: Knowledge-based derivation of document logical structure. Third International Conference on Document Analysis and Recognition **1** (1995) 472–475
12. Rangoni, Y., Belaïd, A.: Document logical structure analysis based on perceptive cycles. Document Analysis Systems **1**(7) (2006) 117–128
13. Küchler, A., Goller, C.: Inductive learning in symbolic domains using structure-driven recurrent neural networks. Lecture Notes in Computer Science (1137) (1996) 183–197
14. Sperduti, A., Starita, A.: Supervised neural networks for the classification of structures. IEEE Transactions on Neural Networks **8**(3) (1997) 714–735
15. Hertz, J., Krogh, A., Palmer, R.G.: Introduction to the theory of neural computation. Addison Wesley (1991)
16. Moody, J., Darken, C.J.: Fast learning in networks of locally-tuned processing units. Neural Computation (1) (1989) 281–294
17. Narendra, K.S., Parthasarathy, K.: Identification and control of dynamical systems using neural networks. IEEE Transactions on Neural Networks **1**(1) (1990) 4–27
18. Hopfield, J.J.: Neural networks and physical systems with emergent collective computational abilities. Proceedings of the National Academy of Sciences of the United States of America **79**(8) (1982) 2554–2558
19. Pineda, F.J.: Dynamics and architecture for neural computation. Journal of Complexity **4**(3) (1988) 216–245
20. Williams, R.J., Zipser, D.: A learning algorithm for continually running fully recurrent neural networks. Neural Computation **1**(2) (1989) 270–280
21. Sperduti, A., Starita, A., Goller, C.: Learning distributed representations for the classification of terms. Proceedings of International Joint Conference on Artificial Intelligence **1**(40) (1995) 509–515
22. Fahlman, S.E., Lebiere, C.: The cascade-correlation learning architecture. Advances in Neural Information Processing Systems **2** (1990) 524–532
23. Côté, M., Lecolinet, E., Cheriet, M., Suen, C.: Automatic reading of cursive scripts using a reading model and perceptual concepts. the Perceptro system. International Journal on Document Analysis and Recognition **1**(1) (1998) 3–17
24. McClelland, J., Rumelhart, D.: An interactive activation model of context effects in letter perception. Psychological Review (88) (1981) 375–407

25. Maddouri, S.S., Amiri, H., Belaïd, A.: Local normalization towards global recognition of arabic handwritten script. Document Analysis and Systems (2000)
26. Vajda, S., Rangoni, Y., Cecotti, H., Belaïd, A.: A fast learning strategy using data selection for feedforward neural networks. International Workshop on Frontiers in Handwriting Recognition (10) (2006)
27. Guyon, I., Elisseeff, A.: An introduction to variable and feature selection. Journal of Machine Learning Research (3) (2003) 1157–1182
28. Nagy, G.: Twenty years of document image analysis in PAMI. Pattern Analysis and Machine Intelligenc **22**(1) (2000) 38–62

Machine Learning for Reading Order Detection in Document Image Understanding

Donato Malerba, Michelangelo Ceci, and Margherita Berardi

Dipartimento di Informatica, Università degli Studi di Bari
via Orabona, 4 - 70126 Bari - Italy
{malerba,ceci,berardi}@di.uniba.it

Summary. Document image understanding refers to logical and semantic analysis of document images in order to extract information understandable to humans and codify it into machine-readable form. Most of the studies on document image understanding have targeted the specific problem of associating layout components with logical labels, while less attention has been paid to the problem of extracting relationships between logical components, such as cross-references. In this chapter, we investigate the problem of detecting the reading order relationship between components of a logical structure. The domain specific knowledge required for this task is automatically acquired from a set of training examples by applying a machine learning method. The input of the learning method is the description of "chains" of layout components defined by the user. The output is a logical theory which defines two predicates, *first_to_read*/1 and *succ_in_reading*/2, useful for consistently reconstructing all chains in the training set. Only spatial information on the page layout is exploited for both single and multiple chain reconstruction. The proposed approach has been evaluated on a set of document images processed by the system WISDOM++.

1 Introduction

Documents are characterized by two important structures: the layout structure and the logical structure. Both are the results of repeatedly dividing the content of a document into increasingly smaller parts, and are typically represented by means of a tree structure. The difference between them is the criteria adopted for structuring the document content: the layout structure is based on the *presentation* of the content, while the logical structure is based on the *human-perceptible meaning* of the content.

The extraction of the layout structures from images of scanned paper documents is a complex process, typically denoted as *document layout analysis*, which involves several steps including preprocessing, page decomposition (or segmentation), classification of segments according to content type (e.g., text, graphics, pictures) and hierarchical organization on the basis of perceptual

D. Malerba et al.: *Machine Learning for Reading Order Detection in Document Image Understanding*, Studies in Computational Intelligence (SCI) **90**, 45–69 (2008)
www.springerlink.com

criteria. *Document image understanding* refers to the process of extracting the logical structure of a document image. This task is strongly application dependent, since the same definition of the logical structure depends on the type of information the user is interested in retrieving in a document.

Most works on document image understanding aim at associating a "logical label" with some components of the layout structure: this corresponds to mapping (part of) the layout structure into the logical structure. Generally, this mapping is based on spatial properties of the layout components, such as absolute positioning with respect to a system of coordinates, relative positioning (e.g., on top, to right), geometrical properties (e.g., height and width), as well as information on the content type (e.g., text, graphics, and picture). Some studies have also advocated the use of textual information of some layout components to base, or at least to refine, the classification of layout components into a set of logical labels.

The main problem for all these approaches remains the large amount of domain specific knowledge required to effectively perform this task. Hand-coding the necessary knowledge according to some formalism, such as block grammars [1], geometric trees [2], and frames [3] is time-consuming and limits the application of a document image understanding system to a set of predefined classes of documents. To alleviate the burden in developing and customizing document image understanding systems, several data mining and machine learning approaches have been proposed with the aim of automatically extracting the required knowledge [4].

In its broader sense, document image understanding cannot be considered synonymous of "logical labeling", since relationships among logical components are also possible and their extraction can be equally important for an application domain. Some examples of relations are the cross reference of a caption to a figure, as well as the cross reference of an affiliation to an author. An important class of relations investigated in this chapter is represented by the *reading order* of some parts of the document. More specifically, we are interested in determining the reading order of most abstract layout components on each page of a multi-page document. Indeed, the spatial order in which the information appears in a paper document may have more to do with optimizing the print process than with reflecting the logical order of the information contained.

Determining the correct reading order can be a crucial problem for several applications. By following the reading order recognized in a document image, it is possible to cluster together text regions labelled with the same logical label into the same textual component (e.g., "introduction", "results", "method" of a scientific paper). Once a single textual component is reconstructed, advanced techniques for text processing can be subsequently applied. For instance, information extraction methods may be applied locally to reconstructed textual components of documents (e.g., sample of the experimental setting studied in the "results" section). Moreover, retrieval of document images on the basis of their textual contents is more effectively realized.

Several papers on reading order detection have already been published in the literature. Their brief description is provided in the next Section. Some are based only on the spatial properties of the layout components, while others also exploit the textual content of parts of documents. Moreover, some methods have been devised for properly ordering layout components (independent of their logical meaning), while others consider the recognition of some logical components, such as "title" and "body", as preliminary to reading order detection. A common aspect of all methods is that they strongly depend on the specific domain and are not "reusable" when the classes of documents or the task at hand change.

As for logical labelling, domain specific knowledge required to effectively determine the reading order can be automatically acquired by means of machine learning methods. In this study we investigate the problem of inducing rules which are used for predicting the proper reading order of layout components detected in document images. The rules are learned from training examples which are sets of ordered layout components described by means of both their spatial properties and their possible logical label. Therefore, no textual information is exploited to understand document images. The ordering of the layout components is defined by the user and does not necessarily reflect the traditional Western-style document encoding rule according to which reading proceeds top-bottom and left-right. For instance, the user can specify a reading order according to which the affiliation of an author immediately follows the author's name, although the two logical components are spatially positioned quite far away on the page layout (e.g., the affiliation is reported at the bottom of the first column of the paper). In multi-page articles, such as those considered in this chapter, ordering is defined at the page level. More precisely, different "chains" of layout components can be defined by the user, when independent pieces of information are represented on the same page (e.g., the end of an article and the beginning of a new one). Chains are mutually exclusive, but not necessarily exhaustive, sets of most abstract layout components in a page, so that their union defines a partial (and not necessarily a total) order on the set of layout objects.

This chapter is organized as follows. In the next section, the background and some related works are reported, while the reading order problem is formally defined in Section 3. The machine learning system applied to the problem of learning from ordered layout components is introduced in Section 4. The representation of training examples as well as the manner in which learned rules are applied to a new document are also illustrated. Some experimental results on a set of multi-page printed documents are reported and commented on in Section 5. Finally, Section 6 concludes and discusses ideas for further studies.

2 Background and Related Works

In the literature there are already several publications on reading order detection. A pioneer work is reported in [5], where multi-column and multi-article documents (e.g., magazine pages) with figures and photographs are handled. Each document page is described as a tree, where each node, except the root, represents a set of adjacent blocks located in the same column, ordered so that the block on the upper location precedes the others. Direct descendants of an internal node are also ordered in sequence according to their locations in the same way that the block to the left and on the top precedes the others. Reading order detection follows a preliminary rough classification of layout components into "title" and "body". Heads are blocks in which there are only a few text lines with large type fonts, while bodies correspond to blocks with several text lines with small type fonts. The reading order is extracted by applying some hand-coded rules which allow the transformation of trees representing layout structures (with associated 'title" and "body" labels) into ordered structures. Once the correct reading order is detected, a further interpretation step is performed to attach some logical labels (e.g., title, abstract, sub-title, paragraph) to each item of the ordered structure.

A similar tree-structured representation of the page layout is adopted in the work by Ishitani [6]. The structure is derived by a recursive XY-cut approach [7], that is, a recursive horizontal/vertical partitioning of the input image. The XY-cut process naturally determines the reading order of the layout components, since for horizontal cuts the top-bottom ordering is applied to the derived sections, while for vertical cuts the right-left (i.e., Japanese style) ordering is applied to the derived columns.

The main problem with this XY-cut approach is that at each recursion step, there are often multiple possible, and possibly conflicting, cuts. In the original algorithm, the widest cut is selected at each recursion. While this strategy works reasonably well for a page segmentation task, it is not always appropriate for a reading order detection task. For this reason, Ishitani proposed a bottom-up approach using three heuristics which take into account local geometric features, text orientation and distance among vertically adjacent layout objects in order to merge some layout objects before performing the XY-cut. As observed by Meunier [8], this aims at reducing the probability of having to face multiple cutting alternatives, but it does not truly prevent them from occurring. For this reason, he proposed to reformulate the problem of recursively cutting a page as an optimization problem, and defined both a scoring function for alternative cuts, and a computationally tractable method for choosing the best partitioning.

A common aspect of all these approaches is that they are based exclusively on the spatial information conveyed by a page layout. On the contrary, Taylor et al. [9], propose the use of linguistic information to define the proper reading order. For instance, to determine whether an article published in a magazine continues on the next page, it is suggested to look for a text, such as 'continued on next page'.

The usage of linguistic information has also been proposed by Aiello et al. [10], who described a document analysis system for logical labelling and reading order extraction of broad classes of documents. Each document object is described by means of both attributes (i.e., aspect ratio, area ratio, font size ratio, font style, content size, number of lines) and spatial relations (defined as extensions of Allen's interval relations [11]). Only objects labelled with some logical labels (title and body) are considered for reading order. More precisely, two distinct reading orders are first detected for the document object types *Title* and *Body*, and then they are combined using a Title-Body connection rule. This rule connects one *Title* with the left-most top-most *Body* object, situated below the *Title*. Each reading order is determined in two steps. Initially, spatial information on the document objects is exploited by a spatial reasoner which solves a constraint-satisfaction problem, where constraints correspond to general document encoding rules (e.g., "in the Western-culture, documents are usually read top-bottom and left-right"). The output of the spatial reasoner is a (cyclic) graph where edges represent instances of the partial ordering relation *BeforeInReading*. A reading order is then defined as a full path in this graph, and is determined by means of an extension of a standard topological sort [12]. Due to the generality of the document encoding rule used by the spatial reasoner, it is likely that one obtains more than one reading order, especially for complex documents with many blocks. For this reason, a natural language processor is used in the second step of the proposed method. The goal is that of disambiguating between different reading orders on the basis of textual information of logical objects. This step works by computing probabilities of sequences of words obtained by joining document objects which are candidates to be followed in reading. The best aspect of this work is the generality of the approach due to the generality of the knowledge adopted in reasoning.

Topological sorting is also exploited in the approach proposed by Breuel [13]. In particular, reading order is defined the basis of text lines segments, which are pairwise compared on the basis of four simple rules in order to determine a partial order. Then a topological sorting algorithm is applied to find at least one global order consistent with this partial order. Columns, paragraphs, and other layout features are determined on the basis of the spatial arrangement of text line segments in reading order. For instance, paragraph boundaries are indicated by relative indentation of consecutive text lines in reading order.

All approaches reported above reflect a clear domain specificity. For instance, the classification of blocks as "title" and "body" is appropriate for magazine articles, but not for administrative documents. Moreover, the document encoding rules appropriate for Western-style documents are different for Japanese papers. Surprisingly, there is no work, to the best of our knowledge, that handles the reading order problem by resorting to machine learning techniques, which can generate the required knowledge from a set of training layout structures whose correct reading order has been provided by the user. In previous works on document image analysis and understanding, we investigated the application of machine learning techniques to several knowledge-based document image processing tasks, such as classification of blocks according to their content type [14], automatic global layout analysis correction [15], classification of documents into a set of pre-defined classes [16], and logical labelling [17]. Experimental results always proved the feasibility of this approach, at least on a small scale, that is, for a few hundred of training document images. Therefore, following this mainstream of research, herein we consider the problem of learning the definition of reading order.

The proposed solution has been tested by processing documents with WISDOM++[1], a knowledge-based document image processing system originally developed to transform multi-page printed documents into XML format. WISDOM++ makes extensive use of knowledge and XML technologies for semantic indexing of paper documents. This is a complex process involving several steps:

1. The image is segmented into basic layout components (basic blocks), which are classified according to the type of content (e.g., text, pictures and graphics).
2. A perceptual organization phase (layout analysis) is performed to detect a tree-like layout structure, which associates the content of a document with a hierarchy of layout components.
3. The first page is classified to identify the membership class (or type) of the multi-page document (e.g. scientific paper or magazine).
4. The layout structure of each page is mapped into the logical structure, which associates the content with a hierarchy of logical components (e.g. title or abstact of a scientific paper).
5. OCR is applied only to those logical components of interest for the application domain (e.g., title).
6. The XML file that represents the layout structure, the logical structure, and the textual content returned by the OCR for some specific logical components is generated.
7. XML documents are stored in a repository for future retrieval purposes.

Four of seven processing steps make use of explicit knowledge expressed in the form of decision trees and rules which are automatically learned by means of two distinct machine learning systems: ITI [18], which returns decision trees useful for block classification (first step), and ATRE [19], which returns rules for layout analysis correction (second step) [15], document image classification (third step) and document image understanding (fourth step) [4]. As explained in Section 4, ATRE is also used to learn the intensional definition of two

[1] http://www.di.uniba.it/~malerba/wisdom++/

predicates, which contribute to determine the reading order chains in a page layout.

3 Problem Definition

In order to formalize the problem we intend to solve, some useful definitions are necessary:

Definition 1. *Partial Order [20]*
Let A be a set of blocks in a document page, a partial order P over A is a relation $P \in A \times A$ such that P is

1. *reflexive* $\forall s \in A \Rightarrow (s, s) \in P$
2. *antisymmetric* $\forall s_1, s_2 \in A$: $(s_1, s_2) \in P \wedge (s_2, s_1) \in P \Leftrightarrow s_1 = s_2$
3. *transitive* $\forall s_1, s_2, s_3 \in A$: $(s_1, s_2) \in P \wedge (s_2, s_3) \in P \Rightarrow (s_1, s_3) \in P$

Definition 2. *Weak Partial Order*
Let A be a set of blocks in a document page, a weak partial order P over A is a relation $P \in A \times A$ such that P is

1. *irreflexive* $\forall s \in A \Rightarrow (s, s) \notin P$
2. *antisymmetric* $\forall s_1, s_2 \in A$: $(s_1, s_2) \in P \wedge (s_2, s_1) \in P \Leftrightarrow s_1 = s_2$
3. *transitive* $\forall s_1, s_2, s_3 \in A$: $(s_1, s_2) \in P \wedge (s_2, s_3) \in P \Rightarrow (s_1, s_3) \in P$

Definition 3. *Total Order*
Let A be a set of blocks in a document page, a partial order T over the set A is a total order iff $\forall s_1, s_2 \in A$: $(s_1, s_2) \in T \vee (s_2, s_1) \in T$

Definition 4. *Complete chain*
Let:

- *A be a set of blocks in a document page,*
- *D be a weak partial order over A*
- *$B = \{a \in A | (\exists b \in A \text{ s.t. } (a, b) \in D \vee (b, a) \in D)\}$ be the subset of elements in A related to any element in A itself.*

If $D \cup \{(a, a) | a \in B\}$ is a total order over B, then D is a complete chain over A

Definition 5. *Chain reduction*
Let D be a complete chain over A
the relation
$C = \{(a, b) \in D | \neg\exists c \in A \text{ s.t. } (a, c) \in D \wedge (c, b) \in D\}$
is the reduction of the chain D over A.

Example 1. Let $A = \{a, b, c, d, e\}$. If $D = \{(a, b), (a, c), (a, d), (b, c), (b, d), (c, d)\}$ is a complete chain over A, then $C = \{(a, b), (b, c), (c, d)\}$ is its reduction (see Figure 1).

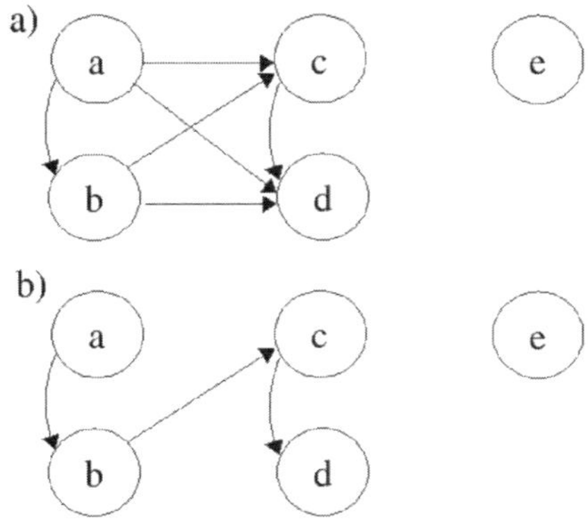

Fig. 1. A complete chain (a) and its reduction (b)

Indeed, for our purposes it is equivalent to deal with complete chains or their reduction. Henceforth, for the sake of simplicity, the term *chain* will denote the reduction of a complete chain.

By resorting to the definitions above, it is possible to formalize the reading order induction problem as follows:

Given :

- A description $DesTP_i$ in the language L of the set of n training pages $TrainingPages = \{TP_i \in \Pi | i = 1..n\}$ (where Π is the set of pages).
- A description $DesTC_i$ in the language L of the set TC_i of chains (over $TP_i \in TrainingPages$) for each $TP_i \in TrainingPages$.

Find :

An intensional definition T in the language L of a chain over a generic page $P \in \Pi$ such that T is complete and consistent with respect to all training chains descriptions $DesTC_i$, $i = 1..n$.

In this problem definition, we refer to the intensional definition T as a first order logic theory. The fact that T is complete and consistent with respect to all training chains descriptions can be formally described as follows:

Definition 6 (Completeness and Consistency).
Let:

- *T be a logic theory describing chains instances expressed in the language L,*
- *E^+ be the set of positive examples for the chains instances ($E^+ = \bigcup_{i=1..n} \bigcup_{TC \in TC_i} TC$),*
- *E^- be the set of negative examples for the chains instances ($E^- = \bigcup_{i=1..n} (TP_i \times TP_i)/E^+$),*

- *$DesE^+$ be the description of E^+ in L,*
- *$DesE^-$ be the description of E^- in L,*

then T is complete and consistent with respect to all training chains descriptions iff $T \models DesE^+ \ \wedge \ T \not\models DesE^-$.

This formalization of the problem permits us to represent and identify distinct reading orders on the same page and avoids including blocks that should not be included in the reading order (e.g. figures or page numbers).

4 Learning Reading Order

In order to learn first order logic theories for reading order identification, in this work we use the learning system ATRE[2] [19]. Indeed, ATRE is particularly suited for the task at hand since the spatial dimension of page layout makes methods developed in inductive logic programming (ILP) [21, 22, 23, 24, 25] the most suitable approaches. This conviction comes from the fact that most of the classical learning systems assume that training data are represented in a single table of a relational database, such that each row (or tuple) represents an independent example (a layout component) and columns correspond to properties of the example (e.g., height of the layout component). This single-table assumption, however, is quite naive for at least three reasons. First, layout components cannot be realistically considered independent observations, because their spatial arrangement is mutually constrained by formatting rules typically used in document editing. Second, spatial relationships between a layout component and a variable number of other components in its neighborhood cannot be properly represented by a fixed number of attributes in a table. Third, different layout components may have different properties (e.g., the property "brightness" is appropriate for half-tone images, but not for textual components), so that properties of the components in the neighborhood cannot be effectively represented by the same set of attributes. Since the single-table assumption limits the representation of relationships (spatial or non) between examples, it also prevents the discovery of this kind of pattern which could be very useful in reading order identification.

To automate the reconstruction of the reading order, WISDOM++ has been opportunely extended in order to:

- Allow the user to define correct reading order chains on training documents through the visual interaction with the system.
- Represent training reading order chains in first order logic formalism which the learning system is able to interpret.
- Run the learning system.
- Apply the learned theories on new (testing) document images.

---
[2] http://www.di.uniba.it/~malerba/software/atre

In the following, we briefly present the learning system ATRE. Subsequently, we present the document descriptions provided by WISDOM++ to the learning system. Lastly, we explain how knowledge acquired by the learning system is exploited for order reconstruction.

4.1 ATRE: The Learning System

ATRE is an ILP system which can learn recursive theories from examples. The learning problem solved by ATRE can be formulated as follows:

Given

- a set of *concepts* C_1, C_2, ..., C_r to be learned
- a set of *observations* O described in a language L_O
- a *background theory* BK
- a *language* of hypotheses L_H
- a user's *preference criterion* PC

Find

A logical theory T expressed in the language L_H and defining the concepts C_1, C_2, ..., C_r, such that T is complete and consistent with respect to O and satisfies the preference criterion PC.

The *completeness* property holds when the theory T explains all observations in O of the r concepts C_1, C_2, ..., C_r, while the *consistency* property holds when the theory T explains no counter-example in O of any concept C_i. The satisfaction of these properties guarantees the correctness of the induced theory with respect to the given observations O. Whether the theory T is correct with respect to additional observations not in O is an extra-logical matter, since no information on the generalization accuracy can be drawn from the training data themselves. In fact, the selection of the “best” theory is always made on the grounds of an inductive bias embedded in some heuristic function or expressed by the user of the learning system (preference criterion).

In the context of the reading order learning, we identified two concepts to be learned, namely $first\_to\_read/1$ and $succ\_in\_reading/2$. The former refers to the first layout component of a chain, while the latter refers to the relation *successor* between two components in a chain. By combining the two concepts it is possible to identify a partial ordering of blocks in a document page.

As to the representation languages, the basic component is the *literal*, which can be of the two distinct forms:

$f(t_1, \ldots, t_n) =$ Value (simple literal)

$f(t_1, \ldots, t_n) \in$ Range (set literal),

where f and g are function symbols called *descriptors*, t_i's are *terms* (constants or variables) and *Range* is a closed interval of possible values taken by f. In the next section descriptors used for the representation of training examples and hypotheses are presented.

4.2 Document Description

In ATRE, training observations are represented by ground multiple-head clauses [26], called *objects*, which have a conjunction of simple literals in the head. The head of an object contains positive and negative examples for the concepts to be learned, while the body contains the description of layout components on the basis of geometrical features (e.g. width, height) and topological relations (e.g. vertical and horizontal alignments) existing among blocks, the type of the content (e.g. text, horizontal line, image) and the logic type of a block (e.g. title or authors of a scientific paper). Terms of literals in objects can only be constants, where different constants represent distinct layout components within a page.

The complete list of descriptors used for this task is reported in Table 1.

Table 1. Descriptors used in the first-order representation of the layout components

Descriptor name	Definition
width(block)	Integer domain
height(block)	Integer domain
x_pos_centre(block)	Integer domain
y_pos_centre(block)	Integer domain
part_of(block1,block2)	Boolean domain: true if block1 contains block2
alignment(block1,block2)	Nominal domain: only_left_col, only_right_col, only_middle_col, only_upper_row, only_lower_row, only_middle_row
to_right(block1,block2)	Boolean domain
on_top(block1,block2)	Boolean domain
type_of(block)	Nominal domain: text, hor_line, image, ver_line, graphic, mixed
<logic_type>(block)	Boolean domain: true if block is labeled as <logic_type> associated to the block
class(doc)	Nominal domain: represents the class associated to the document page
page(doc)	Nominal domain: first, intermediate, last_but_one, last. Represents the page associated to the document page.

The descriptor *<logic_type>(block)* is actually a meta-notation for a class of first order logic predicates, which depend on the application at hand. In particular, *<logic_type>* can be instantiated to the logic labels associated with layout components. In the case of scientific papers, among others, relevant logical labels could be title, author, abstract. This means that $title(block)$, $author(block)$ and $abstract(block)$ are possible descriptors.

In order to explain the semantics of geometrical and topological descriptors, some useful notation is introduced. Let us to consider the reference system whose origin is in the top left corner of the document page as shown in Figure 2. $TL_x(z)$ ($TL_y(z)$) denotes the abscissa (ordinate) of the top left corner of a (rectangular) block z, while $BR_x(z)$ ($BR_y(z)$) denotes the abscissa (ordinate) of the bottom right corner of z. Then the semantics of descriptors in Table 1 is the following:

- ***width(block)*** represents the block width, that is,
  $width(z) = BR_x(z) - TL_x(z)$
- ***height(block)*** represents the block height, that is,
  $height(z) = BR_y(z) - TL_y(z)$
- ***x_pos_centre(block)*** represents the abscissa of the block's centroid, that is, $x\_pos\_centre(z) = (BR_x(z) + TL_x(z))/2$
- ***y_pos_centre(block)*** represents the ordinate of the block's centroid, that is, $y\_pos\_centre(z) = (BR_y(z) + TL_y(z))/2$
- ***part_of(block1,block2)*** represents the fact that the layout component *block1* corresponding to the page contains the layout object *block2*, that is,
  $part\_of(z1, z2) = true \Longleftrightarrow^{def}$
  $TL_x(z1) \leq TL_x(z2) \wedge TL_y(z1) \leq TL_y(z2) \wedge BR_x(z1) \geq BR_x(z2) \wedge BR_y(z1) \geq BR_y(z2)$
- ***alignment(block1,block2)*** represents either horizontal or vertical alignment between two blocks. Six possible nominal values are considered:

  alignment(z1,z2)= only_left_col $\Longleftrightarrow^{def}$
  $abs(TL_x(z1) - TL_x(z2)) \leq \alpha \wedge (TL_y(z1) \leq TL_y(z2))$

  alignment(z1,z2)= only_right_col $\Longleftrightarrow^{def}$
  $abs(BR_x(z1) - BR_x(z2)) \leq \alpha \wedge (TL_y(z1) \leq TL_y(z2))$

  alignment(z1,z2)= only_middle_col $\Longleftrightarrow^{def}$
  $alignment(z1, z2) \neq only\_left\_col \wedge alignment(z1, z2) \neq only\_right\_col \wedge abs(x\_pos\_centre(z1) - x\_pos\_centre(z2)) \leq \alpha \wedge (TL_y(z1) \leq TL_y(z2))$

  alignment(z1,z2)= only_upper_row $\Longleftrightarrow^{def}$
  $abs(TL_y(z1) - TL_y(z2)) \leq \alpha \wedge (TL_x(z1) \leq TL_x(z2))$

  alignment(z1,z2)= only_lower_row $\Longleftrightarrow^{def}$
  $abs(BR_y(z1) - BR_y(z2)) \leq \alpha \wedge (TL_x(z1) \leq TL_x(z2))$

  alignment(z1,z2)= only_middle_row $\Longleftrightarrow^{def}$
  $alignment(z1, z2) \neq only\_upper\_row \wedge alignment(z1, z2) \neq only\_lower\_row \wedge$

$abs(y\_pos\_centre(z1) - y\_pos\_centre(z2)) \leq \alpha \wedge (TL_x(z1) \leq TL_x(z2))$

- ***on_top(block1,block2)*** indicates that *block*1 is above *block*2, that is,
  $on\_top(z1, z2) \Longleftrightarrow^{def}$
  $BR_y(z1) < TL_y(z2) \leq \beta + BR_y(z1) \wedge$
  $(TL_x(z1) \leq x\_pos\_centre(z2) \leq BR_x(z1) \vee$
  $TL_x(z2) \leq x\_pos\_centre(z1) \leq BR_x(z2))$
- ***to_right(block1,block2)*** aims at representing the fact that *block*2 is positioned to the right of *block*1. Its formal definition is:
  $to\_right(z1, z2) \Longleftrightarrow^{def}$
  $BR_x(z1) < TL_x(z2) \leq \gamma + BR_x(z1) \wedge$
  $(TL_y(z1) \leq y\_pos\_centre(z2) \leq BR_y(z1) \vee$
  $TL_y(z2) \leq y\_pos\_centre(z1) \leq BR_y(z2))$.

The last three descriptors are parametrized, that is, their semantics is based on few constants (α, β and γ) whose specification is domain dependent.

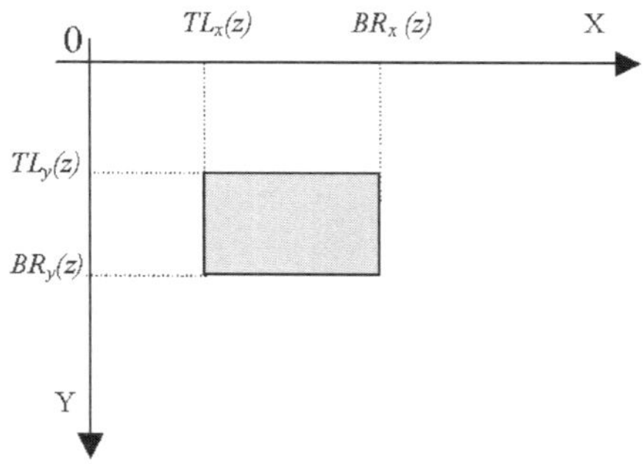

Fig. 2. Reference system used to represent components in the document page. Origin represents the origin in the top left corner of the document page

The description of the document page reported in Figure 3 is reported in the following:

$object('tpami17\_1\text{-}13', [class(p) = tpami,$
$first\_to\_read(0) = true, first\_to\_read(1) = false, ...$
$succ\_in\_reading(0, 1) = true, succ\_in\_reading(1, 2) = true,$
$..., succ\_in\_reading(7, 8) = true,$
$succ\_in\_reading(2, 10) = false, ...,$
$succ\_in\_reading(2, 5) = false],$
$[part\_of(p, 0) = true, ...,$
$height(0) = 83, height(1) = 11, ...$
$width(0) = 514, width(1) = 207, ...,$
$type\_of(0) = text, ..., type\_of(11) = hor\_line,$
$title(0) = true, author(1) = true,$
$affiliation(2) = true, ..., undefined(16) = true, ...$

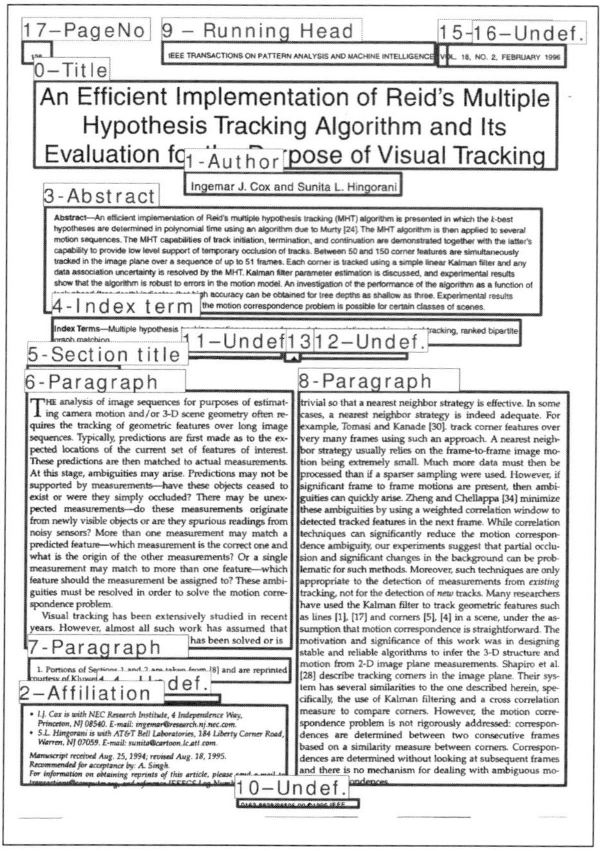

Fig. 3. A document page: For each layout component, both logical labels and constants are shown

$$x\_pos\_centre(0) = 300, x\_pos\_centre(1) = 299, ...,$$
$$y\_pos\_centre(0) = 132, y\_pos\_centre(1) = 192, ...,$$
$$on\_top(9, 0) = true, on\_top(15, 0) = true, ...,$$
$$to\_right(6, 8) = true, to\_right(7, 8) = true, ...$$
$$alignment(16, 8) = only\_right\_col, alignment(17, 5) = only\_left\_col, ...$$
$$alignment(15, 16) = only\_middle\_row,$$
$$class(p) = tpami, page(p) = first]).$$

The constant p denotes the whole page while the remaining integer constants (0, 1, ..., 17) identify distinct layout components. In this example, the block number 0 corresponds to the first block to read ($first\_to\_read(0) = true$), it is a textual component ($type\_of(0) = text$) and it is logically labelled as 'title' ($title(0) = true$). Block number 1 (immediately) follows block 0 in the reading order ($succ\_in\_reading(0, 1) = true$); it is a textual component and it includes information on the authors of the paper ($author(1) = true$).

The expressive power of ATRE is also exploited in order to define background knowledge. In this application domain, the following background knowledge has been defined:

$$at\_page(X) = first \leftarrow part\_of(Y, X) = true, page(Y) = first.$$
$$at\_page(X) = intermediate \leftarrow part\_of(Y, X) = true, page(Y) = intermediate.$$
$$at\_page(X) = last\_but\_one \leftarrow part\_of(Y, X) = true, page(Y) = last\_but\_one.$$
$$at\_page(X) = last \leftarrow part\_of(Y, X) = true, page(Y) = last.$$
$$alignment(X, Y) = both\_rows \leftarrow alignment(X, Y) = only\_lower\_row,$$
$$alignment(X, Y) = only\_upper\_row.$$
$$alignment(X, Y) = both\_columns \leftarrow alignment(X, Y) = only\_left\_col$$
$$alignment(X, Y) = only\_right\_col.$$

The first four rules allow information on the page order to be automatically associated to layout components, since their reading order may depend on the page order. The last two clauses define the alignment by both rows/columns of two layout components.

As explained in the previous section, ATRE learns a logical theory T defining the concepts $first\_to\_read/1$ and $succ\_in\_reading/2$, such that T is complete and consistent with respect to the examples. This means that it is necessary to represent both positive and negative examples and the representation of negative examples for the concept $succ\_in\_reading/2$ poses some feasibility problems due to their quadratic growth. In order to reduce the number of negative examples, we resort to sampling techniques. Indeed, this is a common practice in the presence of unbalanced datasets [27]. In our case, we sampled negative examples by limiting their number to 1000% of the number of positive examples. In this way, it is possible to simplify the learning stage and to have rules that are less specialized and avoid overfitting problems.

In summary, generated descriptions permit us to describe both the layout structure, the logical structure and the reading order chains of a single document page (see Figure 4).

ATRE is particularly indicated for our task since it can identify dependencies among concepts to be learned or even recursion. Examples of rules that ATRE is able to extract are reported in the following:

$$first\_to\_read(X1) = true \leftarrow title(X1) = true,$$
$$x\_pos\_centre(X1) \in [293..341], succ\_in\_reading(X1, X2) = true$$
$$succ\_in\_reading(X2, X1) = true \leftarrow on\_top(X2, X1) = true,$$
$$y\_pos\_centre(X2) \in [542..783]$$

The first rule states that the first block to read is a logical component labelled as title, positioned approximately at the center of the document and followed by another layout component in the reading order. The second rule states that a block $X2$ "follows in reading" another block $X1$ if it is above

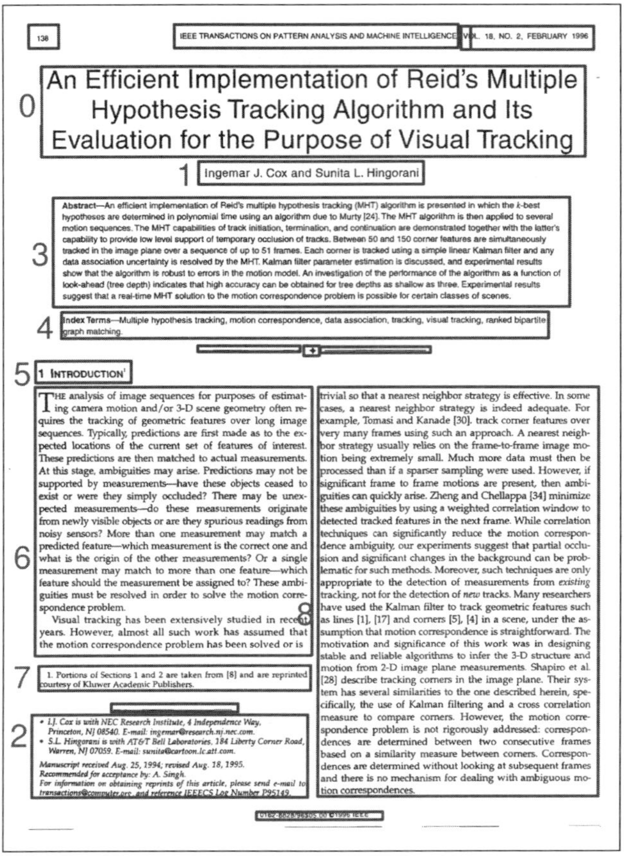
138 IEEE TRANSACTIONS ON PATTERN ANALYSIS AND MACHINE INTELLIGENCE VOL. 18, NO. 2, FEBRUARY 1996

0 An Efficient Implementation of Reid's Multiple Hypothesis Tracking Algorithm and Its Evaluation for the Purpose of Visual Tracking

1 Ingemar J. Cox and Sunita L. Hingorani

3 **Abstract**—An efficient implementation of Reid's multiple hypothesis tracking (MHT) algorithm is presented in which the *k*-best hypotheses are determined in polynomial time using an algorithm due to Murty [24]. The MHT algorithm is then applied to several motion sequences. The MHT capabilities of track initiation, termination, and continuation are demonstrated together with the latter's capability to provide low level support of temporary occlusion of tracks. Between 50 and 150 corner features are simultaneously tracked in the image plane over a sequence of up to 51 frames. Each corner is tracked using a simple linear Kalman filter and any data association uncertainty is resolved by the MHT. Kalman filter parameter estimation is discussed, and experimental results show that the algorithm is robust to errors in the motion model. An investigation of the performance of the algorithm as a function of look-ahead (tree depth) indicates that high accuracy can be obtained for tree depths as shallow as three. Experimental results suggest that a real-time MHT solution to the motion correspondence problem is possible for certain classes of scenes.

4 **Index Terms**—Multiple hypothesis tracking, motion correspondence, data association, tracking, visual tracking, ranked bipartite graph matching.

5 1 INTRODUCTION[1]

6 THE analysis of image sequences for purposes of estimating camera motion and/or 3-D scene geometry often requires the tracking of geometric features over long image sequences. Typically, predictions are first made as to the expected locations of the current set of features of interest. These predictions are then matched to actual measurements. At this stage, ambiguities may arise. Predictions may not be supported by measurements—have these objects ceased to exist or were they simply occluded? There may be unexpected measurements—do these measurements originate from newly visible objects or are they spurious readings from noisy sensors? More than one measurement may match a predicted feature—which measurement is the correct one and what is the origin of the other measurements? Or a single measurement may match to more than one feature—which feature should the measurement be assigned to? These ambiguities must be resolved in order to solve the motion correspondence problem.

Visual tracking has been extensively studied in recent years. However, almost all such work has assumed that the motion correspondence problem has been solved or is

8 trivial so that a nearest neighbor strategy is effective. In some cases, a nearest neighbor strategy is indeed adequate. For example, Tomasi and Kanade [30]. track corner features over very many frames using such an approach. A nearest neighbor strategy usually relies on the frame-to-frame image motion being extremely small. Much more data must then be processed than if a sparser sampling were used. However, if significant frame to frame motions are present, then ambiguities can quickly arise. Zheng and Chellappa [34] minimize these ambiguities by using a weighted correlation window to detected tracked features in the next frame. While correlation techniques can significantly reduce the motion correspondence ambiguity, our experiments suggest that partial occlusion and significant changes in the background can be problematic for such methods. Moreover, such techniques are only appropriate to the detection of measurements from *existing* tracking, not for the detection of *new* tracks. Many researchers have used the Kalman filter to track geometric features such as lines [1], [17] and corners [5], [4] in a scene, under the assumption that motion correspondence is straightforward. The motivation and significance of this work was in designing stable and reliable algorithms to infer the 3-D structure and motion from 2-D image plane measurements. Shapiro et al. [28] describe tracking corners in the image plane. Their system has several similarities to the one described herein, specifically, the use of Kalman filtering and a cross correlation measure to compare corners. However, the motion correspondence problem is not rigorously addressed: correspondences are determined between two consecutive frames based on a similarity measure between corners. Correspondences are determined without looking at subsequent frames and there is no mechanism for dealing with ambiguous motion correspondences.

7 1. Portions of Sections 1 and 2 are taken from [8] and are reprinted courtesy of Kluwer Academic Publishers.

2 • *I.J. Cox is with NEC Research Institute, 4 Independence Way, Princeton, NJ 08540. E-mail: ingemar@research.nj.nec.com.*
• *S.L. Hingorani is with AT&T Bell Laboratories, 184 Liberty Corner Road, Warren, NJ 07059. E-mail: sunita@cartoon.lc.att.com.*

Manuscript received Aug. 25, 1994; revised Aug. 18, 1995.
Recommended for acceptance by: A. Singh.
For information on obtaining reprints of this article, please send e-mail to transactions@computer.org, and reference IEEECS Log Number P95149.

Fig. 4. A document page: the input reading order chain. Sequential numbers indicate the reading order

*X*1 and is positioned in the lower part of a page. Additional rules may state further sufficient conditions for $first\_to\_read$ and $succ\_in\_reading$.

4.3 Application of Learned Rules

Once rules have been learned, they can be applied to new documents in order to generate a set of ground atoms, such as:

$$\{first\_to\_read(0) = true, succ\_in\_reading(0,1) = true, \ldots,$$
$$succ\_in\_reading(4,3) = true, \ldots\}$$

which can be used to reconstruct chains of (possibly logically labelled) layout components. In our approach, we propose two different solutions:

1. Identification of multiple chains of layout components

2. Identification of a single chain of layout components

By applying rules learned by ATRE, it is possible to identify:

- A *directed* graph $G =< V, E >$[3] where V is the set of nodes representing all the layout components found in a document page and edges represent the existence of a *succ_in_reading* relation between two layout components, that is, $E = \{(b_1, b_2) \in V^2 | succ\_in\_reading(b_1, b_2) = true\}$
- A list of initial nodes $I = \{b \in V | first\_to\_read(b) = true\}$

Both approaches make use of G and I in order to identify chains.

Multiple Chains Identification

This approach aims at identifying a (possibly empty) set of chains over the set of logical components in the same document page. It is two-stepped. The first step aims at identifying the heads (first elements) of the possible chains, that is, the set

$$Heads = I \cup \{b_1 \in V | \ \exists b_2 \in V \ (b_1, b_2) \in E \ \wedge \forall b_0 \in V \ (b_0, b_1) \notin E\}.$$

This set contains both nodes for which $first\_to\_read$ is true and nodes which occur as a first argument in a true *succ_in_reading* atom and do not occur as a second argument in any true *succ_in_reading* atom.

Once the set *Heads* has been identified, it is necessary to reconstruct the distinct chains. Intuitively, each chain is the list of nodes forming a path in G which begins with a node in *Heads* and ends with a node without outgoing edges. Formally, an extracted chain $C \subseteq E$ is defined as follows:
$C = \{(b_1, b_2), (b_2, b_3), \ldots, (b_k, b_{k+1})\}$, such that

- $b_1 \in Heads$,
- $\forall i = 1..k : (b_i, b_{i+1}) \in E$ and
- $\forall b \in V \ (b_{k+1}, b) \notin E$.

In order to avoid cyclic paths, we impose that the same node cannot appear more than once in the same chain. The motivation for this constraint is that the same layout component is generally not read more than once by the reader.

Single Chain Identification

The result of the second approach is a single chain. Following the proposal reported in [28], we aim at iteratively evaluating the most promising node to be appended to the resulting chain.

More formally, let $PREF_G : V \times V \rightarrow \{0, 1\}$ be a preference function defined as follows:

[3] G is not a direct acyclic graph (dag) since it could also contain cycles

$$PREF_G(b_1, b_2) = \begin{cases} 1 \text{ if a path connecting } b_1 \text{ and } b_2 \text{ exists in } G \\ 1 \text{ if } b_1 = b_2 \\ 0 \text{ otherwise} \end{cases}$$

Let $\mu : V \rightarrow \mathbb{N}$ be the function defined as follows:

$$\mu(L, G, I, b) = countConnections(L, G, I, b) + outGoing(V/L, b) \\ -inComing(V/L, b)$$

where

- $G =< V, E >$ is the ordered graph
- L is a list of *distinct* nodes in G
- $b \in V/L$ is a candidate node
- $countConnections(L, G, I, b) = |\{d \in L \cup I | PREF_G(d, b) = 1\}|$ counts the number of nodes in $L \cup I$ from which b is reachable.
- $outGoing(V/L, b) = |\{d \in V/L | PREF_G(b, d) = 1\}|$ counts the number of nodes in V/L reachable from b.
- $inComing(V/L, b) = |\{d \in V/L | PREF_G(d, b) = 1\}|$ counts the number of nodes in V/L from which b is reachable.

Algorithm 21 fully specifies the method for the single chain identification. The rationale is that at each step a node is added to the final chain. Such a node is that for which μ is the highest. Higher values of μ are given to nodes which can be reached from I, as well as from other nodes already added to the chain, and have a high (low) number of outgoing (incoming) paths to (from) nodes in V/L. Indeed, the algorithm returns an ordered list of nodes which could be straightforwardly transformed into a chain.

5 Experiments

In order to evaluate the applicability of the proposed approach to reading order identification, we considered a set of multi-page articles published in an international journal. In particular, we considered twenty-four papers, published as either regular or short articles, in the IEEE Transactions on Pattern Analysis and Machine Intelligence (TPAMI), in the January and February issues of 1996. Each paper is a multi-page document; therefore, we processed 211 document images. Each document page corresponds to an RGB 24bit colour image in TIFF format.

Initially, document images are pre-processed by WISDOM++ in order to segment them, perform layout analysis, identify the membership class and map the layout structure of each page into the logical structure. Training examples are then generated by manually specifying the reading order. In all, 211 positive examples and 3,263 negative examples for the concept $fisrt\_to\_read/1$

Algorithm 1 Single chain identification algorithm

```
findChain (G =< V, E >, I)
  Output: L: chain of nodes
L ← ∅;
repeat
   best_mu ← −∞;
   for all b ∈ V/L do
      cc ← countConnections(L, G, I, b);
      inC ← incoming(V/L, b);
      outG ← outGoing(V/L, b);
      if ((cc ≠ 0) AND (inC ≠ 0) AND (outG ≠ 0)) then
         μ ← cc + outG − inC;
         if best_mu < μ then
            best_b ← b;
            best_mu ← μ;
         end if
      end if
   end for
   if (best_mu <> −∞) then
      L.add(best_b);
   end if
until best_mu = −∞
return L
```

and 1,418 positive examples and 15,518 negative examples for the concept *succ_in_reading*/2 are generated.

We evaluated the performance of the proposed approach by means of a 6-fold cross-validation, that is, the dataset is first divided into six *folds* of near-equal size (see Table 2), and then, for every fold, the learner is trained on the remaining folds and tested on them.

When generating descriptions, the following parameters have been set: α=4, β=50 and γ=100. In the task at hand, the following logical labels are considered: *abstract*, *affiliation*, *author*, *biography*, *caption*, *figure*, *formulae*, *index_term*, *reference*, *table*, *page_no*, *paragraph*, *running_head*, *section_title*, *subsection_title*, *title*.

For each learning problem, statistics on precision and recall of the learned logical theory are recorded. In order to evaluate the ordering returned by the proposed approach, we resort to metrics used in information retrieval in order to evaluate the returned ranking of results [29]. For this purpose several metrics have been defined in the literature. Herein we consider the metrics valid for partial orders evaluation.

In particular, we consider the *normalized Spearman footrule distance* which, given two complete lists L and L_1 on a set S (that is, L and L_1 are two different permutations without repetition of all the elements in S), is defined as follows:

Fold	Article ID	Number of pages	Tot number of pages
FOLD1	tpami13	3	36
	tpami11	6	
	tpami04	14	
	tpami16	13	
FOLD2	tpami07	6	38
	tpami24	6	
	tpami17	13	
	tpami01	13	
FOLD3	tpami06	1	33
	tpami19	17	
	tpami23	7	
	tpami02	8	
FOLD4	tpami05	6	35
	tpami18	10	
	tpami22	5	
	tpami03	14	
FOLD5	tpami10	3	33
	tpami20	14	
	tpami21	11	
	tpami08	5	
FOLD5	tpami15	15	36
	tpami09	5	
	tpami14	10	
	tpami12	6	

Table 2. Processed documents

$$F(L, L_1) = \frac{\sum_{b \in S} abs(pos(L, b) - pos(L_1, b))}{|S|^2/2} \tag{1}$$

where the function $pos(L, b)$ returns the position of the element b in the ordered list L.

This measure can be straightforwardly generalized to the case of several lists:

$$F(L, L_1, \ldots, L_k) = 1/k \sum_{i=1 \ldots k} F(L, L_i). \tag{2}$$

Indeed, this measure is specifically designed for total orders and not for partial ones. In order to consider partial orders, we resorted to a variant of this measure (*induced normalized footrule distance*).

$$F(L, L_1, \ldots, L_k) = 1/k \sum_{i=1 \ldots k} F(L|_{L_i}, L_i) \tag{3}$$

where $L|_{L_i}$ is the projection of L on L_i. Since this measure does not take into account the length of single lists, we also adopted the *normalized scaled footrule distance*:

$$F'(L, L_1) = \frac{\sum_{b \in S} abs(pos(L, b)/|L| - pos(L_1, b)/|L_1|)}{|L_1|/2}. \tag{4}$$

Also in this case it is possible to extend the measure to the case of multiple lists:

$$F'(L, L_1, \ldots, L_k) = 1/k \sum_{i=1\ldots k} F'(L|_{L_i}, L_i). \quad (5)$$

In this study, we apply such distance measures to chains. In particular:

- FD=$F(L|_{L_1}, L_1)$ is used in the evaluation of single chain identification.
- SFD=$F'(L|_{L_1}, L_1)$ is used in the evaluation of single chain identification.
- IFD=$F(L, L_1, \ldots, L_k)$ is used in the evaluation of multiple chains identification.
- ISFD=$F'(L, L_1, \ldots, L_k)$ is used in the evaluation of multiple chains identification.

Results reported in Table 3 show that the system has a precision of about 65% and a recall greater than 75%. Some statistics concerning the learned theories are reported in Table 4. It is noteworthy that rules learned for the concept *first_to_read* cover (on average) fewer positive examples than rules learned for the concept *succ_in_reading*. Moreover, by considering the results reported in Table 5, we note that there is no significant difference in terms of recall between the two concepts, while precision is higher for rules concerning the *succ_in_reading* concept. This is mainly due to the specificity of rules learned for the concept *first_to_read* and we can conclude that the concept *first_to_read* appears to be more complex to learn than the concept *succ_in_reading*. This can be explained by the limited number of training examples for this concept (one per page).

	Precision%	**Recall%**
FOLD1	76.60	61.80
FOLD2	73.00	64.90
FOLD3	80.10	67.40
FOLD4	68.00	58.20
FOLD5	76.80	68.40
FOLD6	78.20	62.60
AVG	75.45	63.88

Table 3. Overall Precision and Recall results

Concept	*first_to_read/1*		*succ_in_reading/2*	
	NOC	Training POS exs	NOC	Training POS exs
FOLD1	42	175	162	1226
FOLD2	46	173	145	1194
FOLD3	42	178	149	1141
FOLD4	42	176	114	1171
FOLD5	40	178	166	1185
FOLD6	41	175	177	1173
AVG coverage	4.17		7.77	

Table 4. Number of rules per positive examples

In the following we report some rules learned by ATRE:

Concept	*first_to_read/1*		*succ_in_reading/2*	
	Precision %	Recall%	Precision%	Recall%
FOLD1	75.00	50.00	76.90	64.10
FOLD2	66.70	63.20	74.10	65.20
FOLD3	74.30	78.80	81.00	66.10
FOLD4	69.40	71.40	67.80	56.30
FOLD5	66.70	66.70	78.40	68.70
FOLD6	71.00	61.10	79.40	62.90
AVG	70.52%	65.20%	76.27%	63.88%

Table 5. Precision and Recall results shown per concept to be learned

1. $first\_to\_read(X1) = true \leftarrow x\_pos\_centre(X1) \in [55..177],$
   $y\_pos\_centre(X1) \in [60..121], height(X1) \in [98..138].$
2. $first\_to\_read(X1) = true \leftarrow title(X1) = true,$
   $x\_pos\_centre(X1) \in [293..341], succ\_in\_reading(X1, X2) = true.$
3. $succ\_in\_reading(X2, X1) = true \leftarrow affiliation(X1) = true,$
   $author(X2) = true, height(X1) \in [45..124].$
4. $succ\_in\_reading(X2, X1) = true \leftarrow alignment(X1, X3) = both\_columns,$
   $on\_top(X2, X3) = true, succ\_in\_reading(X1, X3) = true,$
   $height(X1) \in [10..15].$

They can be easily interpreted. For instance, the first rule states that a block at the top of the page, horizontally positioned in the center-left part of the page with a height between 98 an 138 pixels, is the first block to read.

The second rule states that if a block represents the title, is horizontally positioned in the center of the document page and is read before another block, then it is the first to be read. This rule captures concept dependencies. In particular, the predicate $first\_to\_read$ is defined in terms of the predicate $succ\_in\_reading$.

The third rule states that a layout component whose height is between 45 and 124 pixels and labeled as 'affiliation' is read after the logical component 'author'. Since affiliation and author are not close to each other in a typical document page (see Figure 4), this rule would not have been discovered without considering results of the logical structure identification phase.

The fourth rule presents both an example of recursion on the predicate $succ\_in\_reading$ and an example of use of descriptors defined in the background knowledge ($alignment(X1, X3) = both\_columns$).

Experimental results concerning the reconstruction of single/multiple chains are reported in Table 6. We recall that the lower the distance value, the better the reconstruction of the original chain(s). By comparing results in terms of the *footrule distance* measure (IFD vs FD), we note that the reconstruction of multiple chains shows better results than the reconstruction of single chains. Indeed, this result does not take into account the length of the lists. When considering the length of the lists (ISFD vs. SFD), the situation is completely different and the reconstruction of single chains outperforms the reconstruction of multiple chains.

Concept	Multiple chains		Single chain	
	AVG. IFD%	AVG. ISFD%	AVG. FD%	AVG. SFD%
FOLD1	13.18	21.12	47.33	10.17
FOLD2	10.98	18.51	46.32	8.13
FOLD3	1.31	26.91	47.32	17.63
FOLD4	1.32	24.00	49.96	14.51
FOLD5	0.90	22.50	49.31	10.60
FOLD6	0.90	27.65	54.38	12.97
AVG	4.76%	23.45%	49.10%	12.33%

Table 6. Reading order reconstruction results

6 Conclusions

In this chapter, we present a novel approach for automatically determining the reading order in a document image understanding process. Reading order identification is a crucial problem for several applications since it permits us to reconstruct a single textual component to be used in subsequent text processing steps, such as information extraction, information retrieval and text reconstruction for rendering purposes. The proposed approach aims at learning rules which are used for predicting reading order chains of layout components detected in document images. The rules are learned from training examples consisting of sets of ordered layout components described by means of both layout and logical properties. The proposed approach presents two main peculiarities. First, it fully exploits spatial information embedded in the layout structure by resorting to inductive logic programming techniques. Second, it reconstructs reading order chains, which may not necessarily define a total ordering. This last aspect permits us to take into account the case in which independent pieces of information are represented on the same page (e.g., the end of an article and the beginning of a new one) and the case in which some layout components should not be included in the reading order (e.g. images or page numbers).

In the learning phase, rules which identify the first logical component to read and define the successor relation are induced. In the recognition phase such rules are used to reconstruct reading order chains according two different modalities: single vs. multiple chains identification. Results prove that learned rules are quite accurate and that the reconstruction phase significantly depends on the application at hand. In particular, if the user is interested in reconstructing the actual chain (e.g. text reconstruction for rendering purposes), the best solution is in the identification of single chains. On the contrary, when the user is interested in recomposing a text such that sequential components are correctly linked (e.g. in information extraction applications), the most promising solution is the identification of multiple chains.

For future work we intend to consider the entire document (and not the single page) as the analysis unit. This would permit us to reconstruct multiple crossing-pages chains typically found in collections of documents (e.g., conference proceedings or transcriptions of ancient edicts).

References

1. Nagy, G., Seth, S., Viswanathan, M.: A prototype document image analysis system for technical journals. Computer **25**(7) (1992) 10–22
2. Dengel, A., Bleisinger, R., Hoch, R., Fein, F., Hones, F.: From paper to office document standard representation. IEEE Computer **25**(7) (1992) 63–67
3. Wenzel, C., Maus, H.: Leveraging corporate context within knowledge-based document analysis and understanding. IJDAR **3**(4) (2001) 248–260
4. Ceci, M., Berardi, M., Malerba, D.: Relational data mining and ILP for document image understanding. Applied Artificial Intelligence **21**(4-5) (2007) 317–342
5. Tsujimoto, S., Asada, H.: Understanding multi-articled documents. In: in Proceedings of the 10th International Conference on Pattern Recognition. (1990) 551–556
6. Ishitani, Y.: Document transformation system from papers to xml data based on pivot xml document method. In: ICDAR '03: Proceedings of the Seventh International Conference on Document Analysis and Recognition, Washington, DC, USA, IEEE Computer Society (2003) 250
7. Nagy, G., Seth, S.: Hierarchical representation of optically scanned documents. In: Seventh Int'l Conf. Pattern Recognition, IEEE CS Press (1984) 347–349
8. Meunier, J.L.: Optimized xy-cut for determining a page reading order. In: ICDAR '05: Proceedings of the Eighth International Conference on Document Analysis and Recognition, Washington, DC, USA, IEEE Computer Society (2005) 347–351
9. Taylor, S.L., Dahl, D.A., Lipshutz, M., Weir, C., Norton, L.M., Nilson, R., Linebarger, M.: Integrated text and image understanding for document understanding. In: HLT '94: Proceedings of the workshop on Human Language Technology, Morristown, NJ, USA, Association for Computational Linguistics (1994) 421–426
10. Aiello, M., Monz, C., Todoran, L., Worring, M.: Document understanding for a broad class of documents. International Journal on Document Analysis and Recognition-IJDAR **5**(1) (2002) 1–16
11. Allen, J.F.: Maintaining knowledge about temporal intervals. Commun. ACM **26**(11) (1983) 832–843
12. Aiello, M., Smeulders, A.M.W.: Thick 2d relations for document understanding. Inf. Sci. Inf. Comput. Sci. **167**(1-4) (2004) 147–176
13. Breuel, T.M.: High performance document layout analysis. In: Proceedings of the 2003 Symposium on Document Image Understanding (SDIUT '03). (2003)
14. Altamura, O., Esposito, F., Malerba, D.: Transforming paper documents into XML format with WISDOM++. IJDAR **4**(1) (2001) 2–17
15. Malerba, D., Esposito, F., Altamura, O., Ceci, M., Berardi, M.: Correcting the document layout: A machine learning approach. In: ICDAR '03: Proceedings of the Seventh International Conference on Document Analysis and Recognition, Washington, DC, USA, IEEE Computer Society (2003) 97–101
16. Esposito, F., Malerba, D., Lisi, F.A.: Machine learning for intelligent processing of printed documents. J. Intell. Inf. Syst. **14**(2-3) (2000) 175–198
17. Malerba, D., Esposito, F., Lisi, F.A., Altamura, O.: Automated discovery of dependencies between logical components in document image understanding. ICDAR '01: Proceedings of the Sixth International Conference on Document

Analysis and Recognition, Washington, DC, USA, IEEE Computer Society (2001) 174–178
18. Utgoff, P.: An improved algorithm for incremental induction of decision trees. In: Proc. of the Eleventh Int. Conf. on Machine Learning, Morgan Kaufmann (1994)
19. Malerba, D.: Learning recursive theories in the normal ilp setting. Fundamenta Informaticae **57**(1) (2003) 39–77
20. Grimaldi, R.P.: Discrete and Combinatorial Mathematics, an Applied Introduction. Addison Wesley, third edition (1994)
21. Muggleton, S.: Inductive Logic Programming. Academic Press, London (1992)
22. De Raedt, L.: Interactive Theory Revision. Academic Press, London (1992)
23. Lavrač, N., Džeroski, S.: Inductive Logic Programming: techniques and applications. Ellis Horwood, Chichester (1994)
24. Bergadano, F., Gunetti, D.: Inductive Logic Programming: from machine learning to software engineering. The MIT Press, Cambridge, MA (1996)
25. Nienhuys-Cheng, S.W., de Wolf, R.: Foundations of inductive logic programming. Springer, Heidelberg (1997)
26. Levi, G., Sirovich, F.: Generalized and/or graphs. Artif. Intell. **7**(3) (1976) 243–259
27. Mladenić, D., Grobelnik, M.: Feature selection for unbalanced class distribution and naive bayes. In: ICML '99: Proceedings of the Sixteenth International Conference on Machine Learning, Morgan Kaufmann Publishers Inc. (1999) 258–267
28. Cohen, W.W., Schapire, R.E., Singer, Y.: Learning to order things. J. Artif. Intell. Res. (JAIR) **10** (1999) 243–270
29. Dwork, C., Kumar, R., Naor, M., Sivakumar, D.: Rank aggregation methods for the web. In: WWW '01: Proceedings of the 10th international conference on World Wide Web, New York, NY, USA, ACM Press (2001) 613–622

Decision-Based Specification and Comparison of Table Recognition Algorithms

Richard Zanibbi[1], Dorothea Blostein[2], and James R. Cordy[2]

[1] Department of Computer Science, Rochester Institute of Technology, 102 Lomb Memorial Drive, Rochester, New York, USA 14623-5603, rlaz@cs.rit.edu
[2] School of Computing, Queen's University, Kingston, Ontario, Canada, K7L 3N6 {blostein,cordy}@cs.queensu.ca

Summary. Recognition algorithms are difficult to write and difficult to maintain. There is need for better tools to support the creation, debugging, optimization, and comparison of recognition algorithms. We propose an approach that centers on a process-oriented description. The approach is implemented using a new scripting language called RSL (Recognition Strategy Language), which captures the recognition decisions an algorithm makes as it executes. This semi-formal process-oriented description provides a powerful basis for developing and comparing recognition algorithms. Based on this description, we describe new metrics related to the sequence of decisions an algorithm makes during recognition. The capture of intermediate decision outputs and these new process-oriented metrics greatly extend the limited information available from final results and traditional results-oriented metrics such as recall and precision. Using a simple example, we illustrate how these new metrics can be used to understand and improve decisions within a recognition strategy. We believe these new metrics may also be applied in machine learning algorithms that construct optimal decision sequences from sets of decisions and/or strategies.

1 Introduction

Tables are used to summarize vast quantities of information in books, papers, text files, electronic documents, and HTML web pages. Significant efforts have been made towards developing automated and semi-automated methods for searching, extracting, summarizing, and integrating the information they contain. A wide variety of algorithms have been published for detecting tables and analyzing their structure, and a number of surveys on table recognition and processing are available [3, 4, 5, 6, 7, 8].

Algorithms that recognize tables search a space of possible interpretations in order to describe the tables present in an input file. The space of interpretations is defined by some model of table locations and structure whose

R. Zanibbi et al.: *Decision-Based Specification and Comparison of Table Recognition Algorithms*, Studies in Computational Intelligence (SCI) **90**, 71–103 (2008)
www.springerlink.com

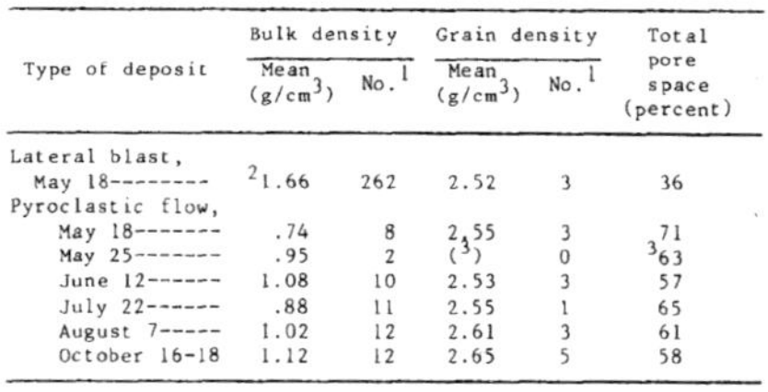

Type of deposit	Bulk density Mean (g/cm[3])	Bulk density No.[1]	Grain density Mean (g/cm[3])	Grain density No.[1]	Total pore space (percent)
Lateral blast,					
May 18--------	[2]1.66	262	2.52	3	36
Pyroclastic flow,					
May 18-------	.74	8	2.55	3	71
May 25-------	.95	2	([3])	0	[3]63
June 12------	1.08	10	2.53	3	57
July 22------	.88	11	2.55	1	65
August 7-----	1.02	12	2.61	3	61
October 16-18	1.12	12	2.65	5	58

a. Original Image

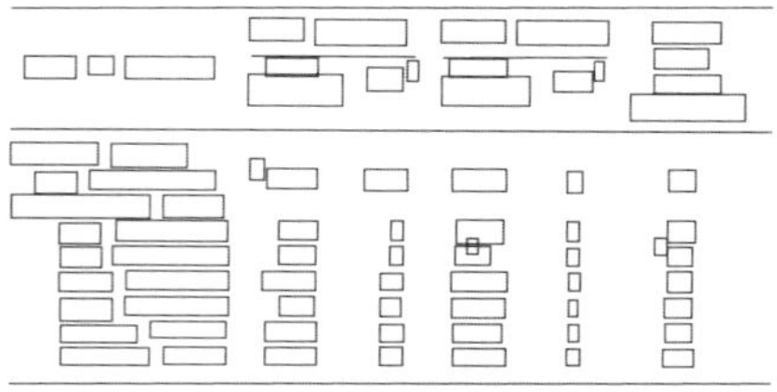

b. Decision 1 (Word Regions → Cells)

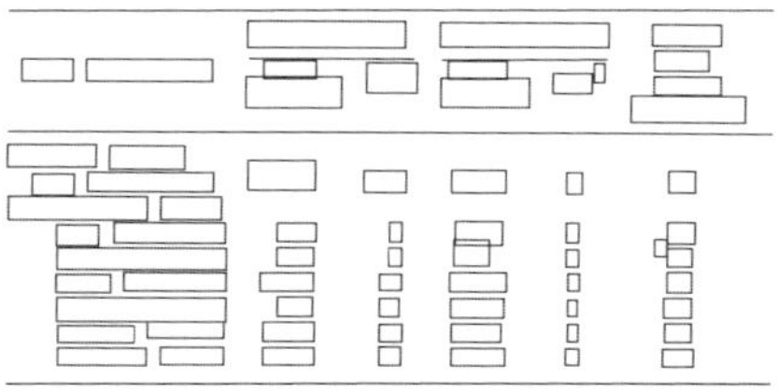

c. Decision 13 (Merge)

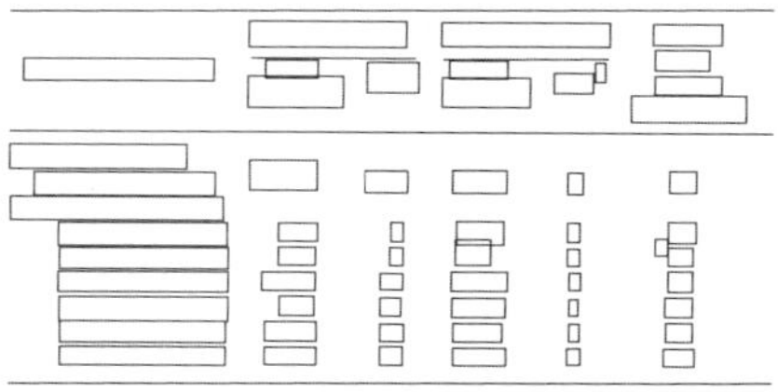

d. Decision 19 (Merge)

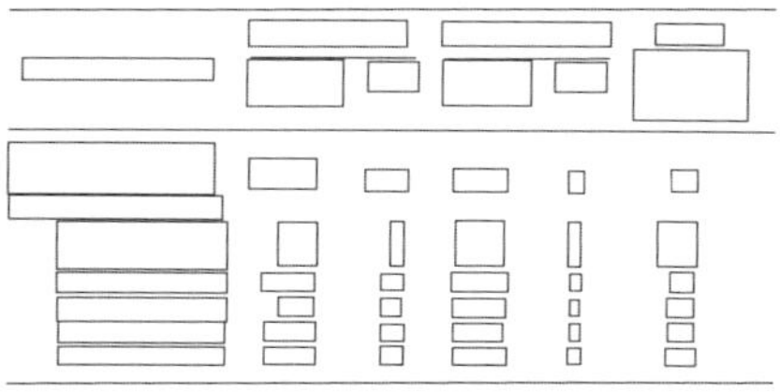

e. Decision 32 (Merge)

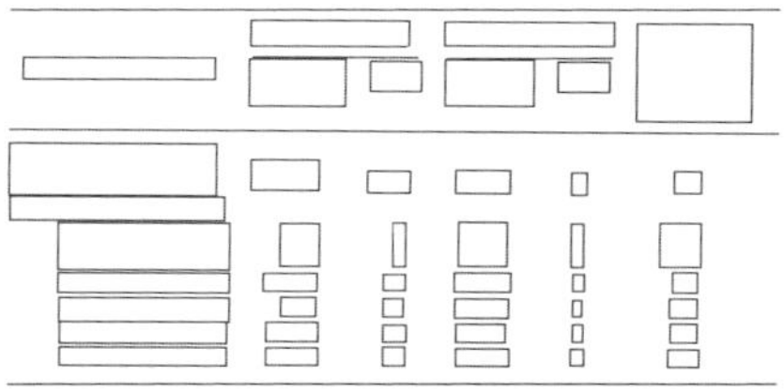

f. Decision 42 (Merge)

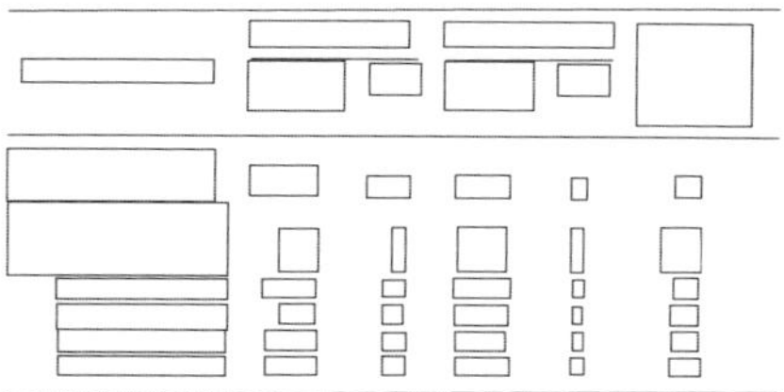

g. Decision 59 (Merge)

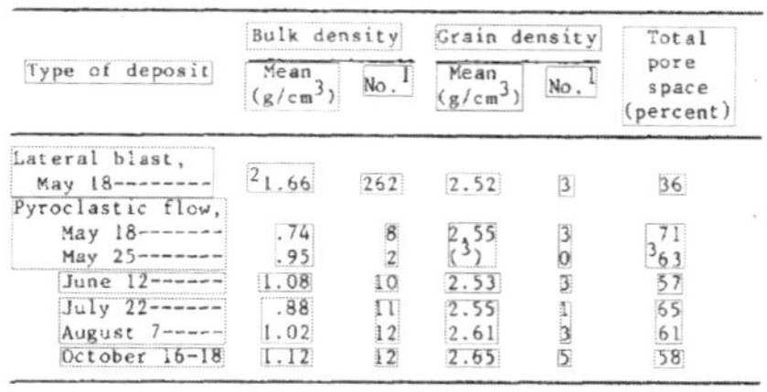

h. Decision 85 (Merge) : Output

Fig. 1. Production of Cell Hypotheses by the Handley Algorithm [1]. (a) Input image from the University of Washington Database [2], page a038. (b) *Line* and *Word* regions are defined manually. All *Word* regions are immediately classified as cells in the first step. (c) to (h) *Cell* region hypotheses are refined through merging (see caption of Figure 9 for further details) operations are invoked by the algorithm while searching. The search for tables and/or structural elements of tables proceeds through a series of decisions that determine which model operations to apply, and which interpretations are worth pursuing or are acceptable for output.

Currently it is difficult to compare table recognition algorithms. One difficulty is that the algorithms often address different problems, and do not use the same model to describe tables as a result [9, 10]. For example, some algorithms assume that all table cells are separated by ruling lines, while in others ruling lines are optional or are assumed to be absent. As another example, some algorithms aim to detect cell locations in an image or text file, while other methods assume cell locations are known, and then attempt to recover the indexing structure of the table from header cells to body cells. Also, algorithms are usually defined informally in the literature, often with no explicit description of the table model or its associated space of possible interpretations. This makes table recognition algorithms hard to replicate, compare, and combine.

We do not believe that it is possible or even desirable for researchers to adopt a standard table model. We propose instead to make table models and their operations more explicit, in order to permit easier identification of common elements in models and recognition techniques. A *decision-based specification* of a table recognition algorithm defines a sequence of decisions that each select from an explicit set of table model operations. Each decision is defined in terms of the input data used to make the decision, the set of model operations from which to select (defined using the inputs), and a function that will decide which operations to select at run-time.

In Section 3 we present the Recognition Strategy Language (RSL [11]), a functional scripting language for creating and executing decision-based specifications. RSL provides a small but general set of decision types used to manipulate interpretations represented by attributed graphs. We then demonstrate how decision-based specifications allow algorithms to be compared through decisions affecting common table model elements. Graph-based summaries of table models used for recognition may be produced through static analysis of RSL programs (see Section 4). After execution, the results of intermediate decisions may be observed and compared in addition to the final results. To illustrate, in Section 6 we compare the detection of table cells in two published algorithms [1, 12] that have been re-implemented in RSL, and then combine the algorithms to produce a better result. Figure 1 illustrates how one of these algorithms creates and discards table cell hypotheses.

Commonly, measures such as recall and precision are used to evaluate recognition algorithms. In the case of cells in table recognition, recall is the percentage of ground-truth cells that are detected correctly, and precision is the percentage of detected table cells that are in the ground truth. Using recall and precision, recognition is characterized only in terms of final results, and not in terms of the *process* of recognition; there is no characterization of whether individual decisions are helpful, detrimental, or irrelevant for accurately recognizing tables within a set of documents.

In conventional performance evaluations there is also no characterization of false negatives: these are valid hypotheses that are created and later rejected by an algorithm. RSL records a complete history of all decision outcomes, which makes it possible to determine which decisions create, reject, and reinstate each generated hypothesis after execution (see Figures 9 and 10). Taking advantage of this additional information, we present *historical recall* and *historical precision*, two new metrics that characterize the complete set of generated hypotheses (e.g. cell locations) in Section 5.

Aside from table recognition, we believe that the decision based-approach to specification, comparison, and evaluation may be usefully applied to many other document recognition and machine learning problems. These include problems in document layout analysis such as page region labeling and interpreting symbol layout within page regions (e.g. in text regions, figures, and diagrammatic notations) where a series of decisions must be made about region types, locations, and relationships, and the problem of classifier combination, where a decision-based specification would support rapid construction and modification of classifier architectures from existing classifiers. Other chapters in this book provide more information about machine learning techniques related to document layout analysis and classifier combination. The decision-based approach provides intermediate interpretations and metrics for decision sequences (e.g. historical recall and precision) that could be used by machine learning algorithms to train or optimize individual decisions within a recognition strategy, or to learn recognition strategies themselves given a set of decisions and/or strategies [13, 14, 15].

We begin our main discussion by introducing table recognition problems as sequential decision-making problems in more detail.

2 Recognizing Tables: A Decision-Making Problem

Researchers have used the term physical structure to refer to explicit properties of an encoding, and logical structure to refer to properties that are implicit in an encoding [5, 16, 17]. Simple examples of physical structure include the number of rows and columns in an image file, the number of characters in each text line of a text file, and the location of a pixel or character within an image or text file (this is usually expressed geometrically, in terms of columns and rows).

Source files for markup language encodings such as HTML and XML have the same physical structure as plain text files, but also represent regions of the file and the relationships between them using text demarcators ('tags'). To recover the information represented by the demarcators requires an inferential process that uses the rules of structure for the appropriate markup language (i.e. we must parse the file using an appropriate grammar). Properties of a file that must be recovered using an inferential process such as this are referred to as logical structure.

Markup language parsers are formal language recognizers that recover implicit logical structure from text encodings. From a pattern recognition perspective they are considered simple because the information used to deduce the tag structure is largely fixed and unambiguous, and parsers assume that input files contain data in the correct format.

Table recognition is a more difficult problem because which information and inferential processes to use for recovering tables are neither fixed nor unambiguous, and are subjects of ongoing study. Table structure is frequently adapted to the idiosyncratic needs of authors, making it difficult and perhaps impossible to make a single, formal definition. Unless strong assumptions about tables in input files are made, there is the additional problem of selecting which model of table structure to apply.

Even if a valid table structure model is used for an input file, the set of possible model instances may be large, and it is often difficult to define inferencing methods that reliably identify table structures due to noise in the file (e.g. smearing in images) and unanticipated properties of the input. For systems recognizing tables in raw images, all table properties are implicit. At the other extreme, for HTML files often table cells, rows, and columns are already represented; however, tags for tables are often used to arrange arbitrary data visually, and determining which encoded tables are real tables is a difficult problem [18, 19].

In our work we have come to view structural pattern recognition problems such as table recognition as sequential decision-making problems: given a model of logical structure to be recovered from a set of input files, what series of decisions about model operations to apply will produce interpretations of logical structure that maximize a given performance criterion? For recognition tasks, the performance criterion will be some combination of metrics for execution time, storage requirements, and accuracy. Similar views of structural recognition problems have been proposed in the machine learning literature [20, 21]. Accuracy metrics are influenced by the chosen source(s) of ground truth; to accommodate this, a problem formulation in which the source of ground truth is an explicit variable may be used (we have proposed one such formulation [10]).

We wish to be able to easily compare, re-sequence, and combine individual decisions about model operations made by different table recognition algorithms. We also wish to be able to easily evaluate interpretations produced by intermediate decisions, and characterize the complete set of hypotheses generated by an algorithm, not just those accepted at a given point in the algorithm's execution.

Currently in the literature, table recognition algorithms are most commonly characterized as a sequence of operations that are implemented in a general-purpose programming language. Less frequently model-driven specifications are employed, in which a representation of the table model is used

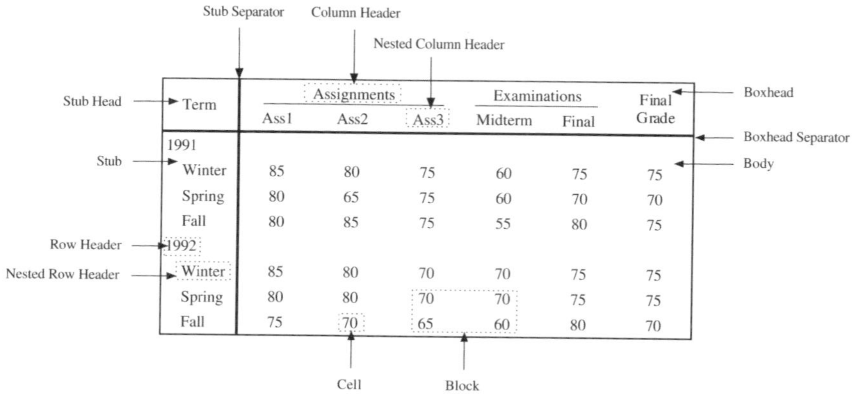

Fig. 2. Table Structure. This example is taken from Wang [26], and uses terminology taken from the Chicago Manual of Style [27]

to 'program' the system (e.g. as an attributed grammar with rules containing associated actions [22, 23, 24]). Given a model specification and an input file, the sequence of decisions to make is determined algorithmically (e.g. by a parser): the model definition is mapped to a decision sequence. The procedural approach has the benefit of being highly flexible, while the model-driven approach has the benefits of concise specification and a level of formality which permits more information about decision-making to be automatically collected.

For our purposes, the primary disadvantage of procedural implementations is that operations for computing input data for decisions, making decisions, and applying model operations are represented in the same way. This makes it difficult to extract individual decisions. A disadvantage of model-driven systems is that their model definitions are usually tied quite tightly to a particular set of recognition techniques, and it can be difficult to modify these systems to accommodate unanticipated requirements and/or new techniques [25].

In Section 3 we present the Recognition Strategy Language (RSL [11]), which provides a way to express recognition algorithms as decision sequences and automatically capture decision outcomes, while maintaining much of the flexibility of procedural implementations. In this chapter we present two procedural table recognition algorithms that have been successfully reimplemented in RSL (see Section 3.3). It may be worth investigating whether there would be benefits to having model-driven systems use a decision-based representation for the output algorithm. A model-driven system might compile a model definition into an RSL program, for example. In this way, decision-based specifications provide an intermediate level of abstraction between procedural and model-driven system specifications.

2.1 Table Structure and Terminology

The most abstract representation of a table is the underlying data set that the table visualizes; often a great deal of domain knowledge about the information represented in a table is needed to recover this. A slightly less abstract representation is the indexing structure, which describes how cells in the body of the table are indexed by header cells in the boxhead, stub, and stubhead of the table. As an example, in Figure 2, we might use a dot notation similar to that of Wang [26] to represent the grade in the top-left corner of the body as ((Term.1991.Winter,Assignments.Ass1),85). This also often requires linguistic and/or domain knowledge to recover as well [28], but to a lesser extent.

Closer to the physical structure of images and text files we have the table grid, which describes the arrangement of cells in rows and columns, and the location of ruling lines and whitespace separators. This is often represented by extending all separators to the boundaries of a table, and then assigning cells and separators to locations within the resulting grid.

The main structural elements of a printed table are illustrated in Figure 2. Cells of the table are organized in rows and columns. Some cells such as the column header 'Assignments' are *spanning cells*, which belong to more than one column and/or row of the table. Cells are further distinguished as *header cells*, which determine how *data cells* in the body of the table are indexed. Header cells which are indexed by other header cells are termed *nested row* and *nested column* headers, as appropriate.

Commonly there are four regions of a table, as illustrated in Figure 2. The *body* contains the data cells, which normally are the values intended to be most easily searched and compared within the table. Header cells appear in the *boxhead* (labeling columns) and *stub* (labeling rows, when present). Sometimes, as in Figure 2, there is a header in the top-left region of the table, called the *stub head*. An adjacent set of cells in the table body is referred to as a *block*.

3 The Recognition Strategy Language: Decision-Based Specification

As a first step towards being able to more easily observe intermediate decisions and combine decisions from different recognition algorithms, we have devised the Recognition Strategy Language (RSL [11]). RSL provides a level of formalization that lies between procedural and model-driven specifications (see Section 2). Based on notes from our survey of table recognition [8], the language allows models of table structure to be defined in terms of three basic decisions for regions in an input file: classification, segmentation, and parsing (binary relationships).

We refer to RSL as a decision-based specification because the basic unit of formalization defines the inputs and acceptable outputs for decisions. An

RSL program defines properties of interpretations, a single set of constant and variable parameters used for making decisions, and the sequencing of the decisions. Decision outcomes are provided at run-time by functions referred to in the decision specifications, but which are defined outside of the RSL program. Functions that provide the decision outcomes may use arbitrary techniques to arrive at a decision.

As currently defined, RSL is a simple functional scripting language influenced by the text-based approach taken in Tcl/Tk [29], with some similarities to an expert system shell [30]. Decisions in RSL are similar to rules in an expert system in that both define transformations of a knowledge representation, though RSL decisions define only possible transformations, and not concrete transformations as in the case with rules. The ability to trace rule applications in an expert system shell is replaced by record-keeping and annotation in RSL, and RSL specifications are sequential, and not declarative (expert system shells often support applying rules using various search algorithms).

RSL strategies may be interpreted for rapid development, or compiled. RSL has been implemented using the TXL language [31], which is a functional language for rapid prototyping that has been used for a wide array of applications including source code transformation, design recovery, and pattern recognition [32, 33].

In the remainder of this section we provide an overview of RSL and a simple example of an RSL program and its execution.

3.1 Transforming Logical Structure Interpretations in RSL

Interpretations of logical structure are represented by attributed directed graphs in RSL. Graph nodes represent geometric regions of an input file (R), and graph edges represent relations on regions. The *contains* relation defines the combination of regions into region segments (S), and additional binary relations may be defined (E). The main elements of an interpretation in RSL are:

- V, the set of expressible geometric locations in an input file, defined using a set of functions. Currently in RSL there are only two functions used to define V: one for bounding boxes, v_{BB} (defined by top-left, bottom-right corners), and another for polylines, v_l (a list of points).
- $R \subseteq (I \times V)$, the set of input file regions used in an interpretation (defines nodes of the graph). I is a set of legal identifiers. Each region in R has a unique identifier, but regions may have identical locations
- $S = \{(S_1 \subseteq R, r_1 \in R), \ldots, (S_n \subseteq R, r_n \in R)\}$, the set of region segments (regions containing other regions). Each set of regions $S_j \subseteq R$ is unique, and defines a region $r_j \in R$: currently in RSL this region is located at the bounding box of the regions in the segment (S_j)
- $C = \{(i_1 \in I, C_1 \subseteq R), \ldots, (i_n \in I, C_n \subseteq R)\}$, the regions associated with each region type identified by $i_j \in I$ (e.g. *Word*, *Cell*, *Row*, and *Column*)

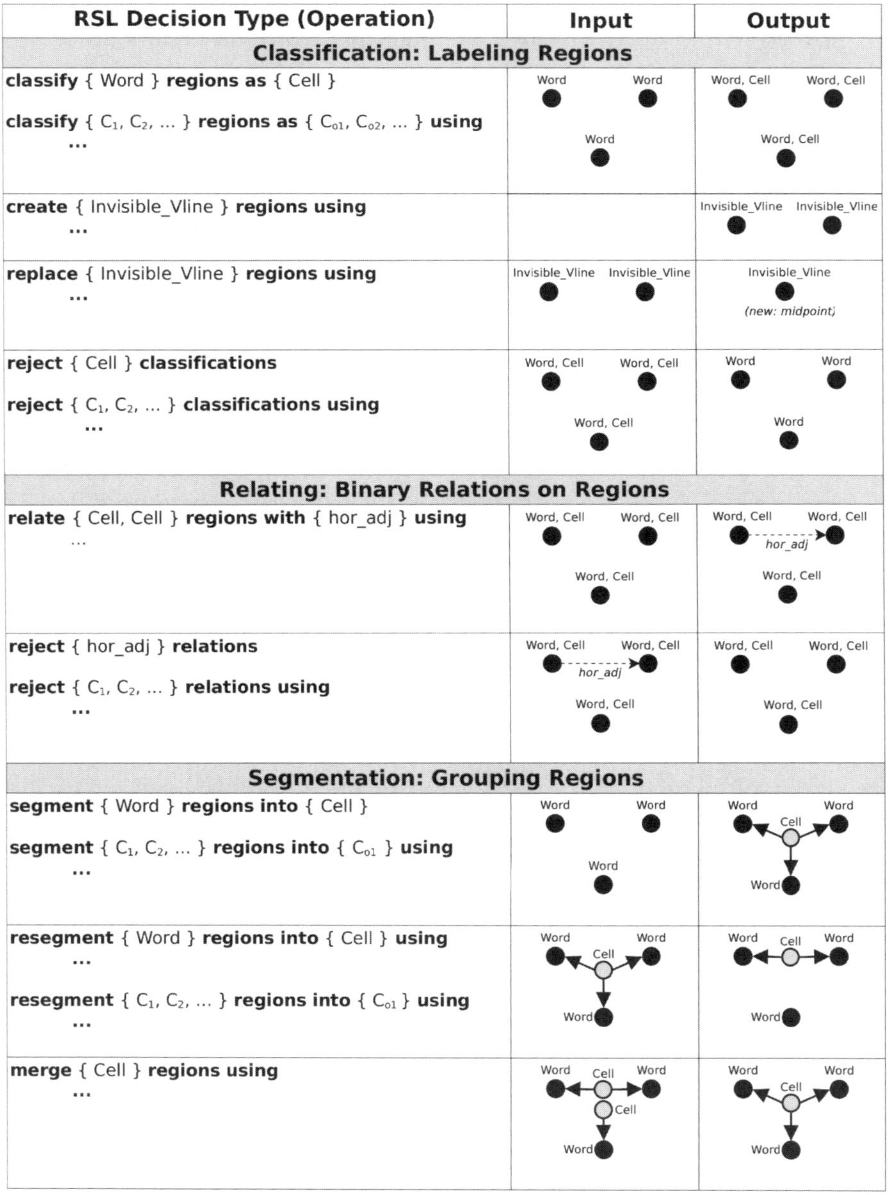

Fig. 3. RSL Decision Types for Transforming Interpretations. For the example inputs and outputs, only accepted hypotheses are shown.

- $E = \{(i_1 \in I, E_1 \subseteq R^2), \ldots, (i_n \in I, E_n \subseteq R^2)\}$, defines binary relations on regions for relation types identified by $i_j \in I$ (e.g. horizontal adjacency)
- A_0, the set of named attributes associated with elements of R, S, and E in the interpretation provided as input (I_0). Attribute values are lists of strings or floating point numbers

Input to an RSL program is an initial interpretation (I_0) that defines the initial sets of regions, segments, region classes, and region relations (R, S, C, and E) and their attributes (A_0). Within A_0, the file described by the interpretation is represented by a single region, with the name of the file provided as an attribute of the region. For example, image files may be represented by a region of type *Image* with a text attribute *FILE_NAME*. Currently only elements in the input interpretation I_0 are permitted to have additional attributes (to avoid side-effects [34]).

The output of an RSL program is a set of accepted interpretations, with each interpretation annotated with a complete history of the model operations that produced it. This allows all intermediate states of an interpretation to be recovered. While RSL supports the selection of multiple alternatives at each decision point [34], here we will consider only the case where each decision selects exactly one alternative, producing a single interpretation as output.

Figure 3 summarizes the available decision types for transforming individual interpretations in RSL. Shown on the left of each row is the first line of an RSL decision specification. Each decision type indicates the interpretation elements that define the set of possible outcomes at run-time. In the case of `create` and `replace`, the alternative outcomes are implicitly defined using the set of all possible input regions (V). Figure 5 illustrates how alternatives are produced for a few decision types.

Examples of input and output interpretations are provided for each of these decision types in Figure 3. Some alternate forms for the decision types are also given. For example, more than one region type may be used to define both the regions to classify and the possible classes in a `classify` operation.

Decisions that generate hypotheses either classify, segment, or relate regions, while the `reject` decision type discards hypotheses. The `replace`, `resegment`, and `merge` operations both assert and reject hypotheses. `replace` returns sets of regions of a given type, replacing each with a new region of the same type. `merge` and `resegment` reject a region type for regions, replacing these with new region segments of the same type. `merge` combines regions and segments into new ones, while `resegment` is more general, and may be used to split segments as well (see examples in Figure 3).

RSL uses a simple method for classification in which all input regions are classified as at most one of the possible output classes (some subset of the region types in C): selecting no class indicates rejection. Segmentation operators simultaneously define segments and assign a type to each segment (altering S and C). Relating operations update the region edge sets in E.

Additional decision types for altering parameters of decision functions and accepting and rejecting interpretations are defined within RSL [34]. Other than to accept all interpretations produced before a strategy completes (see bottom of Figure 4), these are unused in any of the strategies described in this chapter.

3.2 A Simple RSL Strategy

Figure 4 provides an example of a simple RSL strategy, and Figure 5 shows an example execution of the first three decisions of the *main* function. The input interpretation shown in Figure 5 contains three *Word* regions. As discussed in the previous section, the decision type (first line) of each decision operator defines a fixed set of possible outcomes for the associated external decision function. External decision functions and the set of possible alternatives are shown in rounded boxes in Figure 5. Selected alternatives are returned as text records, which are first validated and then used to specify a transformation to apply to the interpretation. RSL records all the decision outcomes, and annotates changes made to the interpretation within the interpretation itself when applying a transformation.

In Figure 4, the `model regions` and `model relations` sections define the set of region and relation types for interpretations used in the strategy (i.e. define the elements of C and E). Any other relations or region types in the input interpretation are left intact, but otherwise ignored by the strategy. Decisions which refer to types not provided in these sections will produce a syntax error. The `recognition parameters` section defines a series of static (constant) and adaptive (variable) parameters for use in the external decision functions. Static parameters are indicated using a prefix 's' (e.g. *sMaxHorDistance* in Figure 4), and adaptive parameters using a prefix 'a' (e.g. *aMaxColSep* in Figure 4). All parameters may be string or floating point number-valued.

Control flow is specified in RSL using *strategy functions*, which define a sequence of decision operations and calls to other strategy functions. Recursion is permitted: see the *recursiveStrategy* in Figure 4. Each strategy function takes the current set of interpretations and parameter values as input, and produces a new set of interpretations as output. For conciseness, this is implicit rather than explicit within an RSL program (i.e. syntactically strategy functions look like procedures, but they are in fact functions).

Following convention, execution begins with the *main* strategy function, which is passed the input interpretation and the initial set of values for the adaptive parameters defined in the `recognition parameters` section.

There is only one form of conditional statement in RSL which acts as a guard for strategy functions. This statement prevents interpretations not selected by an external decision function from being altered by a strategy function, and must appear at the beginning of a strategy function declaration.

For example, in Figure 4 a `for interpretations` statement is used to define the stopping point for the *recursiveStrategy*. The statement given calls an external decision function *adjacencyIncomplete* for each current interpretation, using the *aMaxColSep* and *aMaxHorSep* parameters. *adjacencyIncomplete* is required to specify which of the current interpretations the *recursiveStrategy* may be applied to.

The `observing` keyword is used to add visible types to the interpretations passed to external decision functions; by default, only the interpretation elements that may be altered by the decision are visible (these are called the *scope* types). For example, for classification operations, the visible regions include those associated with the types of input regions to be classified, but none of the possible *output* classes are visible unless they are also an input type or are explicitly added using `observing`. The `create` decision type is unique in that by default *no* region types are visible in the interpretations passed to the decision function: all visible types must be explicitly listed in an `observing` statement.

However, the sets of regions and region segments ($R \cup S$) in the interpretation are always visible to all external decisions. Consider Decision 2 in Figure 5, where the regions associated with *Word* are visible, but *not* their type. As one would expect, only parameters explicitly passed to the external decision function are visible for each decision.

3.3 Handley and Hu et al. Algorithms in RSL

As a proof of concept, we have re-implemented two table structure recognition algorithms in RSL [1, 12]. All external decision functions were implemented using TXL. These algorithms were chosen because they were part of a small set of algorithms described in enough detail to permit replication, and because they exhibited a degree of sophistication. Both algorithms are described procedurally, as sequences of operations.

As a brief summary, both algorithms recover table structure given a list of input *Line* and *Word* regions (note that the Hu et al. algorithm ignores the *Line* regions). The Handley algorithm uses a strictly geometry-based approach, in which *Word* regions are all hypothesized as *Cell* regions, and then merged in a series of steps which makes use of weighted projection profiles and cell adjacency relationships. The Hu et al. algorithm starts by detecting columns based on a hierarchical clustering applied to the horizontal spans of *Word* regions, detects the table boxhead, stub and rows of the table, and finally *Cells* in the body of the table. The Hu algorithm makes use of limited (but powerful) lexical information in its analysis; text in input *Word* regions are classified as alphabetic or alphanumeric. Within RSL, the text of words is represented as attributes of the input *Word* regions (i.e. the word text is defined within A_0).

Reflecting their different intended applications, the Handley algorithm seeks only to recover table cells, while the Hu algorithm returns detected

```
model regions
    Word Cell Header Entry Image
end regions

model relations
    adj_right close_to
end relations

recognition parameters
    sMaxHorDistance 5.0 % mms
    aMaxColSep 20.0 % mms
    aMaxHorSep 25.0 % mms
end parameters

strategy main
    relate { Word } regions with { hor_adj } using
        selectHorizAdjRegions(sMaxHorDistance)

    segment { Word } regions into { Cell } using
        segmentHorizAdjRegions()
        observing { hor_adj } relations

    classify { Cell } regions as { Header, Entry } using
        labelColumnHeaderAndEntries()

    ...
    recursiveStrategy

    accept interpretations
end strategy

strategy recursiveStrategy
    for interpretations using
        adjacencyIncomplete(aMaxColSep,aMaxHorSep)
        observing { Word, Cell } regions { close_to } relations
    ...
    resursiveStrategy
end strategy
```

Fig. 4. A Simple RSL Strategy

rows, columns, and cell indexing structure for the table. Please note that we have modified the Hu algorithm with an extra step to define textlines using a simple projection of *Word* regions onto the Y-axis, defining textlines at gaps greater than a given threshold value. This was necessary because the algorithm was originally specified for detecting table structure within text files.

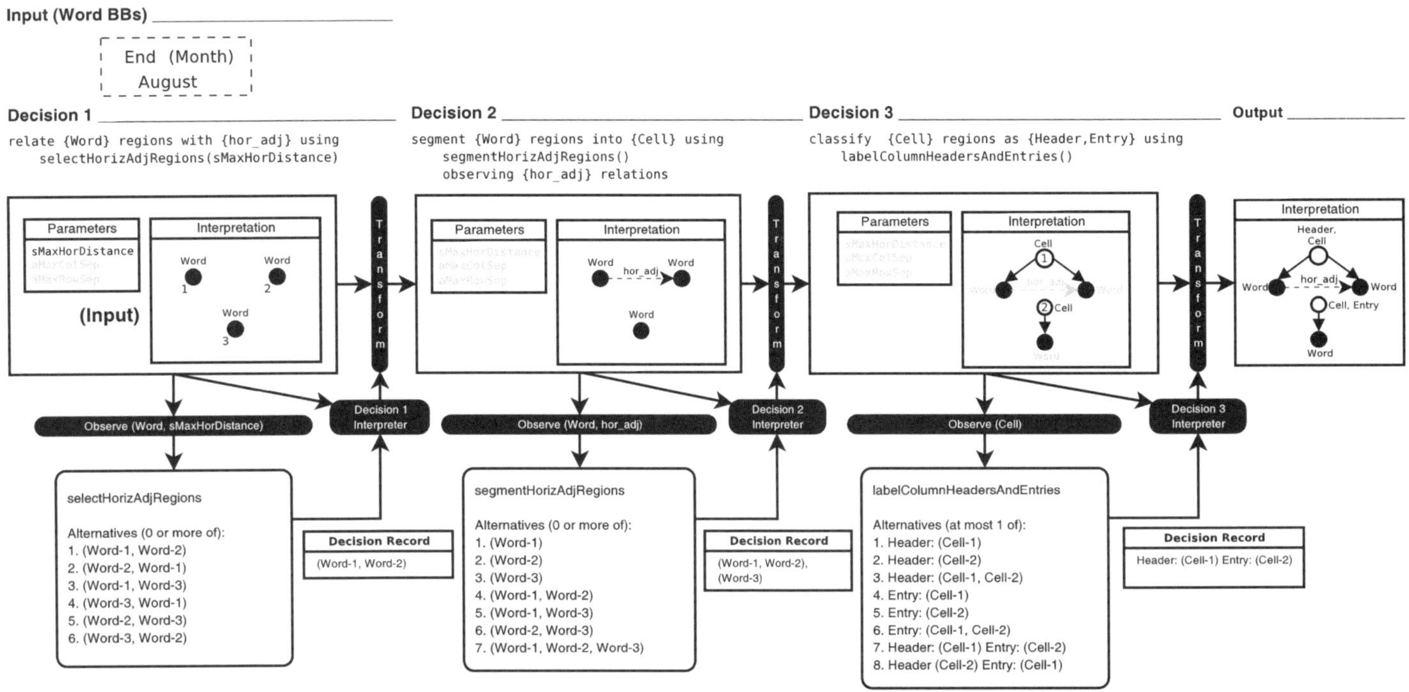

Fig. 5. Execution of the the First Three Decisions for the RSL Strategy Given in Figure 4. Three *Word* regions are provided in the input interpretation, which are 1) related by horizontal adjacency, 2) segmented into *Cell* regions, and 3) classified as *Header* or *Entry* (body) cells. For each decision, parameters and interpretation elements which are not visible to the external decision function are written in light gray. Each external decision function returns a text-based record. A decision interpreter validates chosen alternatives before they are used to transform the current interpretation to produce the input interpretation for the next decision or the output

RSL specifications for the two algorithms are available [34]; the Handley algorithm is specified in roughly 540 lines of RSL, and the Hu algorithm in about 240 lines. These lengths include significant commenting, but not code for the external decision functions. The external functions were implemented in roughly 5000 lines of TXL for the Handley algorithm, and roughly 3000 lines of TXL for the Hu algorithm. Small changes were made to these RSL strategies in order to produce the results shown later in this chapter, in particular renaming common region types to make them more explicit. Some bugs in the implementation of the external decision functions for the Handley algorithm were also corrected, and the performance results provided later in this chapter for the Handley algorithm are better than those reported earlier [34, 35] as a result.

4 Static Analysis of RSL Specifications

Useful insights can be gained from examining static dependencies within an RSL specification, and by comparing static dependencies between algorithms. In this section, we illustrate this process using Figure 6, which provides table model summaries for the Handley and Hu algorithms. First we describe how these table model summaries are derived from a static analysis of the RSL specifications. Next we discuss insights obtained from examining Figure 6, regarding comparisons and contrasts between the Handley and Hu algorithms. This discussion is based on static analysis; dynamic aspects of the algorithms are treated in Section 5.

4.1 Construction of Table Model Summaries

RSL specifications consist of a sequence of decision operations. Each decision operation has an associated set of regions and relation types, a decision function, and decision parameters. Each region or relation type T that may be altered by an RSL decision operator R has four dependencies:

1. On region types used to define the output model operations for R. These are the scope types provided as input to a decision (see Section 3.2)
2. On observed region or relation types (these follow the `observing` keyword)
3. On the decision function used for R
4. On parameters used by the decision function for R

These per-decision dependencies may be used to construct data dependency graphs. These graphs describe how inputs are used to produce outputs, and are commonly used in software engineering and analysis [36]. Once a data

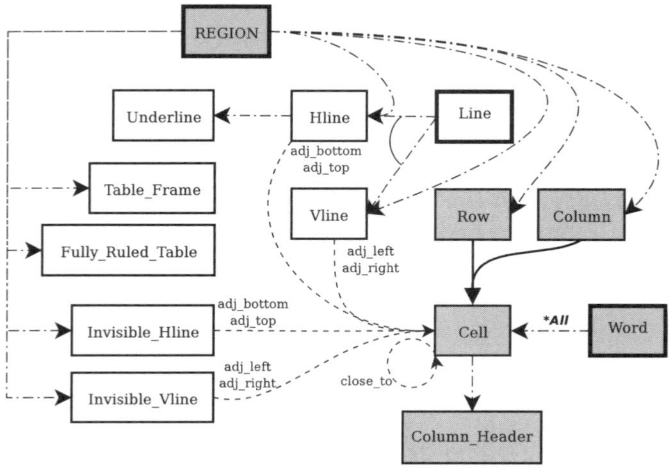

a. Handley

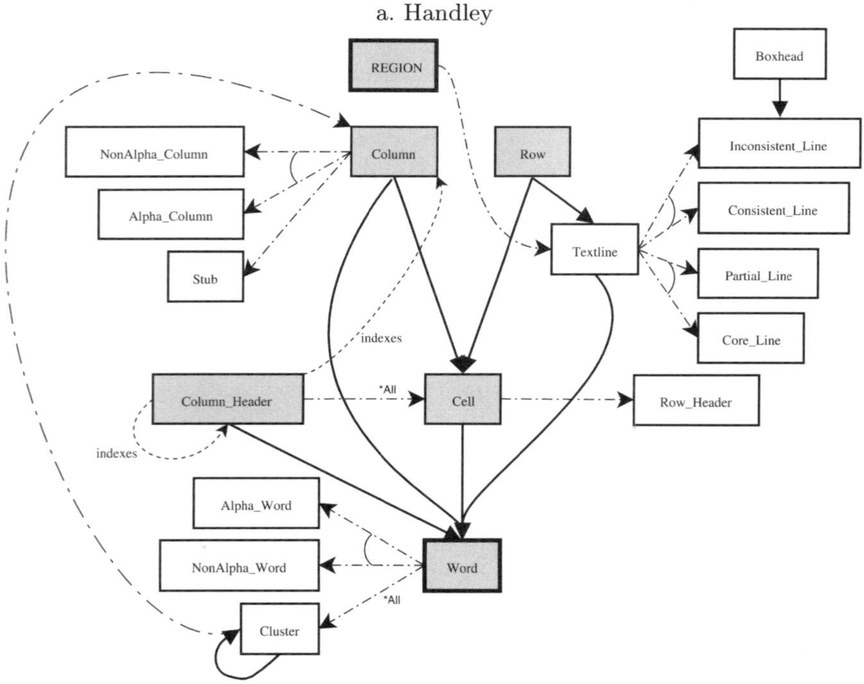

b. Hu et al.

Fig. 6. Table Model Summaries for Two Algorithms: (a) Handley and (b) Hu et al. Boxes represent region types, with thick borders indicating input types (*Word, Line*, and *REGION*). Gray boxes show types common to both algorithms. Labeled dashed lines are relations between region types. Solid lines represent segmentation operations, and dash-dotted lines represent classification operations dependency graph has been constructed, additional analyses of an RSL specification can be made. For example, we can automatically determine the set of decision operations and types associated with a particular decision function. As another example, we can determine which region and relation types may be affected by a given decision parameter. These analyses are analogous to the techniques of backward and forward *program slicing* [37], respectively (please note that slicing requires information about decision sequences not shown in Figure 6).

The summaries of table model structure shown in Figure 6 are simple data dependency graphs which are produced as follows. We compute only dependencies of type 1: dependencies between output types and scope types that define the alternative outcomes. Then we filter all operations that merge and reject regions and relations. The resulting graph summarizes the table model in terms of relationships between scope types and output types. Figure 6 does not represent reject and merge operations because these only modify interpretations within a single type, either removing or combining elements of that type, and we wish primarily to represent relationships between types. Alternative summaries that incorporate merge and reject operations are of course possible.

4.2 Discussion of Table Model Summaries

In Figure 6, boxes represent region types. The three input types (*Word*, *Line*, and *REGION*) are shown in boxes with thick borders. Region types common to both algorithms are shown in gray boxes. Relations between region types are represented using labeled dashed lines (e.g. *indexes* for the Hu algorithm).

The three basic inference types used in RSL are represented using different arrow types: segmentation operations are represented by solid lines, classification by dash-dotted lines, and relations by labeled dashed lines. Segmentation and relation dependencies are drawn as arrows from output types to the scope types on which they depend. For ease of reading, we have reversed the arrow direction for classification operations; arrows representing classification are drawn from scope types (those to be classified) to output types.

To indicate where classifications have more than one output class, connecting arcs are used. For example, in the Handley algorithm, *Line* regions may be (exclusively) classified as horizontal (*Hline*) or vertical (*Vline*) lines, or neither (see Section 3). The annotation **All* is used to indicate labelings, trivial classifications where all regions of the input type have been labeled as the output type. For example, for at least one decision all *Word* regions are labeled as *Cell* regions in the Handley algorithm, and *Cluster* regions in the Hu algorithm. Though neither implemented strategy does so, arcs could also be used to represent segmentation operations which combine multiple region types into a segment (region) type.

The type *REGION* includes the set of all input regions expressible in RSL. Currently this is all bounding boxes and polylines expressible within the input image, as defined by the set V (see Section 3.1). Elements of *REGION* are not promoted directly to a model region type unless a `create` or `replace` operation is used. In the Handley algorithm, many input regions are directly promoted to various types after geometric analyses (e.g. after projecting cells and finding minima in histograms, to define rows and columns). In the Hu algorithm, only *Textline* regions are produced by directly classifying input regions, in the preprocessing step that we added to the algorithm.

The graphs shown in Figure 6 can be interpreted similarly to semantic networks [38]. Segmentation edges correspond roughly to 'has-a' edges, and classification edges correspond roughly to 'is-a' edges, with the remaining edges defining other binary relationships (e.g. adjacency). Unlike a semantic net, non-binary relationships are represented in the graph, using and-or relationships. In this way, each unique set of relationships between scope and output types are represented separately, as an 'or' of 'ands'.

To illustrate the information that can be read directly from Figure 6, consider the Textline regions in the Hu algorithm. The graph edges connecting to the Textline box in Figure 6b tell us the following:

1. *Textline* regions may be segmented into *Row* regions
2. *Word* regions may be segmented into *Textline* regions
3. Image *REGION*s may be classified as a *Textline* region
4. A *Textline* region may be classified as either an *Inconsistent_Line* or *Consistent_Line*, or neither
5. A *Textline* region may be classified as either a *Partial_Line* or *Core_Line*, or neither

Despite their simplicity, these table model summaries provide useful information for analyzing the implemented algorithms. First we discuss the region types which are common and unique to each algorithm. Both algorithms utilize *Word, Cell, Row, Column,* and *Column_Header* regions. However, the *Handley* algorithm takes lines (underlines and ruling lines in the table) into account, and defines spatial relationships that are not used in the *Hu* algorithm. The Hu algorithm on the other hand makes greater use of classification operations, particularly for *Column, Textline,* and *Word* regions. The Hu algorithm also explicitly defines *Boxhead* and *Stub* regions, which the Handley algorithm does not.

Figure 6 also shows interesting differences between the relationships that occur among the common regions. In the Handley algorithm, *Cell* regions are classified as *Column Header* regions, while at some point in the Hu algorithm, all *Column Header* regions are classified as *Cell*s. In the Handley algorithm, *Column* and *Row* regions contain *Cell*s. In contrast, the Hu algorithm composes *Column* and *Row* regions as follows: *Column* regions contain either *Cell* or *Word* regions (but not both), whereas *Row* regions contain either *Cell* or *Textline* regions, but not *Word* regions. The Hu algorithm defines an indexing relation from column headers to *Column*s of headers, while the Handley algorithm has no representation of indexing structure (as the algorithm was not designed to address that problem).

This simple table model summary provides a useful course-grained view of similarities and differences between the table models used by these two algorithms. The relationships provided in the table model summaries are also useful when debugging and during evaluation, as we will see in the next section.

5 Evaluating the Accuracy of Decisions

The three main criteria for evaluating a recognition algorithm are accuracy, speed, and storage requirements. Here, we concern ourselves solely with accuracy but we acknowledge that speed and storage requirements are nearly as important for real-world applications (e.g. for on-line interactive applications), and that some form of trade-off often needs to be made between these three criteria.

Evaluating the accuracy of an algorithm means assessing the ability of the algorithm to produce interpretations that meet the requirements of a given task [10]. This is normally done using test data, for which the ground-truth is known (significant difficulties in defining ground truth for tables have been discussed in the literature [39, 40]). Normally evaluation of recognition algorithms focuses on comparing the final interpretations accepted by algorithms [3, 41, 42, 43, 44, 45].

In this section we use the decision-based approach to evaluate accuracy of recognition by considering the decision process used to produce results. We discuss characterizing individual decisions made by a recognition algorithm as good or bad (Section 5.1), and we augment the traditional measures of recall and precision with the new measures of *historical recall* and *historical precision* (see Section 5.2).

5.1 Evaluating the Accuracy of Individual Decisions

Our goal is to measure the accuracy of individual recognition decisions, and the accuracy of sequences of recognition decisions. This detailed information is useful for planning improvements to a recognition algorithm, and provides a basis for learning algorithms which seek to automatically improve recognition performance.

In our evaluation of decision accuracy, we are concerned only with decisions which affect the comparison with ground truth. It is common for algorithms to hypothesize many objects and relationships that aren't part of the interpretation space used for evaluation. For example, the algorithm comparisons we report in Section 6 use a ground truth that defines the location of cells, but the ground truth does not identify which subset of cells are header cells. The reason we did not record header cells in our ground truth is that one of the algorithms does not aim to identify all header cells. The ground truth, and the comparisons based on ground truth, must be restricted to objects and relationships that are identified by all the algorithms.

In order to define what constitutes a 'good' decision, we first define *correct*, *complete*, and *perfect* decisions. Each decision to apply an interpretation model operation results in asserting and/or rejecting hypotheses. In RSL, each decision record produced by a decision function corresponds to a set of model operations to apply to the current interpretation, chosen from a space of alternative model operations (see Figure 5). Examples for decision types are provided in Section 6.3.

A decision is *correct* if all operations selected generate only valid hypotheses and/or reject only invalid hypotheses. A decision is *complete* if it selects the set of all correct operations in the set of alternatives that *alter* the interpretation (e.g. re-classifying a *Word* as a *Cell* is no more complete than not selecting this redundant operation). A *perfect* decision is both correct and complete; all selected alternatives are correct, and all correct alternatives are in the set. For the case where no alternatives are selected, this is either a perfect decision (when all alternatives are incorrect), or an incomplete decision (when correct alternatives exist).

Using these definitions, 'good' decisions lie somewhere between perfect decisions and those that are totally incorrect. One could characterize a decision using recall and precision metrics (see the next subsection), to give the proportion of correct alternatives selected, and the proportion of selected alternatives that are correct. These metrics could then be thresholded, or used to assign fuzzy membership values for the set of 'good' decisions. More informally, one might consider any decision that is perfect, correct but incomplete, or complete and mostly correct to be a 'good' one.

We could characterize a sequence of decisions (at those decisions for which evaluation information exists) similarly, in terms of distributions of recall and precision for selecting valid model operations. The metrics might also be weighted based on the location of valid operations within the sequence of decision outcomes, to weight earlier decisions more heavily for example.

These are internal performance measures which characterize decisions based on their associated alternative outcomes. While we will only touch on these briefly here, we believe that these may provide a basis for devising table recognition algorithms based on game-theoretic principles such as mini-max optimization [46, 10].

5.2 Historical Recall and Precision

The traditional detection metrics recall and precision measure how similar an algorithm's final interpretation is to the ground truth interpretation. Here we introduce historical versions of these measures [35]. Informally stated, the historical measures give an algorithm credit for correct hypotheses that

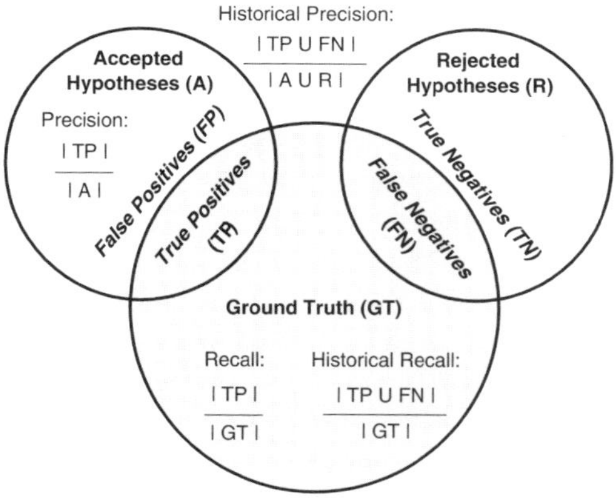

Fig. 7. Recall, Precision, Historical Recall, and Historical Precision

it made somewhere along the way, even if the algorithm later rejected these hypotheses. The historical measures can be evaluated at any point during algorithm execution; this information provides valuable insight into the algorithm's treatment of hypothesis generation and rejection.

Figure 7 illustrates sets of hypotheses and assertions used in our discussion. At a given point in time, the set of generated hypotheses produced by an algorithm (e.g. cell locations) is defined by the union of accepted (A) and rejected (R) hypotheses. We assume that at any given time, every hypothesis is either accepted or rejected, but not both. The validity of individual hypotheses within A and R is determined using GT, a set of ground truth declarations which are taken to be valid (e.g. a set of cell locations taken to be valid). The set of true positives (TP) is defined by the intersection of accepted hypotheses and ground truth ($A \cap GT$). Similarly, the set of false negatives (FN), which consists of ground truth elements that have been proposed and rejected, is defined by the intersection of rejected and ground truth elements ($R \cap GT$).

Also shown in Figure 7 are recall and precision metrics, which describe the ratio of true positives to recognition targets ($|TP|/|GT|$) and accepted hypotheses ($|TP|/|A|$), respectively. *Historical* recall and precision describe the recall and precision of the set of generated hypotheses ($A \cup R$). Together, the true positives and false negatives comprise the set of ground truth elements that have been generated ($TP \cup FN$). Historical recall is the proportion of ground truth hypotheses that have been generated ($|TP \cup FN|/|GT|$), while historical precision is the proportion of generated hypotheses that match ground truth ($|TP \cup FN|/|A \cup R|$). Note that if no hypotheses are rejected (i.e. $R = \{\}$), then the 'conventional' and historical versions of recall and precision are the same. The key difference here is that the historical metrics take rejected hypotheses into account, while the conventional ones do not. For an example of this, compare Figure 11a to Figure 11b; in Figure 11b, cell hypotheses are never rejected.

Conventional and historical recall can be directly compared, as they both describe coverage of the set of ground truth elements. Note that historical recall will always be greater than or equal to recall (refer again to Figure 11). Also, historical recall never decreases during a recognition algorithm's progress, while recall may increase or decrease at any point. The difference between historical and conventional recall is the proportion of recognition targets that have been falsely rejected ($|FN|/|GT|$).

It is harder to relate conventional and historical precision. Precision measures the accuracy of what is accepted as valid, while historical precision measures the accuracy (or *efficiency*) of hypothesis generation. Put another way, historical precision quantifies the accuracy of hypotheses that the algorithm generates and accepts at some point.

6 Decision-Based Comparison of Algorithms in RSL

In this section we will compare the recognition accuracy of our RSL implementations of the Handley and Hu et al. table structure recognition algorithms. We first consider a conventional results-based evaluation, in which the complete sequence of recognition decisions are evaluated as a whole, without reference to rejected hypotheses. We then contrast this with a decision-based comparison, in which the effects of individual decisions may be observed, and rejected hypotheses are taken into account. Using this information, we then design a new strategy which combines the observed strengths of the two algorithms, to produce a better final result. All metrics presented here are in terms of 'external' accuracy, i.e. we compare the state of the interpretation to ground truth after each decision affecting the hypothesis types in question.

Our goal here is not to evaluate these two algorithms in any real sense, but to illustrate decision-based comparisons of algorithms. We will consider only results for a single, reasonably challenging table as input, on which we will try (informally) to optimize recognition. For a real-world application we would train these algorithms by optimizing performance metrics such as conventional and historical recall and precision over a representative sample of the set of tables that we wish to recognize.

Input to both algorithms is a set of *Word* regions with an associated text attribute (set to 'a' for words containing mostly alphabetic characters, and '1' otherwise), and a set of *Line* regions (as seen in Figure 1b). All words in cells were provided as input; any words not within the table (e.g. the table title and footnotes) were not provided.

As can be seen in Figure 6, the Hu algorithm does not pay any attention to the *Line* regions, and leaves them in the produced interpretation untouched. As the classification decisions shown in Figure 6 suggest, the Hu algorithm makes use of the text attribute associated with words; the Handley algorithm makes its analysis based on region geometry and topology alone, and ignores these attributes.

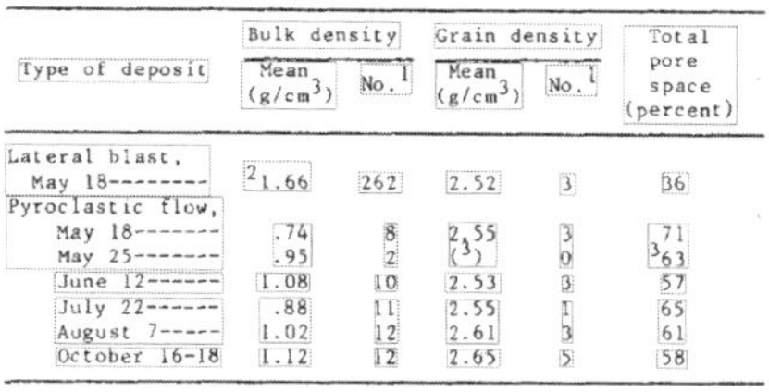

Type of deposit	Bulk density Mean (g/cm3)	No.[1]	Grain density Mean (g/cm3)	No.[1]	Total pore space (percent)
Lateral blast, May 18--------	[2]1.66	262	2.52	3	36
Pyroclastic flow, May 18-------	.74	8	2.55	3	71
May 25-------	.95	2	(3)	0	[3]63
June 12------	1.08	10	2.53	3	57
July 22------	.88	11	2.55	1	65
August 7-----	1.02	12	2.61	3	61
October 16-18	1.12	12	2.65	5	58

a. Handley Cells

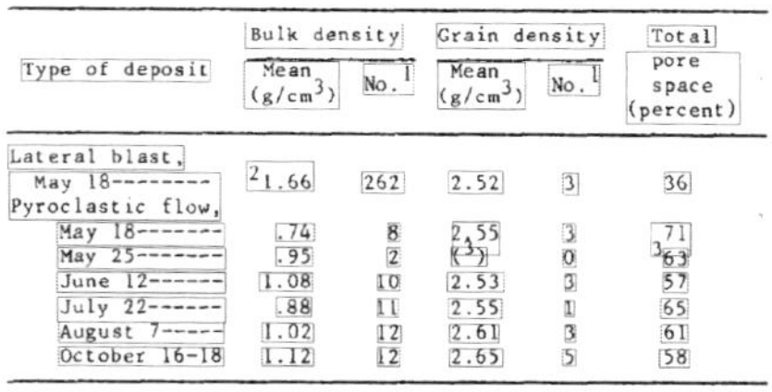

Type of deposit	Bulk density Mean (g/cm3)	No.[1]	Grain density Mean (g/cm3)	No.[1]	Total pore space (percent)
Lateral blast, May 18--------	[2]1.66	262	2.52	3	36
Pyroclastic flow, May 18-------	.74	8	2.55	3	71
May 25-------	.95	2	(3)	0	[3]63
June 12------	1.08	10	2.53	3	57
July 22------	.88	11	2.55	1	65
August 7-----	1.02	12	2.61	3	61
October 16-18	1.12	12	2.65	5	58

b. Hu et al. Cells

	Handley	Hu et al.	Common
TP	25	45	24
FP	13	7	0
S-GT	0	3	0
M-GT	27	4	4
SM-GT	0	0	0
O-GT	0	0	0
FA	0	0	0
Recall	48.1%	86.5%	
Precision	65.8%	86.5%	

c. Cell Hypothesis Sets and Metrics

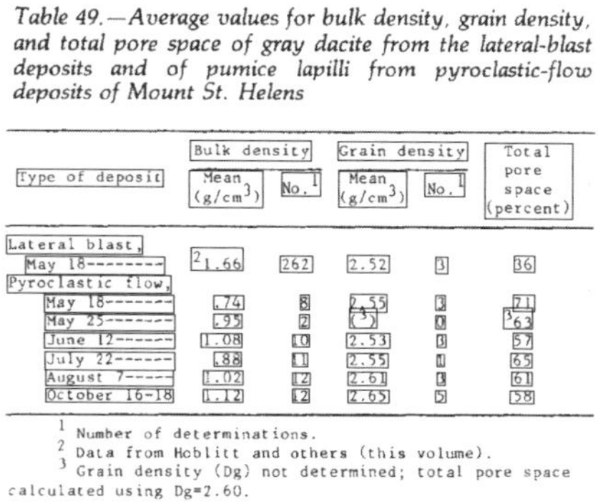

Table 49.—Average values for bulk density, grain density, and total pore space of gray dacite from the lateral-blast deposits and of pumice lapilli from pyroclastic-flow deposits of Mount St. Helens

Type of deposit	Bulk density Mean (g/cm3)	No.[1]	Grain density Mean (g/cm3)	No.[1]	Total pore space (percent)
Lateral blast, May 18--------	[2]1.66	262	2.52	3	36
Pyroclastic flow, May 18-------	.74	8	2.55	3	71
May 25-------	.95	2	(3)	0	[3]63
June 12------	1.08	10	2.53	3	57
July 22------	.88	11	2.55	1	65
August 7-----	1.02	12	2.61	3	61
October 16-18	1.12	12	2.65	5	58

[1] Number of determinations.
[2] Data from Hoblitt and others (this volume).
[3] Grain density (Dg) not determined; total pore space calculated using Dg=2.60.

d. Ground Truth Cells

Fig. 8. Cell Output for UW Database Table (Page a038). Shown in (c) are the cell hypothesis set sizes (*True Positives (TP), False Positives (FP), Split Ground Truth (S-GT), Merged Ground Truth (M-GT), Split-and-Merged Ground Truth (SM-GT), Missing Ground Truth (O-GT), and False Alarms (FA)*) and recall and precision metrics. For each hypothesis set type, the size of the intersection of the Handley and Hu sets is shown in the *Common* column

6.1 Conventional Evaluation

Normally in the table recognition literature when comparing two algorithms, we consider only the final interpretations, for example as shown in Figure 8. Shown are the final cell hypotheses for both algorithms, along with the ground truth interpretation they were evaluated against. Also shown are the sets of true positive, false positive, and errors along with the resulting recall and precision metrics. No errors of omission (cells whose words are entirely missed) or 'false alarms' (cells whose contents do not belong to any ground truth cells) are made, because we provide all words in cells, and only words in cells in the input. There are also no 'spurious' cells (splitting and merging of ground truth producing many-to-many matches with ground truth cells). The approach to error analysis we are using here is based on that of Liang [44]. However, we are presenting errors in terms of ground truth cells here;

for example, the Hu algorithm produces seven false positives, which incorrectly split three ground truth cells and merge four. The Handley algorithm produces thirteen false positives, which incorrectly merge 27 cells.

As can be seen at a glance, both algorithms assign words to the appropriate column. It is the decisions which assign words and cells to rows that have produced the errors in the final interpretations.

In the rightmost column of Figure 8c, we've shown the size of the intersection of each cell hypothesis set type. 24 of the 25 true positives proposed by the Handley algorithm are also proposed by the Hu algorithm; the remaining cell is the 'Total pore space (percent)' column header (see Figure 8d). In Figure 8c we can see that the false positives proposed by the two algorithms are disjoint; the algorithms make different errors.

Figure 8c also describes how ground truth cells are mis-recognized by the two algorithms (i.e. what specific errors are associated with the false positives). All four of the cells incorrectly merged with words from other cells by the Hu algorithm are also over-merged by the Handley algorithm ($M - GT$). Looking at Figures 8a, 8b, and 8d, two of these over-merged cells are located in the leftmost column, and two are located in the table body in the fourth and sixth columns, near the superscripted threes (^{3}). All remaining errors for the Handley algorithm result from merging cells across rows of the table ($M-GT$), while the Hu algorithm also splits ($S - GT$) two cells near the superscripted threes, and splits the rightmost column header ('Total pore space (percent)', as mentioned earlier).

Looking at the recall and precision metrics, the Hu algorithm has higher values for both. From this and our prior analysis, it appears that the Hu algorithm performs better recognition of the cells in this particular table (i.e. makes better decisions) than the Handley algorithm.

6.2 Metrics for Individual Decisions

We now have a reasonably detailed comparison of the outputs of the two algorithms. We also have determined which types of decisions might be studied more closely in order to improve recognition of the table shown in Figure 8: those that affect cells and rows. So our next question is this: which are the decisions that need to be improved, and where are they within our algorithm implementations?

Currently in common practice, answering this question would be dealt with in an ad-hoc way, often by outputting intermediate interpretations at various points, evaluating them, and then trying to determine which parts of the algorithm implementation do not generate or filter cell and row hypotheses as needed. This often produces informal and partial error analyses. We believe that the flaws in this process are in large part due to the effort required to detect errors in decision-making when decisions invoking model operations are not distinguished within the syntax of the implementation language, nor in the output interpretations.

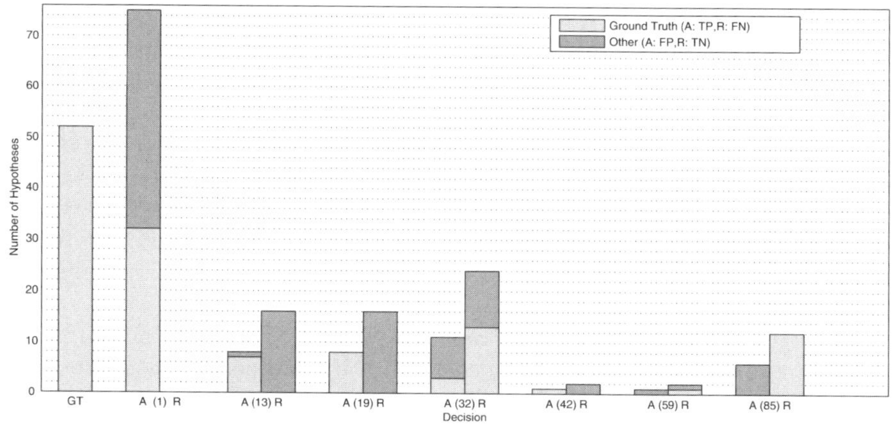

a. Changes in Cell Hypotheses

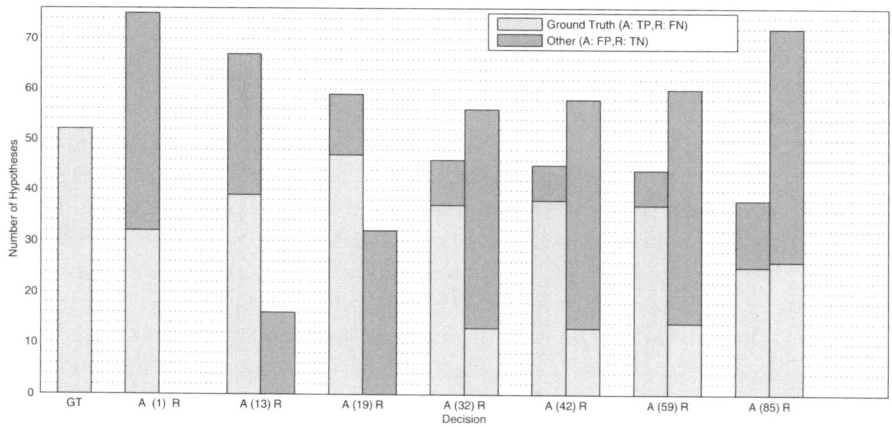

b. Cumulative Cell Hypotheses

Decisions

1 All words classified as cells
13 Merge cells with little horizontal separation which overlap vertically by roughly more than half the height of the taller bounding box
19 Merge cells within columns that overlap vertically by roughly more than half the height of the taller bounding box
32 Merge cells sharing column and row assignments
42 'Total pore space (percent)' header cell detected in boxhead
59 Merge cell in leftmost column alone in its row with the cell below ('Pryoclastic flow,')
85 Merge cells sharing estimated line and whitespace separators for rows and columns

Fig. 9. Detection of Cells Shown in Figure 1 by the Handley Algorithm. *A* represents accepted hypotheses, and *R* represents rejected hypotheses

This is where the benefits of decision-based specification become most apparent. The syntax of a decision-based specification language such as RSL explicitly represents decisions to invoke model operations; algorithms are specified as sequences of possible model operation applications. Further, the outputs of a decision-based language contain the entire history of model operation applications, including *which decision(s) selected them.* Hypotheses are uniquely identified, and their complete history of creation, rejection, and reinstatement are available in the output. Detecting decisions that cause errors and reverting the output to intermediate states is carried out using simple filtering and transformation of the output interpretation, rather than requiring guesswork.

Changes in accepted cell hypotheses for the Handley algorithm are shown in Figure 1, with each change indexed by the decision which produced it (each corresponds to a decision operation in the RSL specification). Decisions concerning cells which had no effect on the interpretation are not shown.

Figure 9 provides additional information about each decision in the Handley algorithm that changed the cell hypotheses. At the bottom of the figure a brief summary for each of these decisions is provided. Figure 9a illustrates the changes made by each decision to the accepted and rejected sets of cell hypotheses. These are shown as the number of newly proposed or rejected cells that are accepted by each decision, and the number of accepted hypotheses which are rejected by each decision. On the far left of the graph the number of cells in ground truth is shown for comparison. For each decision we show the decision number (in parentheses), and the number of cell hypotheses added to the accepted (A) and rejected (R) sets. Ground truth hypotheses are shown in light gray (true positive for A, false negative for R), and other hypotheses are shown in dark gray (false positive for A, true negative for R).

Figure 9b illustrates the sets of accepted and rejected hypotheses after each decision, i.e. the cumulative effects of the changes shown in Figure 9a. Again, light gray is used to illustrate the number of accepted and rejected hypotheses that match ground truth.

The cells produced by decisions in the Hu algorithm are shown in Figures 10a and 10b. Unlike the Handley algorithm, the Hu algorithm revises cell hypotheses only twice for our example table. This is because the Hu algorithm detects rows and columns before cells, and detects boxhead and stub head cells before body and stub cells.

Similar to Figure 9a, Figure 10c illustrates changes in the sets of accepted and rejected cell hypotheses for the Hu algorithm, while Figure 10d illustrates the cumulative effect of the decisions (similar to Figure 9b). Because the Hu algorithm never rejects cell hypotheses, R is always empty.

6.3 Evaluation and Error Analysis for Individual Decisions

Let us briefly try to characterize which are the 'good' decisions shown in Figures 9 and 10. As discussed in Section 5, an *internal* evaluation of decisions

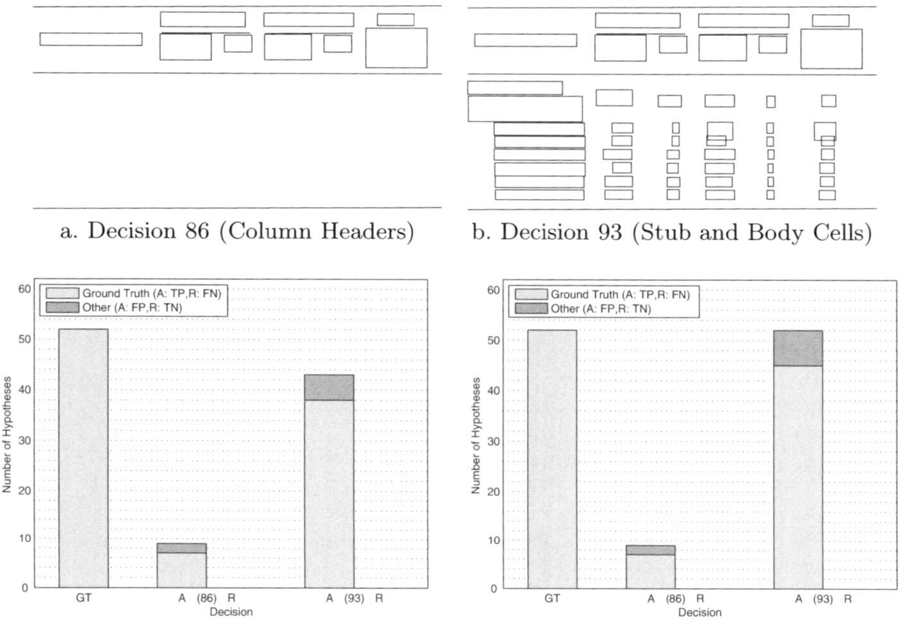

a. Decision 86 (Column Headers)   b. Decision 93 (Stub and Body Cells)

c. Changes   d. Cumulative

Fig. 10. Cell Detection by Hu et al. Over Time. *A* represents accepted hypotheses, and *R* represents rejected hypotheses

is with respect to their associated alternative outcomes, while the type of metrics presented in Figures 9 and 10 are *external* evaluations, which compare interpretations after each decision to goals (i.e. ground truth).

In terms of an internal evaluation, the Handley decision at time 42 is a *perfect* decision which is both correct and complete; that is to say, of all the possible cell merges, only the single valid alternative is selected. All other cells used to define the possible merges are already correct or over-merged cells (as can be seen from the cells shown at time 32 in Figure 1). The Handley decision at time 19 is not a perfect decision, but we will claim that it is quite good, as it is entirely correct (only valid merges are selected), though incomplete: some valid merges within the set of alternatives are not selected, e.g. for some of the column headers.

Both internally and externally, the remaining decisions made by the two algorithms incorrectly merge cells, and lie somewhere between fairly good, with a small number of errors (e.g. the Hu decisions, Handley time 13), and detrimental, producing only errors, as at Handley time 85, when all assertions and rejections are false. Looking at time 85 in Figure 9a, in the left bar we see that all cells accepted by the decision are false positives (dark shading indicates cells not in ground truth), and in the the right bar all cells rejected

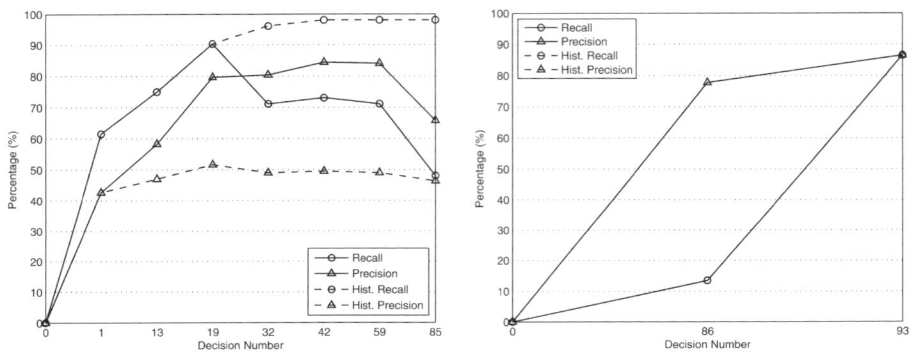

a. Handley b. Hu et al.

Fig. 11. Performance Metrics for Cell Hypotheses

by the decision are false negatives (light shading indicates cells in ground truth).

Now let us identify which decisions in our RSL implementations caused the errors in the final interpretations that we observed in Section 6.1. The Hu algorithm splits a column header cell at Decision 86, and then proposes the cells that have been incorrectly merged and split across rows in the stub and body at Decision 93. We should point out that the errors related to superscripts are caused by our projection-based textline detection addition to the Hu algorithm, which does not take super and sub-scripts into account; the original algorithm was designed for text files, which of course come with textlines already defined. The Handley algorithm completes recognizing the stub head and boxhead cells correctly, using a series of decisions ending at Decision 42, but over-merges cells across rows in the stub and body at Decisions 32, 59, and 85.

These errors may be determined automatically by searching the hypothesis history for false positive cell hypotheses recorded in the output interpretation. Here these errors may be found simply by looking at Figures 1 and 10.

6.4 Evaluating and Improving the Decision-Making Process

In addition to this hypothesis-level view, we can also use historical recall and precision along with conventional recall and precision to give us an external, higher-level view of the decision-making process. Figure 11 presents all four of these metrics for each decision shown in Figures 1 and 10.

Most strikingly, note that our implementation of the Handley algorithm has higher recall after Decision 19 than the recall of the Hu algorithm at any point; if the algorithm had stopped at this point, it would have fared better

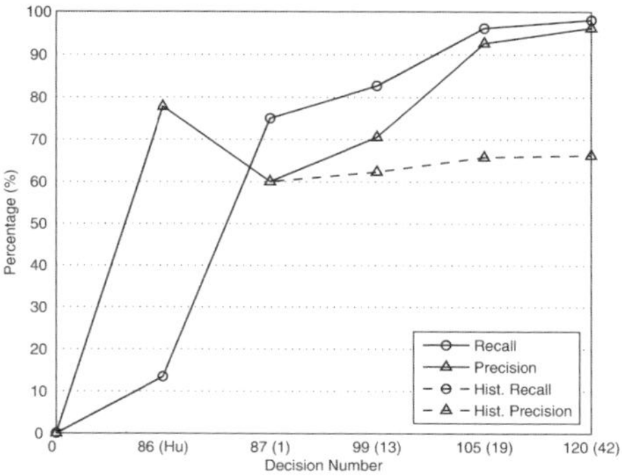

Type of deposit	Bulk density Mean (g/cm³)	No.¹	Grain density Mean (g/cm³)	No.¹	Total pore space (percent)
Lateral blast,					
May 18--------	²1.66	262	2.52	3	36
Pyroclastic flow,					
May 18-------	.74	8	2.55	3	71
May 25-------	.95	2	(³)	0	³63
June 12------	1.08	10	2.53	3	57
July 22------	.88	11	2.55	1	65
August 7-----	1.02	12	2.61	3	61
October 16-18	1.12	12	2.65	5	58

a. Performance Metrics b. Final Result

Fig. 12. Results for Combining Decisions of the Hu et al. and Handley Algorithms. First the Hu algorithm is applied, stopping after its first cell decision (for boxhead and stub head cells). The original decision numbers for the Handley algorithm are shown in brackets in (a). In the output, there are only two false positives, which split one ground truth cell in the rightmost column (^{3}63)

in the conventional evaluation given in Section 6.1. At Decision 19 for the Handley algorithm, the stub and body cells have all been correctly detected with the exception of the cell with a prefix superscript (3) in the last column of the table. Only five cells have not been correctly located: the incorrect body cell, and four of the header cells (these correspond to the twelve false positives at Decision 19, shown as the gray portion of the bar for accepted hypothesis set in Figure 9b).

After Decision 19, the Handley algorithm recall and precision measures start to decrease, and false negatives start being created. At Decision 32, thirteen valid cells are rejected; however, three new ground truth cells (column headers) are produced, causing an increase in historical recall. As mentioned earlier, Decision 42 is a perfect decision, and detects the rightmost column header (increasing all metrics). The final two decisions decrease recall, precision, and historical precision, as they propose invalid merges. The historical precision for the Handley algorithm is uniformly low, partly because of proposing all words as cells initially. This results in many invalid hypotheses being generated. However, the historical recall is very high, and in fact only one ground truth cell is never generated and considered: the '^{3}63' cell in the rightmost column of the body.

The Hu algorithm decisions have fairly high precision, and high final recall (at Decision 93). However, the historical recall is considerably lower than that of the Handley algorithm (it does not generate the ground truth cell missing from the Handley algorithm's hypothesis set, along with other cells). Note that the historical and conventional metrics are identical for the Hu algorithm, because no cell hypotheses are ever rejected.

Can we combine the existing decisions of the Handley and Hu algorithms to produce a result better than either algorithm in isolation? The answer is yes, and the result for one such combination is shown in Figure 12. The combination is produced by first running the Hu algorithm, stopping it after it finishes identifying cells in the boxhead (Decision 86). In the output, we filter all but *Word*, *Line*, and *Cell* region types. We then run the Handley algorithm on the filtered output of the Hu algorithm, after making the following changes to the Handley algorithm:

Decision 1: only words that do not already belong to a cell are classified as cells (this preserves the detected header cells)

Decisions 32, 59, 86: are removed along with their supporting decision sequences (e.g. estimates of row and column structure)

No decision function parameters were altered.

In the combined strategy all the ground truth cells generated by the Handley algorithm are detected, with identical historical and conventional recall after each decision (one cell is still mis-recognized, as for the original algorithms). At the final time, precision is higher than it is for either of the individual algorithms, while the historical precision has been improved relative to the original Handley algorithm. While effective, this particular decision sequence is likely over-fit to this particular table.

7 Conclusion

We have presented decision-based methods for specifying and comparing recognition algorithms. The Recognition Strategy Language (RSL) is a first attempt at formalizing recognition algorithms as sequences of decisions which select among alternative model operations at run-time. Decision-based specifications make a table recognition algorithm's search within the space of possible interpretations defined by a model explicit, and allow more information about an algorithms' decision process to be automatically captured both statically and at run time. We have illustrated the decision-based approach using RSL re-implementations of the Handley [1] and Hu et al. [12] table structure recognition algorithms, and demonstrated by example how we can use the additional information made available to improve recognition by combining decisions from both.

Decision-based specification makes it possible to directly determine which decisions in the recognition process affect a hypotheses. By capturing all hypotheses including those that are rejected, we can measure new metrics on the set of hypotheses generated by an algorithm. We have presented two simple but useful metrics that characterize generated hypotheses in terms of coverage of the set of recognition goals (historical recall) and the accuracy/efficiency of hypothesis generation (historical precision). We believe that other useful metrics may be defined, such as some measure of 'fickleness' that characterizes how frequently hypotheses shift from between the sets of accepted and rejected hypotheses.

In the future we wish to extend RSL to better support feature computations directly within an RSL program; currently all feature computations are defined externally within the functions that make decisions at run-time. To further support studying the combination of recognition strategies, we also wish to explore new decision types that combine and select among decisions and decision sequences within RSL programs. This would allow decision combinations to be produced automatically or semi-automatically, rather than manually as we did in Section 6. Ultimately we would like to use decision-based languages such as RSL for studying the problem of learning recognition strategies [13, 14, 15].

Acknowledgement. We wish to acknowledge Dr. Ching Suen, who provided resources used by the first author to write this chapter while at the Centre for Pattern Recognition and Machine Intelligence (Concordia University, Montreal, Canada), and Joshua Zimler who helped with the design of some figures. This research was supported by the Natural Sciences and Engineering Research Council of Canada.

References

1. Handley, J.: Table analysis for multi-line cell identification. In: Proc. Document Recognition and Retrieval VIII (IS&T/SPIE Electronic Imaging). Volume 4307., San Jose, CA (2001) 34–43
2. Phillips, I., Chen, S., Haralick, R.: CD-ROM document database standard. In: Proc. Second Int'l Conf. Document Analysis and Recognition, Tsukuba Science City, Japan (1993) 478–483
3. Silva, A.e., Jorge, A., Torgo, L.: Design of an end-to-end method to extract information from tables. International Journal on Document Analysis and Recognition **8** (2006) 144–171
4. Embley, D., Hurst, M., Lopresti, D., Nagy, G.: Table-processing paradigms: a research survey. International Journal on Document Analysis and Recognition **8** (2006) 66–86
5. Handley, J.: Document recognition. In: Electronic Imaging Technology. IS&T/SPIE Optical Engineering Press, Bellingham, WA (1999)
6. Lopresti, D., Nagy, G.: Automated table processing: An (opinionated) survey. In: Proc. Third Int'l Workshop on Graphics Recognition, Jaipur, India (1999) 109–134
7. Lopresti, D., Nagy, G.: A tabular survey of automated table processing. In: Lecture Notes in Computer Science. Volume 1941. Springer-Verlag, Berlin (2000) 93–120
8. Zanibbi, R., Blostein, D., Cordy, J.: A survey of table recognition: Models, observations, transformations, and inferences. Int'l J. Document Analysis and Recognition **7**(1) (2004) 1–16

9. Hurst, M.: Towards a theory of tables. International Journal on Document Analysis and Recognition **8** (2006) 123–131
10. Zanibbi, R., Blostein, D., Cordy, J.R.: Recognition tasks are imitation games. In: Lecture Notes in Computer Science. Volume 3686. (2005) 209–218
11. Zanibbi, R., Blostein, D., Cordy, J.R.: The recognition strategy language. In: Proc. Eighth Int'l Conf. Document Analysis and Recognition. (2005) 565–569 Vol. 2
12. Hu, J., Kashi, R., Lopresti, D., Wilfong, G.: Table structure recognition and its evaluation. In: Proc. Document Recognition and Retrieval VIII (IS&T/SPIE Electronic Imaging). Volume 4307., San Jose, CA (2001) 44–55
13. Bottoni, P., Mussio, P., Protti, M.: Metareasoning in the determination of image interpretation strategies. Pattern Recognition Letters **15** (1994) 177–190
14. Draper, B.: Learning control strategies for object recognition. In Ikeuchi, K., Veloso, M., eds.: Symbolic Visual Learning. Oxford Press, New York (1997) 49–76
15. Draper, B., Bins, J., Baek, K.: Adore: Adaptive object recognition. Videre **1**(4) (2000) 86–99 (online journal)
16. Haralick, R.: Document image understanding: Geometric and logical layout. In: Proc. Conf. Computer Vision and Pattern Recognition, Seattle, WA (1994) 385–390
17. Nagy, G.: Twenty years of document image analysis in PAMI. IEEE Trans. Pattern Analysis and Machine Intelligence **22**(1) (2000) 38–62
18. Hurst, M.: Layout and language: Challenges for table understanding on the web. In: Proc. First Int'l Workshop on Web Document Analysis, Seattle, WA (2001) 27–30
19. Wang, Y., Hu, J.: Detecting tables in HTML documents. In: Lecture Notes in Computer Science. Volume 2423., Berlin, Springer-Verlag (2002) 249–260
20. Daumé, H., Langford, J., Marcu, D.: Search-based structured prediction. (unpublished) (2006)
21. Daumeé, H., Marcu, D.: Learning as search optimization: Approximate large margin methods for structured prediction. In: Proc. International Conference on Machine Learning, Bonn (Germany) (2005) 169–176
22. Amano, A., Asada, N., Mukunoki, M., Aoyama, M.: Table form document analysis based on the document structure grammar. International Journal on Document Analysis and Recognition **8** (2006) 201–213
23. Coüasnon, B.: DMOS, a generic document recognition method: Application to table structure analysis in a general and in a specific way. International Journal on Document Analysis and Recognition **8** (2006) 111–122
24. Takasu, A., Satoh, S., Katsura, E.: A document understanding method for database construction of an electronic library. In: Proc. Twelfth Int'l Conf. Pattern Recognition, Jerusalem, Israel (1994) 463–466
25. Bagdanov, A.: Style Characterization of Machine Printed Texts. PhD thesis, University of Amsterdam (2004)
26. Wang, X.: Tabular Abstraction, Editing and Formatting. PhD thesis, University of Waterloo, Canada (1996)
27. Grossman, J., ed.: 12 (Tables). In: Chicago Manual of Style. 14th edn. University of Chicago Press (1993)
28. Hurst, M.: Layout and language: An efficient algorithm for detecting text blocks based on spatial and linguistic evidence. In: Proc. Document Recognition and

Retrieval VIII (IS&T/SPIE Electronic Imaging). Volume 4307., San Jose, CA (2001) 56–67
29. Ousterhout, J.: Tcl and the Tk Toolkit. Addison-Wesley (1994)
30. van Melle, W., Shortliffe, E., Buchanan, B. In: EMYCIN, A Knowledge Engineer's Tool for Constructing Rule-Based Expert Systems. Addison-Wesley (1984) 302–328
31. Cordy, J.: The TXL source transformation language. Science of Computer Programming **61**(3) (2006) 190–210
32. Cordy, J., Dean, T.R. Malton, A., Schneider, K.: Source transformation in software engineering using the TXL transformation system. Journal of Information and Software Technology **44**(13) (2002) 827–837
33. Zanibbi, R., Blostein, D., Cordy, J.R.: Recognizing mathematical expressions using tree transformation. IEEE Trans. Pattern Analysis and Machine Intelligence **24** (2002) 1455–1467
34. Zanibbi, R.: A Language for Specifying and Comparing Table Recognition Strategies. PhD thesis, Queen's University, Kingston (Canada) (2004)
35. Zanibbi, R., Blostein, D., Cordy, J.R.: Historical recall and precision: summarizing generated hypotheses. In: Proc. Eighth Int'l Conference on Document Analysis and Recognition. (2005) 202–206 Vol. 1
36. Horwitz, S., Reps, T.: The use of program dependence graphs in software engineering. In: Proc. 14th International Conference on Software Engineering, New York, NY, USA, ACM Press (1992) 392–411
37. Weiser, M.: Program slicing. In: Proc. Fifth Int'l Conference on Software Engineering, Piscataway, NJ, USA, IEEE Press (1981) 439–449
38. Quillian, M.: Semantic memory. In Minsky, M., ed.: Semantic Information Processing. MIT Press (1968) 216–270
39. Hu, J., Kashi, R., Lopresti, D., Nagy, G., Wilfong, G.: Why table ground-truthing is hard. In: Proc. Sixth Int'l Conf. Document Analysis and Recognition, Seattle, WA (2001) 129–133
40. Lopresti, D.: Exploiting WWW resources in experimental document analysis research. In: Lecture Notes in Computer Science. Volume 2423., Berlin, Springer-Verlag (2002) 532–543
41. Cesarini, F., Marinai, S., Sarti, L., Soga, G.: Trainable table location in document images. In: Proc. Sixteenth Int'l Conf. Pattern Recognition. Volume 3., Québec City, Canada (2002) 236–240
42. Hu, J., Kashi, R., Lopresti, D., Wilfong, G.: Evaluating the performance of table processing algorithms. Int'l J. Document Analysis and Recognition **4**(3) (2002) 140–153
43. Kieninger, T., Dengel, A.: Applying the T-RECS table recognition system to the business letter domain. In: Proc. Sixth Int'l Conf. Document Analysis and Recognition, Seattle, WA (2001) 518–522
44. Liang, J.: Document Structure Analysis and Performance Evaluation. PhD thesis, University of Washington (1999)
45. Lopresti, D., Wilfong, G.: Evaluating document analysis results via graph probing. In: Proc. Sixth Int'l Conf. Document Analysis and Recognition, Seattle,WA (2001) 116–120
46. Colman, A.: Game Theory & its Applications in the Social and Biological Sciences. Butterworth-Heinemann Ltd., London (1995)

Machine Learning for Digital Document Processing: from Layout Analysis to Metadata Extraction

Floriana Esposito, Stefano Ferilli, Teresa M.A. Basile, and Nicola Di Mauro

Università degli Studi di Bari
Dipartimento di Informatica
Via Orabona, 4
70126 Bari - Italy
{esposito,ferilli,basile,ndm}@di.uniba.it

Summary. In the last years, the spread of computers and the Internet caused a significant amount of documents to be available in digital format. Collecting them in digital repositories raised problems that go beyond simple acquisition issues, and cause the need to organize and classify them in order to improve the effectiveness and efficiency of the retrieval procedure. The success of such a process is tightly related to the ability of understanding the semantics of the document components and content. Since the obvious solution of manually creating and maintaining an updated index is clearly infeasible, due to the huge amount of data under consideration, there is a strong interest in methods that can provide solutions for automatically acquiring such a knowledge. This work presents a framework that intensively exploits intelligent techniques to support different tasks of automatic document processing from acquisition to indexing, from categorization to storing and retrieval.

The prototypical version of the system DOMINUS is presented, whose main characteristic is the use of a Machine Learning Server, a suite of different inductive learning methods and systems, among which the more suitable for each specific document processing phase is chosen and applied. The core system is the incremental first-order logic learner INTHELEX. Thanks to incrementality, it can continuously update and refine the learned theories, dynamically extending its knowledge to handle even completely new classes of documents.

Since DOMINUS is general and flexible, it can be embedded as a document management engine into many different Digital Library systems. Experiments in a real-world domain scenario, scientific conference management, confirmed the good performance of the proposed prototype.

1 Introduction & Motivations

In the World Wide Web era, a huge amount of documents in digital format are spread throughout the most diverse Web sites, and a specific research area,

F. Esposito et al.: *Machine Learning for Digital Document Processing: from Layout Analysis to Metadata Extraction*, Studies in Computational Intelligence (SCI) **90**, 105–138 (2008)
www.springerlink.com

focused on principles and techniques for setting up and managing document collections in a digital form, quickly expanded. Usually, these large repositories of digital documents are defined as *Digital Libraries*, intended as distributed collections of textual and/or multimedia documents, whose main goal is the acquisition and the organization of the information contained therein.

During the past years a considerable effort was spent in the development of intelligent techniques in order to automatically transform paper documents into digital format, saving the original layout with the aim of reconstruction. Machine Learning techniques have been applied to attain this goal, and a successful application in preserving cultural heritage material is reported in [1].

Today, most documents are generated, stored and exchanged in a digital format, although it is still necessary to maintain the typing convention of classical *paper* documents. The specific problem we will deal with consists in the application of intelligent techniques to a system for managing a collection of digital documents on the Internet; such a system, aimed at automatically extracting significant information from the documents, is useful to properly store, retrieve and manage them in a *Semantic Web* perspective [2]. Indeed, organizing the documents on the grounds of the knowledge they contain is fundamental for being able to correctly access them according to the user's particular needs. For instance, in the scientific papers domain, in order to identify the subject of a paper and its scientific context, an important role is played by the information available in components such as Title, Authors, Abstract and Bibliographic references. This last component in particular, with respect to others, is a source of problems both because it is placed at the end of the paper, and because it is, in turn, made up of different sub-components containing various kinds of information, to be handled and exploited in different ways.

At the moment we are not aware of techniques able to automatically annotate the layout components of digital documents, without reference to a specific template. We argue that a process is still necessary to identify the significant components of a digital document through three typical phases: Layout Analysis, Document Image Classification and Document Image Understanding. As widely known, *Layout Analysis* consists in the perceptual organization process that aims at identifying the single blocks of a document and at detecting relations among them (*Layout Structure*); then, associating the proper logical role to each component yields the *Document Logical Structure.* Since the logical structure is different according to the kind of document, two steps are in charge of identifying such a structure: *Document Image Classification*, aiming at the categorization of the document (e.g., newspaper, scientific paper, email, technical report, call for papers) and *Document Image Understanding*, aiming at the identification of the significant layout components for that class. Once the class as been defined it is possible to associate to each component a tag that expresses its role (e.g., signature, object, title, author, abstract, footnote, etc.). We propose to apply multistrategy Machine Learning techniques along these phases of document processing where the classical statistical and numerical approaches to classification and learning may fail, being not able to deal with the lack of a strict layout regularity in the variety of documents available online.

The problem of Document Image Processing requires a first-order language representation for two reasons. First, classical attribute-value languages describe a document by means of a fixed set of features, each of which takes a value from a corresponding pre-specified value set; the exploitation of this language in this domain represents a limitation since one cannot know *a priori* how many components make up a generic document. Second, in an attribute-value formalism it is not possible to represent and efficiently handle the relationships among components; the information coming from the topological structure of all components in a document turns out to be very useful in document understanding. For instance, in a scientific paper, it is useful to know that the acknowledgments usually appear above the references section and in the end of the document, or that the affiliation of the authors is reported generally at the beginning of the document, below or on the right of their names.

The continuous flow of new and different documents in a Web repository or in Digital Libraries calls for *incremental* abilities of the system, that must be able to update or revise a faulty knowledge previously acquired for identifying the logical structure of a document. Traditionally, Machine Learning methods that automatically acquire knowledge in developing intelligent systems, require to be provided with a set of training examples, belonging to a defined number of classes, and exploit them altogether in a batch way.

Although sometimes the term *incremental* is used to define some learning method [3, 4, 5, 6], incrementality generally refers to the possibility of adjusting some parameters in the model on the grounds of new observations that become available when the system is already operational. Thus, classical approaches require that the number of classes is defined and fixed since the beginning of the induction step: this prevents the opportunity of dealing with totally new instances, belonging to new classes, that require the ability to incrementally revise a domain theory as soon as new data are encountered. Indeed, Digital Libraries require autonomous or semi-autonomous operation and adaptation to changes in the domain, the context, or the user needs. If any of these changes happens, the classical approach requires that the entire learning process is restarted to produce a model capable of coping with the new scenario. Such requirements suggest that incremental learning, as opposed to classical batch one, is needed whenever either incomplete information is available at the time of initial theory generation, or the nature (and the kinds) of the concepts evolves dynamically. E.g., this is the case of modifications in time of typing style of documents that nevertheless belong to the same class or of the introduction of a completely new class. Incremental processing allows for continuous responsiveness to the changes in the context, can potentially improve efficiency and deals with concept evolution. The incremental setting implicitly assumes that the information (observations) gained at any given moment is incomplete, and thus that any learned theory could be susceptible of changes.

This chapter presents the prototypical version of DOMINUS (DOcument Management INtelligent Universal System): such a system is characterized by the intensive exploitation of intelligent techniques in each step of document processing from acquisition to indexing, from categorization to storing and retrieval. Since it is general and flexible, it can be embedded as a document management engine into many different Digital Library systems. In the following, after a brief description of the architecture of DOMINUS, the results of the layout analysis on digital documents are discussed, with the interesting improvements achieved by using kernel-based approaches and incremental first-order learning techniques: the satisfying results in document layout correction, classification and understanding allow to start an effective structural metadata extraction. Then, the categorization, filing and indexing tasks are described with the results obtained in the effective retrieval of scientific documents. Finally, the application of the system in a real-world domain scenario, scientific conference management, is reported and discussed.

2 The Document Management System Architecture

This Section briefly presents the overall architecture of DOMINUS, reported in Figure 1. A central role is played by the Learning Server, which intervenes during different processing steps in order to continuously adapt the knowledge taking into consideration new experimental evidence and changes in the context. The corresponding process flow performed by the system from the original digital documents acquisition to text extraction and indexing is reported in Figure 2.

The layout analysis process on documents in digital format starts with the application of a pre-processing module, called WINE (Wrapper for the Interpretation of Non-uniform Electronic document formats), that rewrites basic PostScript operators to turn their drawing instructions into objects (see Section 3). It takes as input a digital document and produces (by an intermediate vector format) the initial document's XML basic representation, that describes it as a set of pages made up of basic blocks. Due to the large number of basic blocks discovered by WINE, that often correspond to fragments of words, it is necessary a first aggregation based on blocks overlapping or adjacency, yielding composite blocks corresponding to whole words. The number of blocks after this step is still large, thus a further aggregation (e.g., of words into lines) is needed. Since grouping techniques based on the mean distance between blocks proved unable to correctly handle the case of multi-column documents, such a task was cast to a multiple instance problem (see Section 3.1) and solved exploiting the kernel-based method proposed in [7], implemented in a Learning Server module that is able to generate rewriting rules that suggest how to set some parameters in order to group together

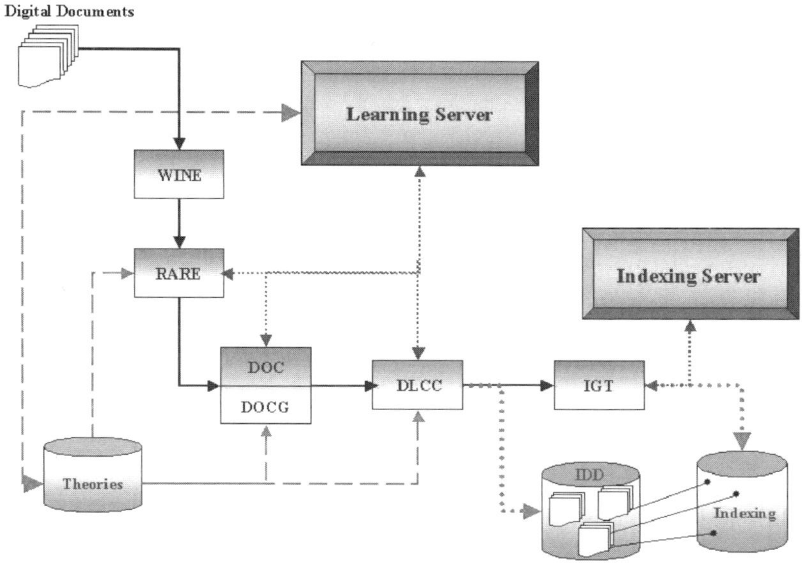

Fig. 1. Document Management System Architecture

word blocks to obtain lines. The inferred rules will be stored in the Theories knowledge base for future exploitation by RARE (Rule Aggregation REwriter) and modification.

Once such a line-block representation is generated, DOC (Document Organization Composer) collects the semantically related blocks into groups by identifying the surrounding frames based on white spaces and the results of the background structure analysis. This is an improvement of the original Breuel's algorithm [8], that finds iteratively the maximal white rectangles in a page: here the process is forced to stop before finding insignificant white spaces such as inter-word or inter-line ones (see Section 3.2).

At the end of this step, some blocks might not be correctly recognized. In such a case a phase of layout correction is needed, that is automatically performed in DOCG (Document Organization Correction Generator) by exploiting embedded rules stored in the Theories knowledge base. Such rules were automatically learned from previous manual corrections collected on some document during the first trials and using the Learning Server.

Once the layout structure has been correctly and definitely identified, a semantic role must be associated to each significant components in order to perform the automatic extraction of the interesting text with the aim of improving document indexing. This step is performed by DLCC (Document and Layout Components Classifier) by firstly associating the document to a class that expresses its type and then associating to every significant layout component a tag expressing its role. Both these steps are performed thanks to theories previously learned and stored in the Theories knowledge base. In

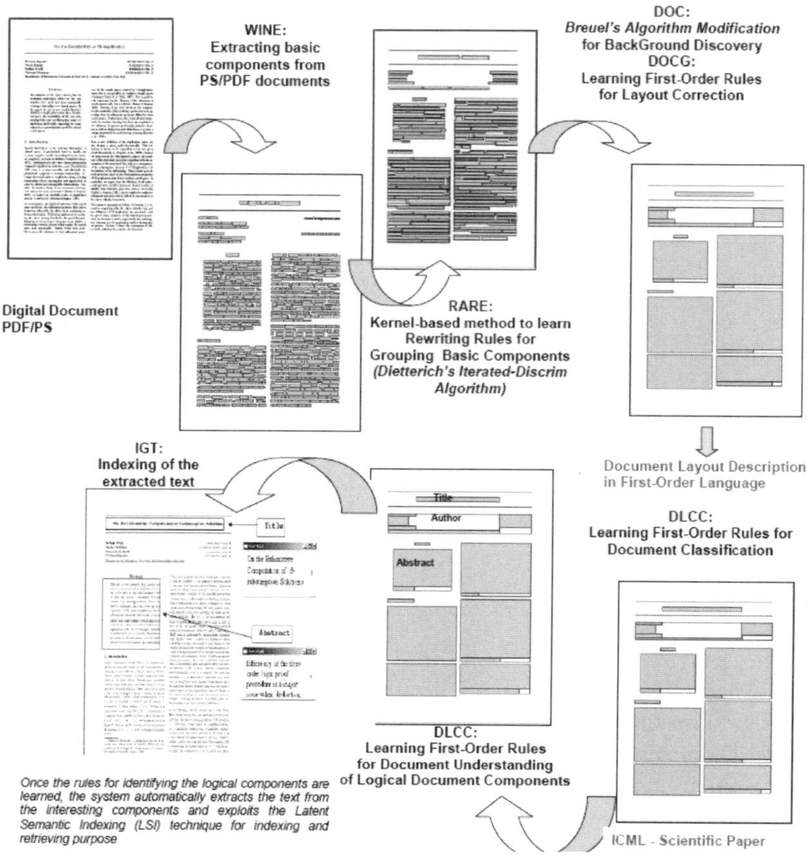

Fig. 2. Document Management System Process Flow

case of failure these theories can be properly updated. The theory revision step is performed by a first-order incremental learning system that runs on the new observations and tries to modify the old theories in the knowledge base. At the end of this step both the original document and its XML representation, enriched with class information and components annotation, is stored in the Internal Document Database, IDD.

Finally, the text is extracted from the significant components and the Indexing Server is called by the IGT (Index Generator for Text) module to manage such information, useful for an effective content-based document retrieval.

3 Layout Structure Recognition

Based on the ODA/ODIF standard, any document can be progressively partitioned into a hierarchy of abstract representations, called its *layout structure*. Here we describe an approach implemented for discovering a full layout hierarchy in digital documents based primarily on layout information.

The layout analysis process starts with a preliminary preprocessing step performed by a module that takes as input a generic digital document and produces a corresponding vectorial description. An algorithm for performing this task is PSTOEDIT [9], but it was discarded because it only applies to PostScript (PS) and Portable Document Format (PDF) documents and returns a description lacking sufficient details for our purposes.

Thus, a module named WINE has been purposely developed. At the moment, it deals with digital documents in PS or PDF formats, that represent the current *de facto* standard for document interchange. The PostScript [10] language is a simple interpretative programming language with powerful graphical capabilities that allow to precisely describe any page. The PDF [11] language is an evolution of PostScript that rapidly gained acceptance as a file format for digital documents. Like PostScript, it is an open standard, enabling integrated solutions from a broad range of vendors. In particular, WINE consists of a rewriting of basic PostScript operators that turns the instructions into objects. For example, the PS instruction to display a text becomes an object describing a text with attributes for the geometry (location on the page) and appearance (font, color, etc.). The output of WINE is a vector format describing the initial digital document as a set of pages, each of which is composed of *basic blocks*. The descriptors used by WINE for representing a document are the following:

box(id,x0,y0,x1,y1,font,size,RGB,row,string): a piece of text in the document, represented by its bounding box;
stroke(id,x0,y0,x1,y1,RGB,thickness): a graphical (horizontal/vertical) line of the document;
fill(id,x0,y0,x1,y1,RGB): a closed area filled with one color;
image(id,x0,y0,x1,y1): a raster image;
page(n,w,h): page information;

where: id is the block identifier; $(x0, y0)$ and $(x1, y1)$ are respectively the upper-left/lower-right coordinates of the block (note that $x0 = x1$ for vertical lines and $y0 = y1$ for horizontal lines); font is the the type font; size represents the text size; RGB is the color of the text, line or area in #rrggbb format; row is the index of the row in which the text appears; string is the text of the document contained in the block; thickness is the thickness of the line; n represents the page number; w and h are the page width and height, respectively. Figure 3 reports an extract of the vectorial transformation of the document.

Such a vectorial representation is translated into an *XML basic representation*, that will be modified as long as the layout analysis process proceeds,

```
box(1, 90.0798, 749.25, 93.1802, 744.0, Times-Roman?, 11, #000000, 744, "J").
box(2, 93.0, 747.75, 96.9849, 744.0, Times-Roman?, 11, #000000, 744, "o").
box(3, 97.1547, 747.75, 101.14, 743.976, Times-Roman?, 11, #000000, 744, "u").
box(4, 101.109, 747.751, 103.763, 744.0, Times-Roman?, 11, #000000, 744, "r").
box(5, 103.852, 747.75, 107.837, 744.0, Times-Roman?, 11, #000000, 744, "n").
box(6, 108.0, 747.75, 111.539, 744.0, Times-Roman?, 11, #000000, 744, "a").
box(7, 111.4, 749.492, 113.615, 744.0, Times-Roman?, 11, #000000, 744, "l").
box(8, 115.5, 747.75, 119.485, 744.0, Times-Roman?, 11, #000000, 744, "o").
box(9, 119.597, 750.0, 122.251, 744.0, Times-Roman?, 11, #000000, 744, "f").
box(10, 124.186, 749.25, 131.272, 744.0, Times-Roman?, 11, #000000, 744, "M").
box(11, 131.25, 747.75, 134.789, 744.0, Times-Roman?, 11, #000000, 744, "a").
box(12, 135.0, 747.75, 138.539, 743.999, Times-Roman?, 11, #000000, 744, "c").
box(13, 138.32, 749.491, 142.304, 744.0, Times-Roman?, 11, #000000, 744, "h").
box(14, 142.362, 749.309, 144.578, 744.0, Times-Roman?, 11, #000000, 744, "i").
box(15, 144.576, 747.75, 148.561, 744.0, Times-Roman?, 11, #000000, 744, "n").
box(16, 148.634, 747.75, 152.173, 744.0, Times-Roman?, 11, #000000, 744, "e").
box(17, 154.066, 749.25, 158.935, 744.0, Times-Roman?, 11, #000000, 744, "L").
box(18, 159.038, 747.75, 162.577, 744.0, Times-Roman?, 11, #000000, 744, "e").
box(19, 162.671, 747.75, 166.209, 744.0, Times-Roman?, 11, #000000, 744, "a").
box(20, 165.953, 747.751, 168.607, 744.0, Times-Roman?, 11, #000000, 744, "r").
box(21, 168.695, 747.75, 172.68, 744.0, Times-Roman?, 11, #000000, 744, "n").
box(22, 172.682, 749.309, 174.898, 744.0, Times-Roman?, 11, #000000, 744, "i").
box(23, 174.896, 747.75, 178.881, 744.0, Times-Roman?, 11, #000000, 744, "n").
box(24, 179.194, 747.75, 183.179, 742.501, Times-Roman?, 11, #000000, 744, "g").
box(25, 184.866, 749.25, 190.182, 744.0, Times-Roman?, 11, #000000, 744, "R").
box(26, 190.249, 747.75, 193.788, 744.0, Times-Roman?, 11, #000000, 744, "e").
box(27, 193.5, 747.75, 196.6, 744.0, Times-Roman?, 11, #000000, 744, "s").
box(28, 196.882, 747.75, 200.421, 744.0, Times-Roman?, 11, #000000, 744, "e").
box(29, 200.25, 747.75, 203.789, 744.0, Times-Roman?, 11, #000000, 744, "a").
box(30, 203.797, 747.751, 206.451, 744.0, Times-Roman?, 11, #000000, 744, "r").
```

Fig. 3. WINE output: Vectorial Transformation of the Document

in order to represent the document by means of increasingly complex aggregations of basic components progressively discovered by the various layout analysis phases.

3.1 A Kernel-Based Method to Group Basic Blocks

The first step in the document layout analysis concerns the identification of rules to automatically shift from the basic digital document description to a higher level one. Indeed, by analyzing the PS or PDF source, the "elementary" blocks that make up the document, identified by WINE, often correspond just to fragments of words (see Figure 3), thus a first aggregation based on their overlapping or adjacency is needed in order to obtain blocks surrounding whole words (*word-blocks*). Successively, a further aggregation starting from the *word-blocks* could be performed to have blocks that group words in lines (*line-blocks*), and finally the *line-blocks* could be merged to build a paragraph (*frames*). As to the grouping of blocks into lines, since techniques based on the mean distance between blocks proved unable to correctly handle cases of multi-column documents, we decided to apply Machine Learning approaches in order to automatically infer rewriting rules that could suggest how to set some parameters in order to group together rectangles (words) to obtain lines. To do this, RARE uses a kernel-based method to learn rewriting rules able to perform the bottom-up construction of the whole document starting from the basic/word blocks up to the lines. Specifically, such a learning task was cast to a Multiple Instance Problem and solved exploiting the kernel-based algorithm

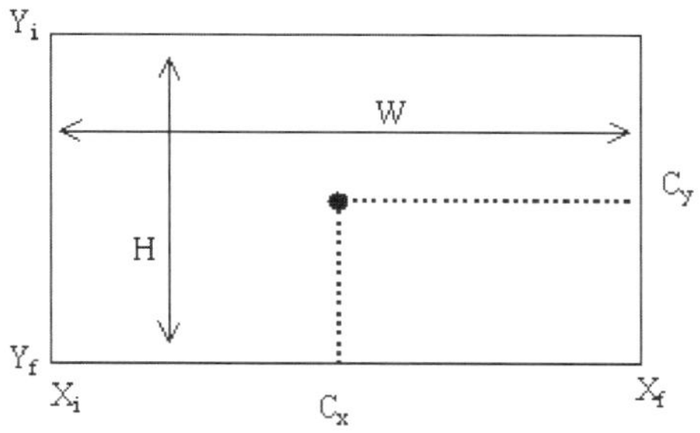

Fig. 4. Block Features

proposed in [7]. In our setting, each elementary block is described by means of a feature-vector of the form:

$$[Block\_Name, Page\_No, X_i, X_f, Y_i, Y_f, C_x, C_y, H, W]$$

made up of parameters interpreted according to the representation in Figure 4, i.e.:

- $Block\_Name$: the identifier of the considered block;
- $Page\_No$: the number of page in which the block is positioned;
- X_i and X_f: the x coordinate values, respectively, for the start and end point of the block;
- Y_i and Y_f: the y coordinate values, respectively, for the start and end point of the block;
- C_x and C_y: the x and y coordinate values, respectively, for the centroid of the block;
- H, W: the distances (height and width) between start and end point of, respectively, x and y coordinate values.

Starting with this description of the elementary blocks, the corresponding example descriptions, from which rewriting rules have to be learned, are built considering each block along with its Close Neighbor blocks: Given a block O_n and the **C**lose **N**eighbor blocks CNO_{nk}, with their own description:

$[O_n, Page\_No, X_{ni}, X_{nf}, Y_{ni}, Y_{nf}, C_{nx}, C_{ny}, H_n, W_n]$,

$[CNO_{nk}, Page\_No, X_{nki}, X_{nkf}, Y_{nki}, Y_{nkf}, C_{nkx}, C_{nky}, H_{nk}, W_{nk}]$

we represent an example E by means of the template $[O_n, CNO_{nk}]$, i.e.:

$[New\_Block\_Name, Page\_No, X_{ni}, X_{nf}, Y_{ni}, Y_{nf}, C_{nx}, C_{ny},$
$X_{nki}, X_{nkf}, Y_{nki}, Y_{nkf}, C_{nkx}, C_{nky}, D_x, D_y]$

where the New_Block_Name is a name for the new block built by appending the names of both the original blocks, the information about the x and y coordinates are the original ones and two new parameters, D_x and D_y, contain the information about the distances between the two blocks.

Fixed a block O_n, the template $[O_n, CNO_{nk}]$ is used to find, among all the word blocks in the document, every instance of close neighbors of the considered block O_n. Such an example (set of instances) will be labelled by an expert as positive for the target concept "the two blocks can be merged" if and only if the blocks O_n and CNO_{nk} are adjacent and belong to the same line in the original document, or as negative otherwise. Figure 5 reports an example of the selected close neighbor blocks for the block $b1$. All the blocks represented with dashed lines could eventually be merged, and hence they will represent the positive instances for the concept *merge*, while dotted lines have been exploited to represent the blocks that could not be merged, and hence will represent the negative instances for the target concept. It is worth noting that not every pair of adjacent blocks has to be considered a positive example since they could belong to different frames in the considered document. Such a situation is reported in Figure 6. Indeed, typical cases in which a block is adjacent to the considered block but actually belongs to another frame are, e.g., when they belong to adjacent columns of a multi-column document (right part of Figure 6) or when they belong to two different frames of the original document (for example, the *Title* and the *Authors* frame - left part of Figure 6).

In such a representation, a block O_n has at least one close neighbor block and at most eight (CNO_{nk} with $k \in \{1, 2, \ldots, 8\}$ or, top-down, from left to right: *top_left_corner, top, top_right_corner, right, bottom_right_corner,*

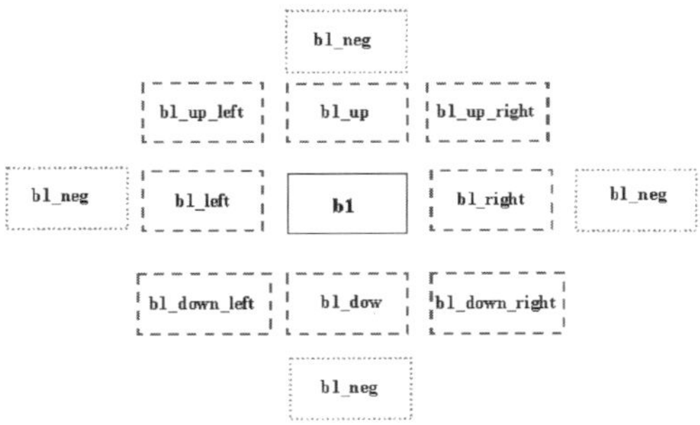

Fig. 5. Close Neighbor blocks for block $b1$

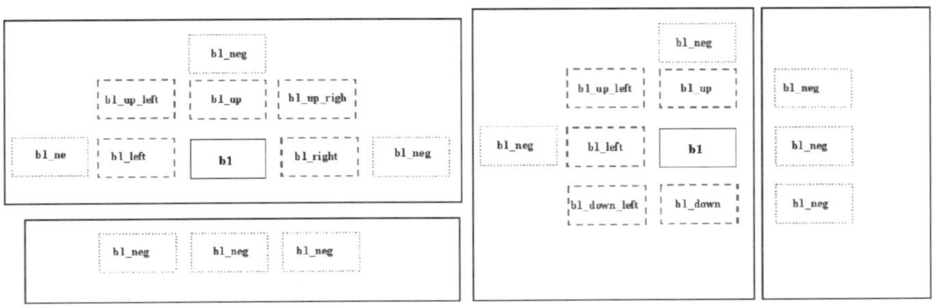

Fig. 6. Selection of positive and negative blocks according to the original document: one-column document on the left, two-columns document on the right

bottom, bottom_left_corner, left); the immediate consequence of the adopted representation is that each single example is actually made up of a bag of instances and, hence, the problem can be clearly cast as a Multiple Instance Problem to be solved by applying the Iterated-Discrim algorithm [7] in order to discover the relevant features and their values to be encoded in rules made up of numerical constraints allowing to automatically set parameters to group together words in lines. In this way, the XML line-level description of the document is obtained, that represents the input to the next step in the layout analysis of the document.

In the following, an example of the representation is provided. Given the representation shown in Figure 5 for the identification of positive and negative blocks, and the template for the example description, a possible representation for the positive example (a set of instances) expressing the description "block b35 can be merged with blocks b36,b34, b24, b43 if and only if such blocks have the reported numeric features (size and position in the document)" is:

```
ex(b35) :-
  istance([b35, b36, 542.8, 548.3, 447.4, 463.3, 553.7, 594.7,
           447.4, 463.3, 545.6, 455.3, 574.2, 455.3, 5.5, 0]).
  istance([b35, b34, 542.8, 548.3, 447.4, 463.3, 529.2, 537.4,
           447.4, 463.3, 545.5, 455.4, 533.3, 455.3, 5.5, 0]).
  istance([b35, b24, 542.8, 548.3, 447.4, 463.3, 496.3, 583.7,
           427.9, 443.8, 545.5, 455.3, 540.1, 435.9, 0, 3.5]).
  istance([b35, b43, 542.8, 548.3, 447.4, 463.3, 538.5, 605.4,
           466.9, 482.8, 545.5, 455.3, 571.9, 474.8, 0, 3.5]).
```

3.2 Discovery of the Background Structure of the Document

The objects that make up a document are spatially organized in *frames*, defined as collections of objects completely surrounded by white space. It is worth noting that there is no exact correspondence between the layout notion of a frame and a logical notion such as a *paragraph*: two columns on a page correspond to two frames, while a paragraph might begin in one column and continue into the next column.

The next step towards the discovery of the document logical structure, after transforming the original digital document into its basic XML representation and grouping the basic blocks into lines, consists in performing the layout analysis of the document by applying an algorithm named DOC, a variant of that reported in [8] for addressing the key problem in geometric layout analysis. DOC analyzes the whitespace and background structure of each page in the document in terms of rectangular covers, and it is efficient and easy to implement.

Once DOC has identified the whitespace structure of the document, thus yielding the background, it is possible to compute its complement, thus

```
<?xml version="1.0" encoding="utf-8" ?>
<document name="bickel06a" numPages="1" xmlns:xsi="http://www.w3.org/2001/XMLSchema-instance" xsi:noNamespaceSchemaLocation="grammar.xsd">
<obstacles>
 <page number="1" x0="0.0" y0="0.0" x1="612.0" y1="792.0" >
  <line number="1" x0="90.0798" y0="42.0" x1="273.291" y1="49.499023" >
   <word number="1" x0="90.0798" y0="42.507996" x1="113.615" y1="48.023987" >
      <box number="1" x0="90.0798" y0="42.75" x1="93.1802" y1="48.0" fontName="Times-Roman?" fontSize="11" fontColor="#000000" >J</box>
      <box number="2" x0="93.0" y0="44.25" x1="96.9849" y1="48.0" fontName="Times-Roman?" fontSize="11" fontColor="#000000" >o</box>
      <box number="3" x0="97.1547" y0="44.25" x1="101.14" y1="48.023987" fontName="Times-Roman?" fontSize="11" fontColor="#000000" >u</box>
      <box number="4" x0="101.109" y0="44.249023" x1="103.763" y1="48.0" fontName="Times-Roman?" fontSize="11" fontColor="#000000" >r</box>
      <box number="5" x0="103.852" y0="44.25" x1="107.837" y1="48.0" fontName="Times-Roman?" fontSize="11" fontColor="#000000" >n</box>
      <box number="6" x0="108.0" y0="44.25" x1="111.539" y1="48.0" fontName="Times-Roman?" fontSize="11" fontColor="#000000" >a</box>
      <box number="7" x0="111.4" y0="42.507996" x1="113.615" y1="48.0" fontName="Times-Roman?" fontSize="11" fontColor="#000000" >l</box>
   </word>
   <word number="2" x0="115.5" y0="42.0" x1="122.251" y1="48.0" >
      <box number="8" x0="115.5" y0="44.25" x1="119.485" y1="48.0" fontName="Times-Roman?" fontSize="11" fontColor="#000000" >o</box>
      <box number="9" x0="119.597" y0="42.0" x1="122.251" y1="48.0" fontName="Times-Roman?" fontSize="11" fontColor="#000000" >f</box>
   </word>
   <word number="3" x0="124.186" y0="42.508972" x1="152.173" y1="48.000977" >
      <box number="10" x0="124.186" y0="42.75" x1="131.272" y1="48.0" fontName="Times-Roman?" fontSize="11" fontColor="#000000" >M</box>
      <box number="11" x0="131.25" y0="44.25" x1="134.789" y1="48.0" fontName="Times-Roman?" fontSize="11" fontColor="#000000" >a</box>
      <box number="12" x0="135.0" y0="44.25" x1="138.539" y1="48.000977" fontName="Times-Roman?" fontSize="11" fontColor="#000000" >c</box>
      <box number="13" x0="138.32" y0="42.508972" x1="142.304" y1="48.0" fontName="Times-Roman?" fontSize="11" fontColor="#000000" >h</box>
      <box number="14" x0="142.362" y0="42.69098" x1="144.578" y1="48.0" fontName="Times-Roman?" fontSize="11" fontColor="#000000" >i</box>
      <box number="15" x0="144.576" y0="44.25" x1="148.561" y1="48.0" fontName="Times-Roman?" fontSize="11" fontColor="#000000" >n</box>
      <box number="16" x0="148.634" y0="44.25" x1="152.173" y1="48.0" fontName="Times-Roman?" fontSize="11" fontColor="#000000" >e</box>
   </word>
   ........
   </line>
   ....
   </page>
   ....
```


Fig. 7. DOC output: XML Representation of the Layout Structure of the Document in Figure 3

obtaining the document content blocks. When computing the complement, two levels of description are generated. The former refers to single blocks filled with the same kind of content, the latter consists in rectangular frames that may be made up of many blocks of the former type. Thus, the overall description of the document includes both kinds of objects, plus information on which frames include which blocks and on the actual spatial relations between frames and between blocks in the same frame (e.g., above, touches, etc.). This allows to maintain both levels of abstraction independently. Figure 7 reports the *XML layout structure* that is the output of DOC. Figure 8 depicts, along with the original document, the graphical representation of the XML generated by a two-column document trough the basic block vectorial transformation and the grouped words/lines representation, obtained by means of a process that is not a merely syntactic transformation from PS/PDF to XML.

It is worth to note that exploiting as-is the algorithm reported in [8] on the basic representation discovered by the WINE tool in real document domains turns out to be unfeasible due to the usually large number of basic blocks discovered. Thus, the preliminary aggregation of basic blocks into words and then of words into lines by means of the above procedure is fundamental for the efficiency and effectiveness of the DOC algorithm. Additionally, some modifications to the algorithm on which DOC is based deserve attention. First of all, any horizontal/vertical line in the layout is considered as a natural separator, and hence is already considered as background (along with all the surrounding white space) before the algorithm starts. Second, any white block whose height or width is below a given threshold is discarded as insignificant (this should avoid returning inter-word or inter-line spaces). Lastly, since the original algorithm tries to find iteratively the maximal white rectangles, taking it to its natural end and then computing the complement would result again in the original basic blocks coming from the previous steps and provided as input. This would be useless, and hence raised the problem of identifying a stop criterion to end this process.

Such a criterion was empirically established as the moment in which the area of the new white rectangle retrieved, $W(R)$, represents a percentage of the total white area in the document page, $W(D)$, less than a given threshold δ, i.e.:
Let $A(D)$ be the area of the document page under consideration, $A(R_i)$, $i = 1, \ldots, n$ be the areas of the blocks identified thus far in the page, and $W(D) = A(D) - \sum_{i=1,\ldots,n} A(R_i)$ be the total white area in the page (computed as the difference between the total page area and the area of the blocks in the page), then the stop criterion is established as:

$$\frac{W(R)}{W(D)} < \delta$$

The empirical study was performed applying the algorithm in full on a set of 100 documents of three different categories, and it took into account the values of three variables in each step of the algorithm: number of new white rectangles (black line in Figure 9) normalized between 0 and 1, ratio of the last white area retrieved with respect to the total white area of the current page of the document (bold line in Figure 9), ratio of the white area retrieved so far with respect to the total white area of the current page of the document (dashed line in Figure 9). The ratio of the white area retrieved, the dashed

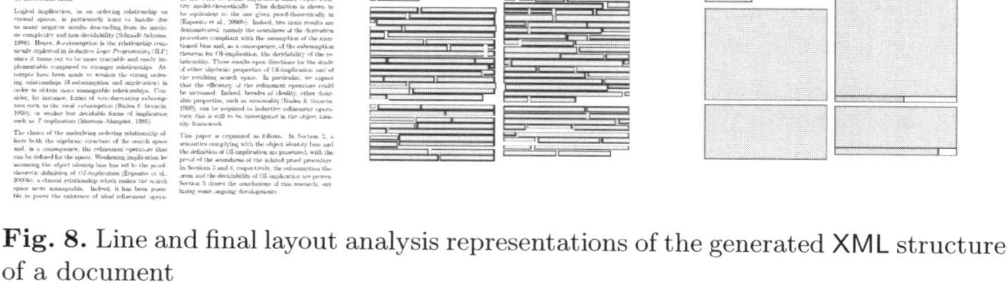

Fig. 8. Line and final layout analysis representations of the generated XML structure of a document

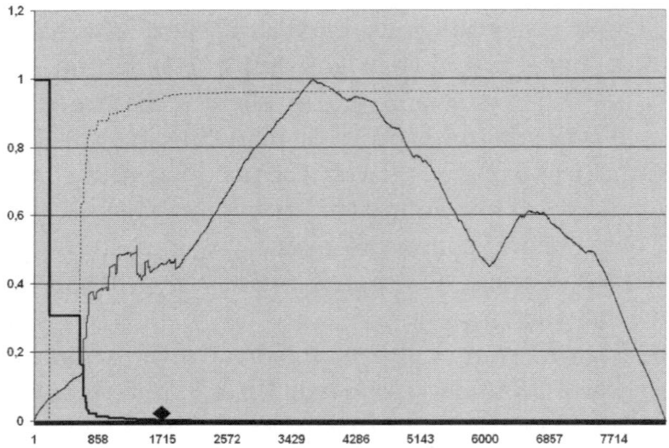

Fig. 9. Stop Criterion Analysis

line, is never equal to 1 (the algorithm does not find all the white area), but it becomes stable before reaching 1/4 of the total steps of the algorithm. Such a consideration is generally valid for all the documents except for those having a scattered appearance. Such a point, highlighted in the figure with a black diamond, is the best stop point for the algorithm since before it the layout is not sufficiently detailed, while after it useless white spaces are found, as shown with the black line in the graphic. Indeed, this is the point in which all the useful white spaces in the document, e.g. those between columns and sections, have been identified. Such a consideration is confirmed by analyzing the trend of the ratio of the last white area retrieved with respect to the total white area in the current page of the document (bold line), that decreases up to 0 in such a point. This suggests to stop executing the algorithm just there. It is worth noting that this value is reached very early, and before the size of the structure containing the blocks waiting to be processed starts growing dramatically, thus saving lots of time and space resources.

4 Structural Metadata Extraction

The organization of the document collection and the extraction of the interesting text is a fundamental issue for a more efficient storage and retrieval process in a digital library. To perform such tasks, one has to firstly identify the correct type the document belongs to (e.g. understand whether the document is a magazine, or a book, or a scientific paper) in order to file it in the corresponding record. Then, the significant components of the document have to be identified in order to extract from them the information needed to categorize it. Since carrying out manually such a process is unfeasible due to the huge amount of documents, our proposal is the use of a concept learning system to infer rules able to correctly classify the document type along with its significant components. The inborn complexity of the document domain, and the need to express relations among components, suggests the exploitation of symbolic first-order logic as a powerful representation language to handle such a situation. Furthermore, based on the belief that in typical digital libraries on the Internet new documents continuously become available over time and are to be integrated in the collection, we consider incrementality as a fundamental requirement for the techniques to be adopted. Even more difficult, it could be the case that not only single definitions turn out to be faulty and need revision, but whole new document classes are to be included in the collection as soon as the first document for them becomes available. This represents a problem for most existing systems, that require not only all the information on the application domain to be available when the learning process starts, but also the set of classes for which they must learn definitions to be completely defined since the beginning.

These considerations, among others about the learning systems available in the literature, led to the exploitation of INTHELEX (INcremental THEory Learner from EXamples) [12], whose most characterizing features are its incremental nature, the reduced need of a deep background knowledge, the exploitation of negative information and the peculiar bias on the generalization model, which reduces the search space and does not limit the expressive power of the adopted representation language.

4.1 The Learning System

INTHELEX is an Inductive Logic Programming [13] system that learns hierarchical logic theories from positive and negative examples. It is fully incremental (in addition to the possibility of refining previously generated hypotheses/definitions, learning can also start from an empty theory), and adopts Datalog^{OI} [14] as a representation language: based on the Object Identity assumption (different symbols must denote different objects), it ensures effectiveness of the descriptions and efficiency of their handling, while preserving the expressive power of the unrestricted case. It can learn simultaneously multiple concepts/classes, possibly related to each other; it can retain all the processed examples, so to guarantee validity of the learned theories on all of them.

INTHELEX has a closed loop architecture (i.e., feedback on performance is used to activate the theory revision phase [15]). The learning cycle it performs, depicted in Figure 10, can be described as follows. A set of examples of the concepts to be learned, possibly selected by an expert, is provided by the environment. This set can be subdivided into three subsets (training, tuning, and test set) according to the way in which examples are exploited during the learning process. Specifically, training examples, previously classified by the expert, are stored in the base of processed examples, and exploited to obtain an initial theory that is able to explain them. In INTHELEX, such a

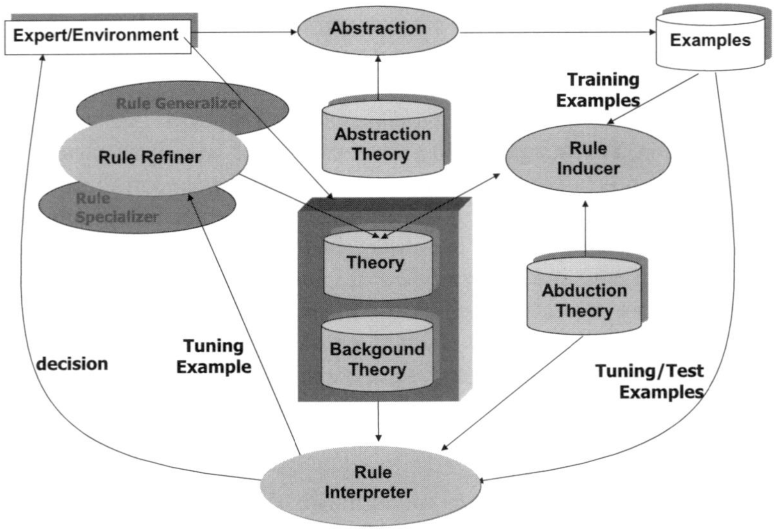

Fig. 10. Learning System Architecture

theory can also be provided by the expert, or even be empty. Subsequently, the validity of the theory against new available tuning/test examples, also stored in the example base as long as they are processed, is checked against the set of inductive hypotheses, producing a decision that is compared to the correct one. Test examples are exploited just to check the predictive capabilities of the theory, intended as its behavior on new observations, without causing a refinement of the theory in the case of incorrectness. Conversely, in case of incorrectness on a tuning example, the cause of the wrong decision can be located and the proper kind of correction chosen, firing the theory revision process. In this way, tuning examples are exploited incrementally to modify incorrect theories according to a data-driven strategy.

Specifically, INTHELEX incorporates two inductive refinement operators to revise the theory, one for generalizing definitions that reject positive examples, and the other for specializing definitions that explain negative examples. If an example is positive and not covered, the system first tries to generalize one of the available definitions of the concept the example refers to, so that the resulting revised theory covers the new example and is consistent with all the past negative examples. If such a generalization is found, then it replaces the chosen definition in the theory, or else a new clause is chosen to compute generalization. If no definition can be generalized in a consistent way, the system checks whether the example itself can represent a new alternative (consistent) definition of the concept. If so, such a definition is added to the theory, or else the example itself is added as an exception. If the example is negative and covered, specialization is needed. Among the theory definitions that concur in covering the example, INTHELEX tries to specialize one by adding to it one or more conditions which characterize all the past positive examples and can discriminate them from the current negative one. In case of failure, the system tries to add the negation of a condition, that is able to discriminate the negative example from all the past positive ones. If this fails too, the negative example is added to the theory as an exception. New incoming observations are always checked against the exceptions before applying the rules that define the concept they refer to.

Another peculiarity in INTHELEX is the embedding of multistrategy operators that may help in solving the theory revision problem by pre-processing the incoming information. It was operated according to the theoretical framework for integrating different learning strategies known as Inferential Theory of Learning [16]. Deduction refers to the possibility of better representing the examples and, consequently, the inferred theories. INTHELEX exploits deduction to recognize known concepts that are implicit in the examples description and explicitly add them to the descriptions. The system can be provided with a *Background Knowledge*, supposed to be correct and hence not modifiable, containing (complete or partial) concept definitions to be exploited during deduction. Differently from abstraction (see next), all the specific information used by deduction is left in the example description. Hence, it is preserved in the learning process until other evidence reveals it is not significant for the concept definition, which is a more cautious behavior. Abduction was defined by Peirce as hypothesizing facts that, together with a given theory, could explain a given observation, and aims at completing possibly partial information in the examples (adding more details). According to the framework proposed in [17], this can be done by exploiting a set of *abducibles* (concepts about which assumptions can be made, that carry all the incompleteness of the domain: if it were possible to complete their definitions then the theory would be correctly described) and a set of *integrity constraints* (each corresponding to a combination of conditions that is not allowed to occur, that provide indirect information about abducibles). Abstraction is a pervasive activity in human perception and reasoning, and aims at removing superfluous details from the description of both the examples and the theory. Thus, the exploitation of abstraction results in the shift from the language in which the theory is described to a higher level one. According to the framework proposed in [18], in INTHELEX abstraction takes place by means of a set of operators that replace a number of components by a compound object, or decrease the granularity of a set of values, or ignore whole objects or just part of their features, or neglect the number of occurrences of some kind of object.

4.2 Representation Language

In order to work, the learning system must be provided with a suitable first-order logic representation of the documents. Thus, once the layout components of a document are automatically discovered as explained in Section 3, the next step concerns the automatic description of the pages, blocks and

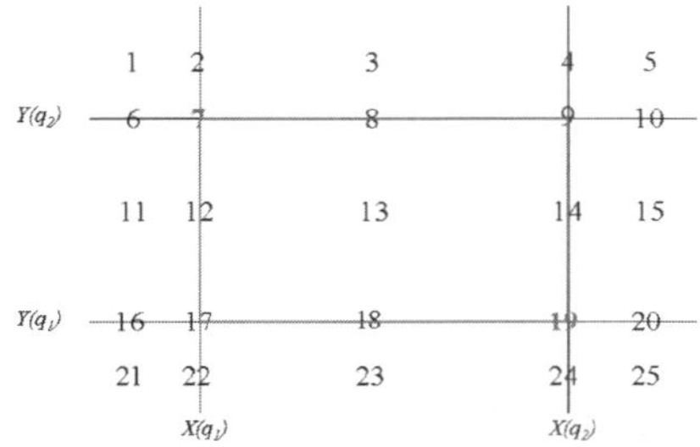

Fig. 11. Representation Plans according to [19]

frames according to their size, spatial [19] and inclusion relations. Dealing with multi-page documents, the document description must be enriched with *page information* such as: page number and position (whether it is at the beginning, in the middle or at the end of the document, and specifically whether it is the last one), total number of pages in the document. As pointed out, the automatic process results in a set of content rectangles recognized in each page. Such rectangles are described by means of their size (height and width), their type (text, graphic, line) and their horizontal and vertical position in the document. Furthermore, the algebraic relations $\subset$ and $\supset$ are exploited to express the inclusion between frames and pages, e.g. $contain(page_i, frame_j)$, and between blocks and frames, e.g. $contain(frame_j, block_k)$.

Another possible relation between rectangles is the spatial one. Given a rectangle r, one can ideally divide the plan containing it in 25 parts (see Figure 11), and describe the relative position between the other rectangles and r in terms of the plans they occupy with respect to r. Such a technique is applied to every block belonging to a same frame and to all the adjacent frames, where a rectangle is adjacent to another rectangle r if it is the nearest rectangle to r in some plan. Additionally, such a kind of representation of the plans allows also to express in the example description the topological relations [20, 19], such as closeness, intersection and overlapping between rectangles. However, the topological information can be deduced by the spatial relationships, and thus it can be included by the system during the learning process by means of deduction and abstraction. For instance, the following fragment of background knowledge could be provided to the system to infer the topological relations between two blocks or frames:

```
top_alignment(B1,B2):-
        occupy_plane_9(B1,B2), not(occupy_plane_4(B1,B2)).
top_alignment(B1,B2):-
       occupy_plane_10(B1,B2), not(occupy_plane_5(B1,B2)).
bottom_alignment(B1, B2) :-
      occupy_plane_19(B1, B2), not(occupy_plane_24(B1, B2)).
bottom_alignment(B1, B2) :-
      occupy_plane_20(B1, B2), not(occupy_plane_25(B1, B2)).
```

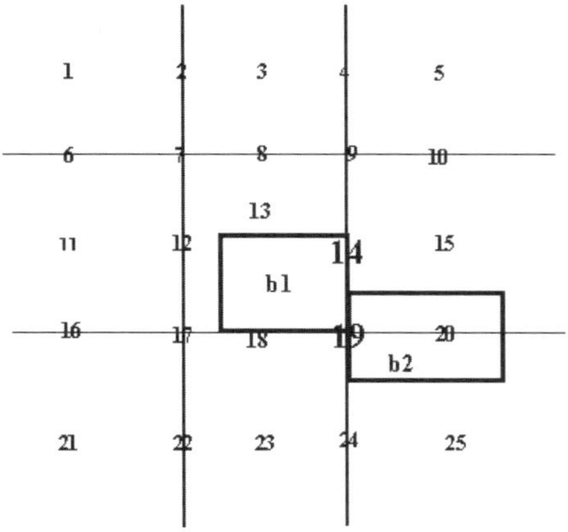

Fig. 12. Block representation

```
left_alignment(B1,B2):-
       occupy_plane_17(B1,B2), not(occupy_plane_16(B1,B2)).
left_alignment(B1,B2):-
       occupy_plane_22(B1,B2), not(occupy_plane_21(B1,B2)).
right_alignment(B1, B2) :-
      occupy_plane_19(B1, B2), not(occupy_plane_20(B1, B2)).
right_alignment(B1, B2) :-
      occupy_plane_24(B1, B2), not(occupy_plane_25(B1, B2)).
touch(B1,B2):-
       occupy_plane_14(B1,B2), not(occupy_plane_13(B1,B2)).
touch(B1,B2) :-
       occupy_plane_17(B1,B2), not(occupy_plane_13(B1,B2)).
touch(B1,B2) :-
       occupy_plane_18(B1,B2), not(occupy_plane_13(B1,B2)).
touch(B1,B2):-
       occupy_plane_19(B1,B2), not(occupy_plane_13(B1,B2)).
```

Thus, given the representation of the two blocks reported in Figure 12 where block B2 occupies plans 14, 15, 19, 24, 25 while block B1 occupies plans 13, 14, 18, 19, and having in common the plans 14 and 19, the initial representation will be made up, among other descriptors, by:

```
....., occupy_plane_14(b2, b1), occupy_plane_19(b2, b1), ......
```

and the system is able to recognize the topological relations above reported giving the following:

```
..., touch(b2,b1), bottom_alignment(b2,b1),....
```

In this language unary predicate symbols, called attributes, are used to describe properties of a single layout component (e.g. height and length), while n-ary predicate symbols, called relations, are used to express spatial relationships between layout components. A complete list of attributes and relations is reported in Table 1.

Page Descriptors

page_number(d,p): p is the number of current page in document d
last_page(p): true if page p is the last page of the document
in_first_pages(p): true if page p belongs to the first n pages of the document ($n < 1/3$ total number of the pages in the document)
in_middle_pages(p): true if page p is in the middle n pages of the document ($1/3 < n < 2/3$ total number of the pages in the document)
in_last_pages_pagine(p): true if page p belongs to the last n pages of the document ($n > 2/3$ total number of pages in the document)
number_of_pages(d, n): n is the total number of pages in document d
page_width(p,w): w is the page width (a value normalized in [0,1])
page_height(p,h): h is the page height (a value normalized in [0,1])

Frame/Block Descriptors

frame(p,f): f is a frame of page p
block(p,b): b is a block of page p
type(b,t): t is the type of the block content (text, graphic, mixed, empty, vertical_line, horizontal_line, oblique_line)
width(b,w): w is the block width in pixels
height(b,h): h is the block height in pixels
x_coordinate_rectangle(r,x): x is the horizontal coordinate of the start point of the rectangle (frame or block) r
y_coordinate_rectangle(r,y): y is the vertical coordinate of the start point of the rectangle (frame or block) r

Topological Relation Descriptors

belong(b, f): block b belongs to frame f
pos_upper(p, r): rectangle r is positioned in the upper part of page p
pos_middle(p, r): the rectangle r is vertically positioned in the middle part of page p
pos_lower(p, r): the rectangle r is positioned in the lower part of page p
pos_left(p, r): the rectangle r is positioned in the left part of page p
pos_center(p, r): the rectangle r is horizontally positioned in the center part of page p
pos_right(p, r): the rectangle r is positioned in the right part of page p
touch(b1,b2): block b1 touches block b2 and *vice versa*
on_top(b1,b2): block b1 is positioned on block b2
to_right(b1,b2): block b1 is positioned on the right of block b2
top_alignment(b1, b2): block b1 is over block b2
bottom_alignment(b1, b2): block b1 is under block b2
left_alignment(b1, b2): block b1 is on the left of block b2
right_alignment(b1, b2): block b1 is on the right of block b2

Table 1. Attributes/Relations used to describe the documents

4.3 Layout Correction

Due to the fixed stop threshold (see section 3.2), it might happen that after the layout analysis step some blocks are not correctly recognized, i.e. background areas are considered as content ones and/or vice versa. In such a case a phase of layout correction would be desirable. A first correction of the automatically recognized layout can be performed by allowing the system user to manually force further forward steps, or to go some step backward, in the algorithm with respect to the stop threshold. This is possible since the system maintains three structures that keep track of: all white rectangles found (W), all black rectangles found (B) and all rectangles that it has not analyzed yet (N: if no threshold is given all the rectangles are analyzed and N will be empty at the end of processing). Hence, when the user is not satisfied by the discovered layout because some background is missing, he can decide to go forward, and the system will extract and process further rectangles from N. Conversely, if the user notes that the system has found insignificant background pieces, he can decide to go back and the system will correspondingly move blocks between W, B and N.

However, such a solution is not always effective in case of lost significant background rectangles (e.g., small areas that represent the cut point between two frames), since they could be very small and hence it would be necessary to perform many forward steps (during which the system would probably restore insignificant white rectangles) before being able to retrieve them. Even worse, the system could be completely unable to retrieve the needed background just because it is too small to satisfy the constraints.

To solve both problems, DOCG, a module to improve the analysis performed by DOC, was implemented. It uses machine learning techniques to automatically infer rules for recognizing interesting background rectangles among those discarded or not yet analyzed by the layout analysis algorithm, according to their size and position with respect to the surrounding blocks. Specifically, we first processed a number of documents, then presented to the user all the blocks in the N structure and asked him to force as background some rectangles that the system had erroneously discarded (even if such rectangles do not satisfy the constraints), and to remove insignificant rectangles erroneously considered as background by the system. These blocks were then considered as examples for the learning system in order to infer rules to automatically perform this task during future layout analysis processes. Again, due to the need of expressing many relationships among blocks in order to represent these situations, a first-order description language was required, and INTHELEX was exploited as a learning system. Specifically, each example described the block to be forced plus all the blocks around it, along with their size and position in the document, both before and after the manual correction.

Classification

After detecting the document layout structure, a logic role can be associated to some of its components. In fact, the role played by a layout component represents meta-information that could be exploited to tag the document and help its filing and management within the digital library. The logical components can be arranged in another hierarchical structure, which is called logical structure. The logical structure is the result of repeatedly dividing the content of a document into increasingly smaller parts, on the basis of the human-perceptible meaning of the content. The leaves of the logical structure are the basic logical components, such as authors and title of a magazine article or the date and signature in a commercial letter. The heading of an article encompasses the title and the author and is therefore an example of composite logical component. Composite logical components are internal nodes of the logical structure. The root of the logical structure is the document class. The problem of finding the logical structure of a document can be cast as the problem of associating some layout components with a corresponding logical component.

The first component that can be tagged is the document itself, according to the class it belongs to (*document image classification* step). Indeed, in general many kinds of documents with different layout structures can be present in one library, and they have to be exploited in different ways according to their type. In turn, the type of a document is typically reflected by the layout structure of its first page: e.g., humans can immediately distinguish a bill from an advertisement or a letter or a (newspaper or scientific) article without actually reading their content, but just based on their visual appearance.

For this reason, we decided to apply machine learning to infer rules that allow to automatically classify new incoming documents according to their first-page layout, in order to determine how to file them in the digital repository and what kind of processing they should undergo next. This step turns out to be very significant in a digital library, where a lot of different layout structures for the documents, either belonging to different classes or even to the same class, can be encountered. Again, the diverse and complex relationships that hold between the layout components of a document suggested the use of a first-order representation language and learning system. Additionally, the possibility of continuously extending the repository with new classes of documents or with modifications of the existing ones asked for incremental abilities that INTHELEX provides.

Classification of multi-page documents is performed by matching the layout structure of their first page against the automatically learned models of classes of documents. These models capture the invariant properties of the images layout structures of documents belonging to the same class. They consist of rules expressed in a first-order logic language, so that the document classification problem can be reformulated as a matching test between a logic formula that represents a model and another logic formula that describes the image/layout properties of the first page. The choice of a first-order logic language fulfils the requirements of flexibility and generality.

Understanding

Once the class of a document has been identified on the basis of its first page layout, its logical components that are present in any page can be located and tagged by matching the layout structure of each page against models of logical components. Indeed, if we assume that it is possible to identify logical components by using only layout information, just as humans do, these models capture the invariant layout properties of the logical components of documents in the same class.

This is the task of the *document image understanding* phase, that must necessarily follow document image classification since the kind of logical components that can be expected in a document strongly depends on the document class (e.g., in a commercial letter one expects to find a sender, possibly a logo, an address, an object, a body, a date and a signature, whereas in a scientific paper one could be interested in its title, authors and their affiliations, abstract and bibliographic references). Once again, they are expressed as rules in a first-order logic language. However, differently from document classification, the document understanding problem cannot be effectively reformulated as a simple matching test between logic formulae. The association of the logical description of pages with logical components requires a full-fledged theorem prover, since it is typical that one component is defined and identified in relation to another one.

5 Categorization, Filing and Indexing

One of the most important tasks in digital library management concerns the categorization of documents. Effectiveness in performing such a task represents the success factor in the retrieval process, in order to identify documents that are really interesting for the users. Indeed, a problem of most existing word-based retrieval systems consists in their ineffectiveness in finding interesting documents when the users exploit different words than those by which the information they seek has been indexed. This is due to a number of tricky features that are typical of natural language: different writers use different words to describe the same idea (*synonymy*), thus a person issuing a query in a search engine might use different words than those that appear in an interesting document, and could not retrieve the document; one word can have multiple meanings (*polysemy*), so a searcher can get uninteresting documents concerning the alternate meanings. To face such a problem, the Latent Semantic Indexing (LSI) technique [21] has been adopted, that tries to overcome the weaknesses of term-matching based retrieval by treating the unreliability of observed term-document association data as a statistical problem. Indeed,

LSI assumes that there exists some underlying latent semantic structure in the data, that is partially obscured by the randomness of word choice with respect to the retrieval phase, and that can be estimated by means of statistical techniques.

As a weighting function, a combination of the local and global relevance of terms has been adopted in the following way:

$$w_{ij} = L(i,j) * G(i)$$

where $L(i,j)$ represents the local relevance of the term i in the document j and $G(i)$ represents the global value of the term i. A good way to relate such values is represented by the log entropy function, where:

$$L(i,j) = \log(tf_{ij} + 1)$$

$$G(i) = 1 - \sum_{j=1,\ldots,N} \frac{p_{ij} * \log(p_{ij})}{\log(N)}$$

Here, N represents the number of documents and $p_{ij} = \frac{tf_{ij}}{gf_i}$, where tf_{ij} is the local relevance for each term (the frequency of the term i in the document j, TF) and gf_i is the global relevance for each term (the frequency of the term i in the whole set of documents, IDF). This way, the logarithmic value of the local factor $L(i,j)$ mitigates the effects due to large variations in term frequencies, while the entropy function of the global factor $G(i)$ mitigates the noise that could be present in the documents.

The success of the retrieval step turns out to be strictly related to the choice of the parameter k that represents the best new rank, lower than the original one, to reduce the matrix. In our system, it is set as the minimum number of documents needed to cover the whole set of terms. As to the retrieval phase, the following weighting function was applied to each element of the query vector in order to create, for the query too, the correspondence between the local and global factor:

$$q_{ij} = (0.5 + \frac{0.5 * tf_i}{maxtf}) * \log \frac{N}{n}$$

where tf_i is the frequency of term i in the query, $maxtf$ is the maximum value among all the frequencies, N represents the total number of documents and n is the number of documents in which term i appears. In our system the *cosine similarity* function [22] was exploited to perform the comparison between the query vector and each document vector. Documents that show a high degree of similarity according to the value computed are those interesting for the user query.

However, the large amount of items that a document management system has to deal with, and the continuous flow of new documents that could be added to the initial database, require an incremental methodology to update the initial LSI matrix. Indeed, applying from scratch at each update the LSI method, taking into account both the old (already analyzed) and the new documents, would become computationally inefficient. Two techniques have been developed in the literature to update (i.e., add new terms and/or documents to) an existing LSI generated database: Folding-In [23] and SVD-Updating [24]. An analysis on the performance of both techniques shows that SVD-Updating is more suitable to be exploited in a digital library environment. Indeed, the former is a much simpler alternative that uses the existing LSI matrix to represent new information but yields poor-quality updated matrices, since the semantic correlation information contained in the new documents/terms is not exploited by the updated semantic space. The latter represents a trade-off between the former and the recomputation from scratch.

6 Exploitation and Evaluation

The system for automated digital documents processing was evaluated in each step, from document acquisition to document indexing for categorization and information retrieval purposes. Since the system can be embedded as a document management engine into many different domain-specific applications, in this section we focus on the Conference Management scenario. As we will see DOMINUS can usefully support some of the more critical and knowledge-intensive tasks involved by the organization of a scientific conference, such as the assignment of the submitted papers to suitable reviewers.

6.1 Scientific Conference Management Scenario

Organizing scientific conferences is a complex and multi-faceted activity that often requires the use of a Web-based management system to make some tasks a little easier to carry out, such as the job of reviewing papers. Some of the features typically provided by these packages are: submission of abstracts and papers by Authors; submission of reviews by the Program Committee Members (PCMs); download of papers by the Program Committee (PC); handling of reviewers preferences and bidding; Web-based assignment of papers to PCMs for review; review progress tracking; Web-based PC meeting; notification of acceptance/rejection; sending e-mails for notifications.

Let us now present a possible scenario. An Author connects to the Internet and opens the submission page, where (after registering, or after logging in if already registered) he can browse his hard disk and submit a paper by choosing the corresponding file in one of the accepted formats. After uploading, the paper undergoes the following processing steps. The layout analysis algorithm is applied, in order to single out its layout components. Then, it is translated into a first-order logic description and classified by a proper module according to the theory learned so far for the acceptable submission layout standards. A single conference can allow different layout standards for the submitted papers (e.g., full paper, poster, demo) and it can be the case that many conferences have to be managed at the same time. Depending on the identified class, a further step consists in locating and labelling the layout components of interest for that class (e.g., title, author, abstract and references in a full paper). The text contained in each of such components is read, stored and used to automatically file the submission record (e.g., by filling its title, authors and abstract fields). If the system is unable to carry out any of these steps, such an event is notified to the Conference administrators, that can manually fix the problem and let the system complete its task. Such manual corrections are logged and used by the incremental learning component to refine the available classification/labeling theories in order to improve their performance on future submissions. Nevertheless, this is done off-line, and the updated theory replaces the old one only after the learning step is successfully completed, thus allowing further submissions in the meantime. Alternatively, the corrections can be logged and exploited all at once to refine the theory when the system performance falls below a given threshold.

The next step, which is currently under investigation, concerns the automatic categorization of the paper content on the grounds of the text it contains. This allows to match the paper topic against the reviewers' expertise, in order to find the best associations for the final assignment. Specifically, we exploit the text in the title, abstract and bibliographic references, assuming that they concentrate the subject and research field the paper is concerned with. This requires a pre-processing step that extracts the meaningful content from each reference (ignoring, e.g., page numbers, place and editors). Furthermore, the paper topics discovered in the indexing phase are matched with the conference topics with the aim of supporting the conference scheduling.

6.2 Experimental Results

In the above scenario, the first step concerns document image classification and understanding of the documents submitted by the Authors. In order to evaluate the system on this phase, experiments were carried out on a fragment of 353 documents coming from our digital library, made up of documents of the last ten years available in online repositories (i.e., publishers' online sites, authors' home pages, the Scientific Literature Digital Library CiteSeer, our submissions, etc.) interesting for our research topics. The resulting dataset is made up of four classes of documents: the Springer-Verlag Lecture Notes in Computer Science (LNCS) series, the Elsevier journals style (ELSEVIER), the Machine Learning Journal (MLJ) and the Journal of Machine Learning Research (JMLR). Specifically, 70 papers were formatted according to the LNCS style (proofs and initial submission of the papers), 61 according to the ELSEVIER style, 122 according to the MLJ (editorials, Kluwer Academy and Springer Science publishers) style and 100 according to the JMLR style.

Journal of Machine Learning Research 3 (2003) 1107–1135 Submitted 4/02; Published 2/03

Matching Words and Pictures

Kobus Barnard KOBUS@CS.ARIZONA.EDU
Computer Science Department
University of Arizona
Tucson, AZ 85721-0077, USA

Pinar Duygulu DUYGULU@CENG.METU.EDU.TR
Department of Computer Engineering
Middle East Technical University
Ankara, Turkey

David Forsyth DAF@CS.BERKELEY.EDU
Computer Science Division
University of California
Berkeley, CA 94720-1776, USA

Nando de Freitas NANDO@CS.UBC.CA
Department of Computer Science
University of British Columbia
Vancouver, B.C. V6T 1Z4, Canada

David M. Blei BLEI@CS.BERKELEY.EDU
Computer Science Division
University of California
Berkeley, CA 94720-1776, USA

Michael I. Jordan JORDAN@CS.BERKELEY.EDU
Computer Science Division and Department of Statistics
University of California
Berkeley, CA 94720-1776, USA

Editors: Jaz Kandola, Thomas Hofmann, Tomaso Poggio and John Shawe-Taylor

Abstract

We present a new approach for modeling multi-modal data sets, focusing on the specific case of segmented images with associated text. Learning the joint distribution of image regions and words has many applications. We consider in detail predicting words associated with whole images (auto-annotation) and corresponding to particular image regions (region naming). Auto-annotation might help organize and access large collections of images. Region naming is a model of object recognition as a process of translating image regions to words, much as one might translate from one language to another. Learning the relationships between image regions and semantic correlates (words) is an interesting example of multi-modal data mining, particularly because it is typically hard to apply data mining techniques to collections of images. We develop a number of models for the joint distribution of image regions and words, including several which explicitly learn the correspondence between regions and words. We study multi-modal and correspondence extensions to Hofmann's hierarchical clustering/aspect model, a translation model adapted from statistical machine translation (Brown et al.), and a multi-modal extension to mixture of latent Dirichlet allocation (MoM-LDA). All models are assessed using a large collection of annotated images of real

©2003 Kobus Barnard, Pinar Duygulu, David Forsyth, Nando de Freitas, David Blei and Michael Jordan.

Journal of Machine Learning Research 3 (2003) 1209-1227 Submitted 5/02; Published 3/03

Extensions to Metric-Based Model Selection

Yoshua Bengio BENGIOY@IRO.UMONTREAL.CA
Nicolas Chapados CHAPADOS@IRO.UMONTREAL.CA
Dept. IRO, Université de Montréal
C.P. 6128, Montreal, Québec, H3C 3J7, Canada

Editors: Isabelle Guyon and André Elisseeff

Abstract

Metric-based methods have recently been introduced for model selection and regularization, often yielding very significant improvements over the alternatives tried (including cross-validation). All these methods require unlabeled data over which to compare functions and detect gross differences in behavior away from the training points. We introduce three new extensions of the metric model selection methods and apply them to feature selection. The first extension takes advantage of the particular case of time-series data in which the task involves prediction with a horizon h. The idea is to use at t the h unlabeled examples that precede t for model selection. The second extension takes advantage of the different error distributions of cross-validation and the metric methods: cross-validation tends to have a larger variance and is unbiased. A hybrid combining the two model selection methods is rarely beaten by any of the two methods. The third extension deals with the case when unlabeled data is not available at all, using an estimated input density. Experiments are described to study these extensions in the context of capacity control and feature subset selection.

Keywords: Metric-based Methods, Model Selection

1. Model Selection and Regularization

Supervised learning algorithms take a finite set of input/output training pairs $\{(x_1,y_1), \ldots, (x_m,y_m)\}$, sampled (usually independently) from an unknown joint distribution $P(X,Y)$, and attempt to infer a function $g \in G$ that minimizes the expected value of the loss $L(g(X),Y)$ (also called the *generalization error*). In many cases one faces the dilemma that if G is too "rich", which often occurs if the dimension of X is too large, then the average training set loss (*training error*) will be low but the expected loss may be large (*overfitting*), and vice-versa if G is not "rich" enough (*underfitting*).

In many cases one can define a collection of increasingly complex function classes $G_0 \subset G_1 \subset \cdots \subset G$ (although some methods studied here work as well with a partial order). In this paper we focus on the case in which the function classes differ by the dimensionality of the input X. *Model selection* methods attempt to choose one of these function classes to avoid both overfitting and underfitting. Variable subset selection is a form of model selection in which one has to select both the number of input variables and the particular choice of these variables (but note that the two decisions may be profitably decoupled). One approach to model selection is based on *complexity penalization* (Vapnik, 1982, 1998, Rissanen, 1986, Foster and George, 1994): one first applies the learning algorithm to each of the function classes (on the whole training set), yielding a sequence of hypothesis functions $g_0, g_1, \ldots$. Then one of them is chosen based on a criterion that combines average training loss and a measure of complexity (which usually depends only on the function

©2003 Yoshua Bengio and Nicolas Chapados.

Fig. 13. Two first pages from the JMLR class

It is worth to note that even documents in the same class might fall in different layout standards, according to the period of publication, since the publisher layout style may have changed in time. Thus, the changes in spatial organization of the first page might affect the classification step (see Figure 13).

This calls for the incremental abilities of the incremental system that must generate different concept definitions at the same time. Indeed, the system is able, at any moment, to learn the layout description of a new class of document style preserving the correct definition of the others. In this way a global theory is built, containing the definitions of different document styles, that could be used for many conferences.

Each document was described according to the features reported in Section 4.2, and was considered as a positive example for the class it belongs to, and as a negative example for all the other classes to be learned. The system performance was evaluated according to a 10-fold cross validation methodology, ensuring that the training and test sets contained the same percentage of positive and negative examples. Furthermore, the system was provided with background knowledge expressing topological relations (see Section 4.2), and abstraction operators were used to discretize numeric values concerning size and position into intervals expressed by symbolic descriptors. In the following, an example of the abstraction rules for rectangles width discretization is given.

```
width_very_small(X):-
    rectangle_width(X, Y), Y >= 0, Y =< 0.023.
```

```
width_small(X):-
    rectangle_width(X, Y), Y > 0.023, Y =< 0.047.
width_medium_small(X):-
    rectangle_width(X, Y), Y >= 0.047, Y =< 0.125.
width_medium(X):-
    rectangle_width(X, Y), Y > 0.125, Y =< 0.203.
```

A first experiment was run to infer the document classification rules; good results were obtained in terms of runtime, predictive accuracy, number of theory revisions (Rev = total revisions, Rev+ = revisions performed on positive examples only, Rev- = revisions performed on negative examples). Furthermore, in order to evaluate the theory revision rate, some additional measures were considered: the global percentage of revisions Rev on the whole training set (RevRate), the percentage of revisions Rev- on the positive examples (RevRate+) and the percentage of revisions Rev- on the negative examples (RevRate-)), as reported in Table 2. The lowest accuracy and poorest performance was obtained on MLJ, that reflects the variety of corresponding paper formats and typing styles.

Table 2. Learning System Performance: inferring rules for paper class identification

Class	Rev	Rev+	Rev-	RevRate	RevRate+	RevRate-	Time (s.)	Acc. %
LNCS	16	11.7	4.3	0.05	0.18	0.02	662.88	97.2
MLJ	28.2	18.7	9.5	0.08	0.17	0.04	2974.87	93.5
ELSEVIER	13.6	11.2	2.4	0.04	0.20	0.01	303.85	98.9
JMLR	12.7	10	2.7	0.04	0.11	0.01	1961.66	98.2

As to the revision rate, Figure 14 sketches the system performance with respect to revisions and accuracy on the training phase in the classification step in one fold (the nearest to the average reported in Table 2). The curve represents the trend in accuracy as long as new examples are analyzed, while the cuts represent the revision points. These points become very sparse as the number of analyzed examples increases and the accuracy curve, after a first phase in which many revisions have to be performed to restore the theory correctness, tends to increase towards a stable condition. The results concerning class MLJ are perfectly consistent with the composition of the selected sample; the variety of typing conventions and formats underlying the documents requires to extend the training set.

Once the classification step has been completed the image document understanding phase starts. The second experiment was performed on the *title*, *authors*, *abstract* and *references* layout components of documents belonging to the LNCS class. This class was chosen since it represents the layout standard of papers submitted to the 18th Conference on Industrial & Engineering Applications of Artificial Intelligence & Expert Systems (IEA/AIE 2005) which

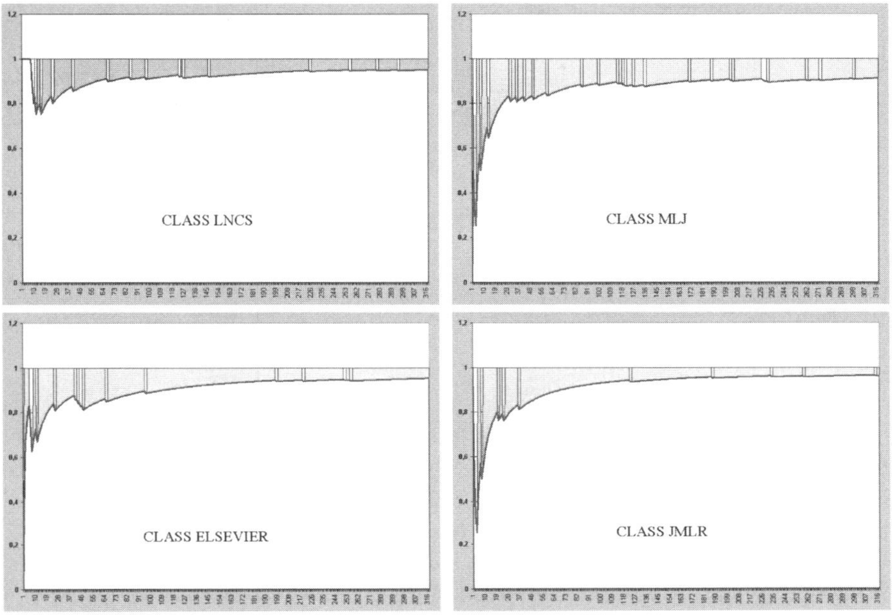

Fig. 14. Accuracy and revision rate of the learning system on tuning phase

has been used as a real testbed. In Table 3 the averaged results of the 10 folds are reported, that can be considering satisfying from both the accuracy and the time consuming point of view.

Table 3. Learning System Performance: inferring rules for components label identification

Label	Rev	Rev+	Rev-	RevRate	RevRate+	RevRate-	Time (s.)	Acc. %
Title	16.5	13.7	2.8	0.06	0.22	0.01	217.60	95.3
Abstract	10.5	9.4	1.1	0.04	0.15	0.01	104.07	96.2
Author	14.6	11.1	3.5	0.05	0.17	0.02	146.48	98.6
Ref	15.4	10.6	4.8	0.06	0.17	0.02	150.93	97.4

A very hard task in the organization of Scientific Conferences is the reviewers assignment; due to the many constraints, manually performing such a task is very tedious and difficult, and does not guarantee the best results. The proposed document management system can assist the conference program chair both in indexing and retrieving the documents and their associated topics, although not explicitly reported by the paper authors. In the following we present an experiment carried out on the above reported dataset consisting of

264 papers submitted to the IEA/AIE 2005 conference, whose Call for Papers included 34 topics of interest.

Firstly, the layout of each paper in digital format was automatically analyzed in order to recognize the significant components. In particular, the abstract and title were considered the most representative of the document subject, and hence the corresponding text was extracted to apply the LSI technique. The words contained therein were stemmed according to the technique proposed by Porter [25], resulting in a total of 2832 word stems. Then, the same procedure was applied to index the reviewers expertise according to the titles of their papers appearing in the DBLP Computer Science Bibliography repository (http://www.informatik.uni-trier.de/~ley/db/), resulting in 2204 stems.

In both cases, the LSI parameters were set in such a way that all the conference topics were covered as different concepts. The experiment consisted first in performing 34 queries, each corresponding to one conference topic, both on papers and on reviewers, and then in associating respectively to each paper/reviewer the first l results of the LSI queries. The results obtained on document topic recognition showed that considering 88 documents per query is enough to cover the whole set of documents. However, considering just 30 documents per query, 257 out of 264 documents (97.3%) were already assigned to at least one topic. This is an acceptable trade-off since the remaining 7 documents can be easily assigned by hand. Moreover, 30 documents are a good choice to assure the equidistribution over the document. Interestingly, more than half of the documents (54.7%) concern $2 \div 4$ topics so confirming the extremely specialized nature of the conference and the high correlation between the topics. The results, compared to the conference program chair indications, showed a 79% accuracy on average. Setting $l = 10$, the automatic assignment of the topics to the reviewers resulted in 65% accuracy compared to the suggestions of the conference program chair.

Lastly, the expert system GRAPE (Global Review Assignment Processing Engine) [26] has the task of automatically assigning the papers to reviewers taking into account specific knowledge (i.e., conflicts, nationality, common interest, etc.). The final assignments were considered very useful suggestions by the experts so confirming the goodness of the indexing process and of the topic associations.

7 Related Work

Image Document analysis refers to algorithms and techniques developed in order to obtain a computer-readable description of a scanned document [27].

While an impressive amount of contribution has been presented applied to scanned image documents, only recently a few works have faced the problem of handling digital document formats such as PDF and PS. Most of them aim at extracting (some part of) the document content by means of a syntactic parsing of the PDF [28, 29, 30] or at discovering the background by means of statistical analyses applied to the numerical features of the documents and its components [31]. A further step towards digital document analysis as opposed (but complementary) to document image analysis is represented by the work reported in [32]; here, a method is proposed for the extraction of the logical structure from PDF files by examining the visual appearance and geometric position of text and image blocks distributed over the whole document and exploiting the information on line spacing and font usage in order to bridge the semantic gap between the document image and its content. In this work, the PDF file is firstly decomposed by syntactically parsing it, then grouping words into lines (by means of APIs provided by the Acrobat Exchange viewer), lines in bins (based on their point size, font name and their coordinates in the page) and finally bins in blocks. Successively, relationships (greater/lesser status) among two blocks are discovered by analyzing their features (font name and point size) and labels of the discovered blocks are identified by applying (and possibly modifying after new blocks are evaluated) a set of rules purposely codified by a domain expert for the class/tags at hand.

Recently, some works [33, 34] proposed a strategy that mixes the layout extraction methods from digital documents with the most widely used document analysis techniques. The approach consists into three steps:

- parsing syntactically the PDF file to extract the document primitives (text, image or graphics);
- recognizing and grouping homogeneous entities among the extracted primitives;
- extracting the logical layout structure by means of text entities labelling (e.g., title, author, body) and document modelling in which the entities are *projected* in a document model.

Here, for each class of documents an expert provides a model, representing its grammar, i.e. a hierarchy of logical entities.

A similar approach which uses grammars to annotate document components is proposed in [35]. Here, based on the model provided by an expert, a set of possible roles is assigned to each layout object. Then, they are collected into more complex objects until the logical structure is produced.

All the approaches reported above perform geometric layout analysis by means of a syntactic parsing of the document. Then, the mapping between the geometric and logical structure is supported by using a template of the document, a grammar representing its layout, or an expert system whose knowledge base must be provided by a human expert.

8 Conclusion

The huge amount of documents available in digital form and the flourishing of digital repositories raise problems about document management, concerned with effectiveness and efficiency of their successive retrieval, that cannot be faced by manual techniques. This paper proposed DOMINUS, an intelligent system characterized by the intensive application of Machine Learning techniques as a support to all phases of automated document processing, from document acquisition to document understanding and indexing. The core of DOMINUS is the Learning Server, a suite of different inductive learning methods and systems, among which the more suitable for the specific document processing phase is chosen and applied. The most interesting is INTHELEX, a proprietary incremental learning system able to handle structural descriptions and to automatically revise first-order theories.

Experiments in the real-world domain of automatic Scientific Conference Management have been presented and discussed, showing the validity of the proposed approach.

Several future work directions are planned for the proposed system. First of all, the automatic processing of bibliographic references, that can improve the identification of the document subject and context. Secondly, the use of ontologies in text processing in order to improve the effectiveness of content-based retrieval.

References

1. Esposito, F., Malerba, D., Semeraro, G., Ferilli, S., Altamura, O., Basile, T.M.A., Berardi, M., Ceci, M., Mauro, N.D.: Machine learning methods for automatically processing historical documents: From paper acquisition to XML transformation. In: Proceedings of the First International Workshop on Document Image Analysis for Libraries (DIAL 2004). (2004) 328–335
2. Berners Lee, T., Hendler, J., Lassila, O.: The semantic web. Scientific American **284**(5) (2001) 34–43
3. Utgoff, P.E.: Incremental induction of decision trees. Machine Learning **4**(2) (1989) 161–186
4. Cauwenberghs, G., Poggio, T.: Incremental and decremental support vector machine learning. In: Advances in Neural Information Processing Systems (NIPS 2000). Volume 13., Cambridge, MA, USA, MIT Press (2000) 409–415
5. Solomonoff, R.: Progress in incremental machine learning. In: NIPS Workshop on Universal Learning Algorithms and Optimal Search, Dec. 14, 2002, Whistler, B.C., Canada, 27 pp. (2003)
6. Wong, W., Fu, A.: Incremental document clustering for web page classification. In: IEEE 2000 Int. Conf. on Info. Society in the 21st century: emerging technologies and new challenges (IS2000), Nov 5-8, 2000, Japan. (2000)
7. Dietterich, T.G., Lathrop, R.H., Lozano-Perez, T.: Solving the multiple instance problem with axis-parallel rectangles. Artificial Intelligence **89**(1-2) (1997) 31–71
8. Breuel, T.M.: Two geometric algorithms for layout analysis. In: Workshop on Document Analysis Systems. (2002)
9. Glunz, W.: pstoedit - a tool converting postscript and PDF files into various vector graphic formats (2007) (http://www.pstoedit.net).

10. Adobe Systems Inc.: PostScript language reference manual – 2nd ed. Addison Wesley (1990)
11. Adobe Systems Inc.: PDF Reference version 1.3 – 2nd ed. Addison Wesley (2000)
12. Esposito, F., Ferilli, S., Fanizzi, N., Basile, T.M., Di Mauro, N.: Incremental multistrategy learning for document processing. Applied Artificial Intelligence: An Internationa Journal **17**(8/9) (2003) 859–883
13. Muggleton, S., Raedt, L.D.: Inductive logic programming: Theory and methods. Journal of Logic Programming **19/20** (1994) 629–679
14. Semeraro, G., Esposito, F., Malerba, D., Fanizzi, N., Ferilli, S.: A logic framework for the incremental inductive synthesis of datalog theories. In Fuchs, N., ed.: Proceedings of the 7th International Workshop on Logic Program Synthesis and Transformation. Volume 1463 of LNCS., Springer (1998) 300–321
15. Becker, J.: Inductive learning of decision rules with exceptions: Methodology and experimentation. Master's thesis, Dept. of Computer Science, University of Illinois at Urbana-Champaign, Urbana, Illinois (1985) B.S. diss., UIUCDCS-F-85-945
16. Michalski, R.: Inferential theory of learning. developing foundations for multistrategy learning. In Michalski, R., Tecuci, G., eds.: Machine Learning. A Multistrategy Approach. Volume IV. Morgan Kaufmann (1994) 3–61
17. Kakas, A., Mancarella, P.: On the relation of truth maintenance and abduction. In: Proceedings of the 1st Pacific Rim International Conference on Artificial Intelligence, Nagoya, Japan (1990)
18. Zucker, J.D.: Semantic abstraction for concept representation and learning. In Michalski, R.S., Saitta, L., eds.: Proceedings of the 4th International Workshop on Multistrategy Learning. (1998) 157–164
19. Papadias, D., Theodoridis, Y.: Spatial relations, minimum bounding rectangles, and spatial data structures. International Journal of Geographical Information Science **11**(2) (1997) 111–138
20. Egenhofer, M.: Reasoning about binary topological relations. In Gunther, O., Schek, H.J., eds.: Second Symposium on Large Spatial Databases. Volume 525 of Lecture Notes in Computer Science., Springer (1991) 143–160
21. Deerwester, S.C., Dumais, S.T., Landauer, T.K., Furnas, G.W., Harshman, R.A.: Indexing by Latent Semantic Analysis. Journal of the American Society of Information Science **41**(6) (1990) 391–407
22. Baeza-Yates, R.A., Ribeiro-Neto, B.A.: Modern Information Retrieval. ACM Press / Addison-Wesley (1999)
23. Berry, M.W., Dumais, S.T., O'Brien, G.W.: Using linear algebra for intelligent information retrieval. SIAM Rev. **37**(4) (1995) 573–595
24. O'Brien, G.W.: Information management tools for updating an SVD-encoded indexing scheme. Technical Report UT-CS-94-258, University of Tennessee (1994)
25. Porter, M.F.: An algorithm for suffix stripping. In Karen, J.S., Willet, P., eds.: Readings in information retrieval. Morgan Kaufmann Publishers Inc., San Francisco, CA, USA (1997) 313–316
26. Di Mauro, N., Basile, T.M.A., Ferilli, S.: GRAPE: An expert review assignment component for scientific conference management systems. In: Innovations in Applied Artificial Intelligence: 18th International Conference on Industrial and Engineering Applications of Artificial Intelligence and Expert Systems (IEA/AIE 2005). Volume 3533 of Lecture Notes in Computer Science., Springer Verlag (2005) 789–798

27. Nagy, G.: Twenty years of document image analysis in PAMI. IEEE Transactions on Pattern Analysis and Machine Intelligence **22**(1) (2000) 38–62
28. Futrelle, R.P., Shao, M., Cieslik, C., Grimes, A.E.: Extraction, layout analysis and classification of diagrams in PDF documents. In: Proceedings of Seventh International Conference on Document Analysis and Recognition (ICDAR 2003). (2003) 1007–1014
29. Chao, H.: Graphics extraction in PDF document. In Kanungo, T., Smith, E.H.B., Hu, J., Kantor, P.B., eds.: Proceedings of SPIE - The International Society for Optical Engineering. Volume 5010. (2003) 317–325
30. Ramel, J.Y., Crucianu, M., Vincent, N., Faure, C.: Detection, extraction and representation of tables. In: Proceedings of the Seventh International Conference on Document Analysis and Recognition (ICDAR 2003), Washington, DC, USA, IEEE Computer Society (2003) 374–378
31. Chao, H., Fan, J.: Layout and content extraction for pdf documents. In: Document Analysis Systems VI, Proceeding of the Sixth International Workshop (DAS 2004). Volume 3163 of Lecture Notes in Computer Science., Springer Verlag (2004) 213–224
32. Lovegrove, W.S., Brailsford, D.F.: Document analysis of PDF files: methods, results and implications. Electronic Publishing – Origination, Dissemination and Design **8**(2-3) (1995) 207–220
33. Hadjar, K., Rigamonti, M., Lalanne, D., Ingold, R.: Xed: A new tool for extracting hidden structures from electronic documents. In: DIAL '04: Proceedings of the First International Workshop on Document Image Analysis for Libraries (DIAL'04), Washington, DC, USA, IEEE Computer Society (2004) 212
34. Rigamonti, M., Bloechle, J.L., Hadjar, K., Lalanne, D., Ingold, R.: Towards a canonical and structured representation of PDF documents through reverse engineering. In: ICDAR '05: Proceedings of the Eighth International Conference on Document Analysis and Recognition, Washington, DC, USA, IEEE Computer Society (2005) 1050–1055
35. Anjewierden, A.: AIDAS: Incremental logical structure discovery in pdf documents. In: Proceedings of Sixth International Conference on Document Analysis and Recognition (ICDAR 2001). (2001) 374–378

Classification and Learning Methods for Character Recognition: Advances and Remaining Problems

Cheng-Lin Liu[1] and Hiromichi Fujisawa[2]

[1] National Laboratory of Pattern Recognition (NLPR), Institute of Automation, Chinese Academy of Sciences, P.O. Box 2728, Beijing 100080, P.R. China
liucl@nlpr.ia.ac.cn

[2] Central Research Laboratory, Hitachi Ltd., 1-280 Higashi-koigakubo, Kokubunji-shi, Tokyo 185-8601, Japan
hiromichi.fujisawa.sb@hitachi.com

Summary. Pattern classification methods based on learning-from-examples have been widely applied to character recognition from the 1990s and have brought forth significant improvements of recognition accuracies. This kind of methods include statistical methods, artificial neural networks, support vector machines, multiple classifier combination, etc. In this chapter, we briefly review the learning-based classification methods that have been successfully applied to character recognition, with a special section devoted to the classification of large category set. We then discuss the characteristics of these methods, and discuss the remaining problems in character recognition that can be potentially solved by machine learning methods.

1 Introduction

The methods popularly used in the early stage of OCR (optical character recognition) research and development are template matching and structural analysis [1]. An intermediate approach between them is feature analysis, also referred to as feature matching. The templates or prototypes in these early methods were either designed artificially, selected or averaged from few samples. As the number of samples increases, these simple methods are insufficient to accommodate the shape variability of samples, and so, are not able to yield high recognition accuracy. To take full advantage of large sample data, the character recognition community turned attention to learning-based classification methods, especially artificial neural networks (ANNs) from the late 1980s and the 1990s. Due to the close connection between ANNs and statistical pattern recognition, statistical classification methods are also considered seriously from then. Meanwhile, the research activities in pattern recognition and machine learning communities are becoming close to each other. New

C.-L. Liu and H. Fujisawa: *Classification and Learning Methods for Character Recognition: Advances and Remaining Problems*, Studies in Computational Intelligence (SCI) **90**, 139–161 (2008)
www.springerlink.com

learning methods, especially support vector machines (SVMs, and more generally, kernel methods) and ensemble methods (multiple classifier systems), are now actively studied and applied in pattern recognition.

Learning methods[3] have benefited character recognition tremendously: they release engineers from painful job of template selection and tuning, and the recognition accuracies have been improved significantly because of learning from large sample data. Some excellent results have been reported by, e.g. [2, 3, 4]. While enjoying the benefits of learning-from-examples, we are aware that the problem is far from being solved: the recognition accuracies of either machine-printed characters on degraded image or freely handwritten characters are insufficient; the existing learning methods do not work well on large category set; huge sample data and ever-increasing data; recognition errors cannot be eliminated even if we reject a large percentage of samples, etc. The solution of these remaining problems should still rely on learning: to better utilize knowledge and samples.

In this chapter, we first give a brief survey of classification methods in character recognition. A special section is devoted to the classification of large category set. We then discuss the strengths and weaknesses of these methods, identify the needs of improved performance in character recognition, and suggest some research directions of pattern classification that can help meet with these needs. We will focus on the classification of isolated (segmented) characters, though classification methods are also important for other tasks like layout analysis and segmentation (see [5]). The classification of characters is also important for segmentation, when over-segmentation-based or character-model-based word/string recognition schemes are adopted. When we discuss classification, it is assumed that pre-processing and feature extraction procedures have been performed appropriately.

2 Brief Survey of Classification Methods

The classification methods for character recognition can be roughly categorized into feature-vector-based methods and structural methods. Feature-vector-based methods are prevailing, especially in off-line character recognition, because of their simple implementation and low computational complexity. Whereas a feature vector can be easily extracted from character images, for structural methods, the extraction of components or strokes are rather difficult. Meanwhile, there is not an off-the-shelf method for learning structural models from examples. Hence, we mainly discuss feature-vector-based methods, including statistical classification methods, ANNs, SVMs, and multiple classifier combination. A comprehensive survey of classification methods has been given by Jain et al. [6]. Statistical methods and ANNs are systematically

[3] We refer to learning when classifier design is concerned, and refer to classification when the task of recognition is concerned

treated by Fukunaga [7] and Bishop [8], respectively. The textbook of Duda et al. [9] emphasizes statistical methods, but covers other methods as well. In the following, we briefly review the methods that have been successfully applied to character recognition.

2.1 Statistical Methods

Statistical classification methods are rooted in the Bayes decision rule. In the case of 0-1 loss, the input pattern is classified to the class of maximum a posteriori (MAP) probability, which is computed by the Bayes formula from the a priori probability (usually assumed equal for defined classes) and the conditional probability density. Statistical classifiers are divided into parametric ones and non-parametric ones depending on the probability density estimation approach. Parametric classifiers assume for each class a known form of density function, usually a Gaussian function, with unknown parameters estimated on training samples by maximum likelihood (ML). Non-parametric classifiers approximate arbitrary density functions by interpolating the local densities of training samples (Parzen window), or estimate the a posteriori probabilities directly from samples (k-nearest neighbor (k-NN)). Non-parametric methods are expensive in both storage space and execution, however. Though parametric methods assume restrictive density functions, they perform fairly well for practical problems.

When assuming Gaussian density and equal a priori probabilities, the Bayesian discriminant function is equivalent to a quadratic discriminant function (QDF), which is often taken as a standard classifier in benchmarking. When further assuming that the Gaussian density functions of all classes share a common covariance matrix, the QDF is reduced to a linear discriminant function (LDF). If more restrictively, the conditional density function is spherical Gaussian with equal variance, the discriminant function is reduced to the Euclidean distance from class mean, which was often taken in early feature matching methods. The QDF does not necessarily outperform the LDF because it has as many parameters as square of feature dimensionality, and so, is sensitive to the training sample size. The regularized discriminant analysis (RDA) method [10] alleviates this problem by smoothing the covariance matrices. On the other hand, Kimura et al. replace the minor eigenvalues of covariance matrix of each class with a constant [11]. The resulting modified quadratic discriminant function (MQDF) involves less parameters and lower computation than the QDF, and results in improved generalization accuracy (accuracy on un-trained samples). The MQDF is popularly used, especially in handwritten Chinese/Japanese character recognition. An improvement of MQDF with elaborate parameter estimation is called as modified Bayes discriminant function (MBDF) [12]. Another method that is often referred is the projection distance [13], in which the distance of input pattern from a linear subspace of each class serves a reasonable discriminant function. An improvement, called modified projection distance (MPD), has a functional form similar to the MQDF [14].

Other than regularizing Gaussian density, the Gaussian mixture model (mixture of Gaussians) can model multi-modal distributions. For high-dimensional feature space as is the case of character recognition, however, it does not generalize well. Using a mixture of low-dimensional linear subspaces lowers the complexity of Gaussian mixture while maintaining the multi-modal nature, and the classification performance can be largely improved, as have been demonstrated in handwritten numeral recognition [15, 16].

Under the umbrella of statistical pattern recognition are also feature selection and transformation methods. Feature transformation can reduce the dimensionality of feature space and often improve the classification accuracy. Principal component analysis (PCA) and Fisher discriminant analysis (FDA) are two linear subspace learning methods that have been popularly used. PCA is effective mainly in the recognition of small character set, whereas for large character set, FDA is more efficient [12, 17]. Heteroscedastic discriminant analysis and nonlinear dimensionality reduction have been actively studied in pattern recognition, but are rarely applied to pratical character recognition.

Feature selection is also an active research field in pattern recognition and machine learning. It benefits recognition when there is a large number of features containing redundant and/or noisy ones. Extracting various types of features followed by feature subset selection may yield higher performance than classification on the original feature set or an artificially selected subset. Promising results of handwritten digit recognition using feature subset selection have been reported in [18].

Both parametric and non-parametric classifiers estimate the density parameters of each class independently without considering the separability of different classes. Some methods have been proposed to improve the classification accuracy by modifying the parameters according to the recognition errors on training samples made by the ML classifier, like the LDA method [19] and the mirror image learning method [20]. Parameter optimization methods by error minimization will be reviewed in the context of neural networks.

2.2 Artificial Neural Networks

The connecting weights of artificial neural networks (ANNs) are adjustable to fit an objective of functional approximation, e.g. minimum regression error. Feedforward neural networks, including single-layer perceptron (SLP), multilayer perceptron (MLP), radial basis function (RBF) network, higher-order neural network (HONNs), etc., have been widely applied to pattern recognition. Usually, each output node of the network corresponds to a class, and the maximum output gives the decision of classification. The connecting weights are usually adjusted to minimize the square error between the outputs and target values on training samples (supervised learning). The minimum square error training algorithm for MLP is referred to as back-propagation (BP) in particular. /indexback-propagation

The descriptions of supervised learning, SLP, MLP, and RBF network can be found in most neural networks textbooks. The output of SLP can be viewed as a linear discriminant function, with the weights estimated by error minimization instead of maximum lieklihood (ML), and so, the SLP often gives higher classification accuracy than the LDF with ML estimation. The MLP is flexible to approximate nonlinear functions and capable of separating patterns of complicated distributions. This power makes it a popular tool for pattern classification. Many works in character recognition have taken the MLP as a standard classifier or benchmark. The generalization performance of MLP can be improved by weight decay, local connection (local receptive fields), weight sharing, structure selection and stopping by cross-validation, etc. A network using local connection and shared weights, called convolutional neural network, has reported great success in character recognition [2, 21]. It directly works on character image and the hidden nodes with local connection can be viewed as trainable feature extractors. For feature vector-based classification, using a modular network for each class can also improve the accuracy [22].

The RBF network has one hidden layer of Gaussian functions, which are combined linearly by the output nodes. In early stage, the parameters of RBF networks were usually estimated in two phases: Gaussian parameter estimation by clustering and weight learning by error minimization. Since the clustering procedure does not consider the speparability of patterns, the Gaussian parameters learned this way do not lead to good classification performance. A substantial improvement is to adjust all the parameters simultaneously by error minimization [8]. This makes the RBF network competitive with the MLP in classification accuracy.

The HONN is also referred to as functional-link network [23], polynomial network, or polynomial classifier [24]. Its output is a weighted combination of pattern features and their polynomial expansions. For high-dimensional features, the number of (even 2nd-order) polynomial terms is extremely large. This complexity can be reduced by dimensionality reduction before polynomial expansion [25] or polynomial term selection [26]. A recently proposed class-specific feature polynomial classifier (CFPC) improves the classification accuracy by polynomial expansion on class-specific linear subspaces [27].

Some unsupervised learning methods have also been applied to pattern recognition, among them are competitive learning for vector quantization (VQ, can be used for learning prototypes for each class) and auto-association network (an application to character recognition can be seen in [28]). On the other hand, Zhang et al. learn mixtures of linear subspaces using neural networks for classification [29].

The learning vector quantization (LVQ) algorithm of Kohonen [30] learns class prototypes with the aim of separating the samples of different classes. LVQ is a supervised learning method and can give higher classification accuracy than VQ. We view VQ and LVQ as neural-like methods because like neural networks, the parameters (prototypes) are adjusted in online mode (stochastic gradient descent, iteratively on training samples). Some improvements of LVQ learn prototypes by minimizing classification or regression error instead of heuristic adjustment [31].

The discriminative learning quadratic discriminant function (DLQDF) [32] can also be viewed as a neural-like classifier. The DLQDF inherits the structure and initial parameters from the MQDF, but the parameters are optimized on training samples by minimizing the classification error by stochastic gradient descent. In experiments of handwritten numeral recognition, the DLQDF was shown to outperform most statistical and neural classifiers.

2.3 Kernel Methods

Kernel methods, including support vector machines (SVMs) [33, 34] primarily and kernel PCA, kernel FDA, etc., are receiving increasing attention and have shown superior performance in pattern recognition. Kernel methods use a kernel function to represent the inner product of two patterns in expanded nonlinear feature space (possibly of infinite dimensionality). Both training and classification are performed via the kernel function without explicit access of the nonlinear space. An SVM is a binary classifier with discriminant function being the weighted combination of kernel functions over all training samples. The weights (coefficients) are learned by quadratic programming (QP) with the aim of maximizing the margin in feature space. After learning, the samples of non-zero weights are called support vectors (SVs), which are stored and used in classification. The maximal margin criterion of SVM learning leads to good generalization performance, but the resulting large number of SVs brings about heavy storage and computation in classification.

For multi-class classification, binary SVMs can be combined in two ways: one-versus-all (one-against-others) or one-versus-one (pairwise). The pairwise combination scheme was shown to outperform one-versus-all when using linear kernel [35]. When nonlinear kernels are used, the one-versus-all scheme performs sufficiently. In recent years, many results of character recognition using SVM classification have been reported, mostly for small category set problems like numeral recognition. The results (e.g. [4]) show that SVMs indeed yield higher accuracies than statistical and neural classifiers, but the storage and computation of large number of SVs are expensive. A strategy to alleviate the computation cost is to use a statistical or neural classifier for selecting two candidate classes, which are then discriminated by SVM [36]. Dong et al. used a one-versus-all scheme for large set Chinese character recognition with fast training [37]. They speed up the recognition by using a coarse classifier for candidate selection, but cannot avoid the problem of storing large number of SVs.

2.4 Multiple Classifier Systems

Combining multiple classifiers has been long pursued for improving the accuracy of single classifiers [38, 39]. Rahman et al. give a survey of combination methods in character recognition, including various structures of classifier organization [40]. Moreover, other chapters of this book are dedicated to this subject. Parallel (horizontal) combination is more often adopted for high accuracy, while sequential (cascaded, vertical) combination is mainly used for accelerating large category set classification. According to the information level of classifier outputs, the decision fusion methods for parallel combination are categorized into abstract-level, rank-level, and measurement-level combination. Measurement-level combination takes full advantage of output information, and many fusion methods have been proposed to it [41, 42, 43]. Some character recognition results using multiple classifiers combined at different levels are reported by Suen and Lam [44].

The classification performance of multiple classifiers not only depends on the fusion strategy, but also relies on the complementariness (also referred to as independence or diversity) of the classifiers. Complementariness can be yielded by varying training samples, pattern features, classifier structure, learning methods, etc. In recently years, methods for generating multiple classifiers (called an ensemble) by exploring the diversity of training samples based on a given feature representation are receiving high attention, among them are the Bagging [45] and the Boosting [46]. For character recognition, combining classifiers based on different pre-processing and feature extraction techniques is effective. Yet another effective method uses a single classifier to classify multiple deformations (called perturbations or virtual test samples) of the input pattern and combine the decisions on multiple deformations [47, 48]. The deformations of training samples can also be used to train the classifier for improving the generalization performance [48, 21].

3 Strategies for Large Category Set

Unlike numerals and English letters that have only tens of classes, the character sets of some oriental languages, like Chinese, Japanese, and Korean, have thousands of daily-used characters. A standard of Chinese, GB2312-80, contains 3,755 characters in the level-1 set and 3,008 characters in the level-2 set, 6,763 in total. A general-purpose Chinese character recognition system needs to deal with an even larger set because those not-often-used characters should be recognized as well.

For classifying a large category set, many classifiers become infeasible because either the training time or the classification time becomes unacceptably long. Classifiers based on discriminative supervised learning (called discriminative classifiers hereof), like ANNs and SVMs, are rarely used to directly classify a large category set. Two divide-and-conquer schemes are often used to accelerate classification. In one scheme, a simple and fast classifier is used to select a dynamic subset from the whole class set such that the input pattern belongs to the subset with high probability. In another scheme, the class set is divided into static (possibly overlapping) clusters and the input pattern is assigned to one or several clusters, whose unification gives the subset of classes for further discrimination. A hierarchical classification method using both schemes was reported in [49]. Tree classifiers were ever pursued for fast classification of large character set (e.g. [50]) but the accumulation of error along hierarchies makes them insufficient in accuracy, especially for recognizing handwritten characters.

In divide-and-conquer schemes, the second-stage classifier for discriminating a subset of classes (called fine classifier) can be a quadratic classifier or a discriminative classifier. The main advantage of quadratic classifiers is that the parameters of each class are estimated independently using the samples of one class only. The training time is hence linear with the number of classes (NoC). Successful quadratic classifiers include the MQDF of Kimura et al. [11, 12] and some modifications of Mahalanobis distance, which have lower complexity and yield higher accuracy than the original QDF. A further improvement is the compound discriminant functions [51, 14], which discriminate pairs of confusing classes without extra parameters compared to the baseline quadratic classifier. The asymmetric Mahalanobis distance of Kato et al. [52] yields superior recognition accuracy, though with higher complexity than the MQDF.

The training time for a discriminative classifier is square of the NoC since the total number of samples is linear with the NoC, and each sample is used for training the parameters of all classes. To alleviate this problem for large category set, neural networks are usually trained with a subset of samples. Fu and Xu designed probabilistic decision-based neural networks for discriminating groups of classes divided by clustering [53], with each network trained with the samples of the classes in a group. Kimura et al. design an MLP for each of confusing classes, which are determined from the classification on training samples using a statistical classifier [54]. Each MLP discriminates one target class from some rivals that are confused to the target class by the statistical classifier. In classification, an MLP is activated only when its target class is the top-rank class given by the statistical classifier. Saruta et al. design an MLP for each class, but the MLP is trained with the samples of a few classes only [55].

Training SVMs with all samples for Chinese character recognition has been attempted by Dong et al., who designed a fast training algorithm [37]. Though the training with all samples is now feasible due to the increasing power of computers, reducing the complexities of training, storage and classification is concerned for practical applications.

As a discriminative classifier, the LVQ classifier has moderate complexity for large category set [17, 31]. Fukumoto et al. has used a generalized LVQ (GLVQ) algorithm for discriminatively adjusting the class means of quadratic classifiers for large character set recognition [56]. The DLQDF [32], discriminatively adjusting all the parameters of quadratic classifier, provides more accurate classification than LVQ, but its training is very computationally expensive for large category set. By introducing hierarchical rival class search for acceleration, the training of DLQDF on large category set is feasible [57]. Compared to the ML-based MQDF, however, the DLQDF improves the accuracy of handwritten Chinese character recognition only slightly [57, 58].

The mirror image learning method of Wakabayashi et al. [20], for adjusting the covariance parameters of quadratic classifier, was recently applied to handwritten Chinese character recognition with success [59]. Running quadratic classification and modifying covariance matrices for five cycles on training samples, the accuracy of MQDF on test samples was improved from 98.15% to 98.38%. Using compound quadratic discriminant functions for pair discrimination, the test accuracy was further improved to 98.50%.

Feature dimensionality reduction also plays an important role in large character set recognition, since it reduces the classifier complexity (both parameter storage and computation) and possibly, improves the classification accuracy. The Fisher discriminant analysis (FDA) has shown success in many recognition systems [12, 58], though it assumes equal covariance for all classes and tends to blur the difference between nearby classes. Previous hetorescedastic discriminant analysis (HDA) methods are computationally formidable for large category set. A new HDA method was proposed recently and applies effectively to Chinese character recognition [60].

A feature subspace learning method by error minimization, called discriminative feature extraction (DFE) [61], has been tried to improve the accuracy of Chinese character recognition [17, 62, 63, 57]. DFE optimizes the subspace vectors and classifier parameters simultaneously by stochastic gradient descent. With a classifier of single prototype per class, the optimization for thousands of classes is computationally feasible, and the simultaneous optimization of class prototypes and subspace can be viewed as a combination of LVQ and DFE. Using a quadratic classifier on the feature subspace learned by DFE with a prototype classifier, the accuracy of handwritten Chinese character recognition is improved significantly compared to classification on FDA subspace [57].

4 Comparison of Classification Methods

In this section we collect some character recognition results reported in the literature for comparing the performance of the classification methods reviewed above, and we discuss the characteristics of these methods regarding their impacts on practical applications.

4.1 Performance Comparison

The various experiments of character recognition differ in many factors such as the sample data, pre-processing technique, feature representation, classifier structure and learning algorithm. It is hard to assess the performance of a special classification or learning method from the recognition accuracies reported by different works since the other factors are variable. Only a few works have compared different classification/learning methods based on the same feature data.

For handwritten character recognition, more experiments have been reported for off-line recognition than for on-line recognition. Regarding the target of recognition, the 10 Arabic numerals are most often tested, while Chinese characters or Japanese Kanji characters are often tested in large character set recognition. The numeral databases that have been widely tested include the CENPARMI, NIST Special Database 19 (SD19), MNIST, etc. The NIST SD19 contains huge number of character images, but researchers often use different partitions of data for training and testing, unlike that the CENPARMI and MNIST databases are partitioned into standard training and test sets.

Performance on Handwritten Numerals

We first collect some high recognition accuracies reported on standard numeral databases, then summarize some results of classification on common feature data.

The CENPARMI database contains 4,000 training samples and 2,000 test samples. Early works using structural analysis hardly reached 95% of correct recognition on this test set [64]. In recently years, it is easy to achieve a recognition rate over 98% by extracting statistical features and training classifiers. Suen et al. reported a correct rate 98.85% by training neural networks on 450,000 samples [3]. By training with the standard 4,000 samples, correct rates over 99% have been given by polynomial classifier (PC) and SVMs with efficient image normalization and feature extraction [4, 65].

The MNIST database contains 60,000 training samples and 10,000 test samples. Each sample was normalized to a gray-scale image of 20×20 pixels, which is located in a 28×28 plane. The pixel values of normalized image are used as feature values, on which different classifiers and learning algorithms can be fairly compared. LeCun et al. collected a number of test accuracies given by various classifiers [2]. A high accuracy, 99.30%, was given by a boosted convolutional neural network (CNN) trained with distorted data. Simard et al. improved both the distorted sample generation and the implementation of CNN and resulted in a test accuracy 99.60% [21]. Instead of the trainable feature extractors in CNN, extracting heuristically discriminating features also lead to high accuracies. Without training with distorted samples, Teow and Loe obtained a test accuracy 99.57% by extracting local structure features and classification using triowise linear SVMs [66]. On 200D gradient direction feature, Liu et al. obtained a test accuracy 99.58% by SVM classification, 99.42% by polynomial classifier, and over 99% by many other classifiers [4].

On the MNIST database, training classifiers without feature extraction show worst performance. Since image pre-processing and feature extraction are both important to character recognition, a better scheme to compare classifiers is to train them on a common discriminating feature representation. Holmström et al. compared various statistical and neural classifiers on PCA features extracted from normalized images [67]. However, the PCA feature does not perform satisfactorily. In the comparison studies of Liu et al. [68, 4], the features used, chaincode and gradient direction features, are widely recognized and well-performing in practice. Their results show that parametric statistical classifiers (especially the MQDF) generalize better than neural classifiers when training with small sample data, while neural classifiers outperforms when training with large sample data. The SVM classifier with RBF kernel mostly gives the highest accuracy. The best neural classifier was shown to be the polynomial classifier (PC), which is far less complex in storage and execution than SVMs. And the RBF network mostly outperforms the MLP when training all its parameters discriminatively.

A citation of error rates from [4] is shown in Table 1, where "4-grad" and "8-grad" stand for 4-orientation and 8-direction gradient features, respectively; and "SVC-poly" and "SVC-rbf" denotes one-versus-all support vector classifiers with polynomial kernel and RBF kernel, respectively. In this table, the RBF network is shown to be inferior to the MLP on the MNIST dataset, but on many other datasets, the RBF network outperforms the MLP [4].

Table 1. A citation of error rates (%) on the MNIST test set

Feature	pixel	PCA	4-grad	8-grad
k-NN	3.66	3.01	1.26	0.97
MLP	1.91	1.84	0.84	0.60
RBF	2.53	2.21	0.92	0.69
PC	1.64	N/A	0.83	0.58
SVC-poly	1.69	1.43	0.76	0.55
SVC-rbf	1.41	1.24	0.67	0.42

Performance on Large Character Sets

In the area of Chinese/Japanese character recognition, a public handprinted (constrained handwriting) database ETL9B has been widely tested. Various classification methods have been proposed, but they have never been compared on a common feature representation of samples.

The ETL9B database contains 200 samples for each of 3,036 classes, including 2,965 Kanji and 71 hiragana characters. Early works often used 100 samples of odd number from each class for training and the even-numbered samples for testing, and focused on image normalization and feature extraction for improving the performance of feature matching. Nonlinear normalization based on line density equalization [69, 70] and edge direction feature extraction are now widely accepted. Using the class means of training samples as prototypes, the recognition accuracy on test samples was hardly over 95%. On this sample partitioning scheme, Saruta et al. achieved a correct rate of 95.84% by using class-modular neural networks for fine classification [55]. Using FDA for dimensionality reduction and GLVQ for optimizing the class means, Fukumoto et al. reported a correct rate of 97.22% for Euclidean distance, 98.30% for projection distance (PD) and 98.41% for modified PD (MPD) [56]. The PD and MPD classifiers have comparable complexity with the MQDF, however.

High accuracies have been reported on ETL9B by using quadratic classifiers and SVMs. Nakajima et al. used 160 samples per class for training and the remaining 40 samples for testing, and reported a correct rate 98.90% using MPD and compound MPD [14]. Dong et al. tested on a partially different set of 40 samples per class, and reported a correct rate 99.00% by using SVMs trained on enhanced samples for fine classification [37]. Kimura et al. tested on 40 samples per class in rotation and reported average rate 99.15% by using modified Bayes discriminant function on enhanced training samples [12]. Suzuki et al. [51] and Kato et al. [52] tested on 20 samples per class in rotation, and both used partial inclination detection for improving normalization. Using compound Mahalanobis distance for fine classification, Suzuki et al. improved the recognition rate from 99.08% to 99.31%. Kato et al. reported a correct rate 99.42% by using asymmetric Mahalanobis distance for fine classification.

Some works reported results on ETL9B as well as databases of handwritten Chinese characters, say, HCL2000 [58] and CASIA [57]. The Chinese databases are not available for free use, however. From the reported results, the Chinese samples turn out to be more difficult to recognize than the samples of ETL9B. Based on nonlinear normalization and gradient direction feature extraction, the accuracies on ETL9B (with samples partitioned as [14]) are as high as 99.33% and 99.39%, while the accuracies on HCL2000 and CASIA databases are 98.56% and 98.43%, respectively. The underlying classification methods are DLQDF+compound quadratic discrminant [58] and DFE+DLQDF [57], respectively.

4.2 Statistical vs. Discriminative Classifiers

We refer to statistical classifiers as those based on parametric or non-parametric density estimation, and discriminative classifiers as those based on minimum (regression or classification) error training. Discriminative classifiers include neural networks and SVMs, for which the parameters of one class are trained on the samples of all classes or selected confusing classes. For statistical classifiers, the parameters of one class are estimated from the samples of its own class only. Non-parametric classifiers like Parzen window method and k-NN rule are not practical for real-time applications, and so, are not considered in the following discussions. We compare the characteristics of statistical and discriminative classifiers in the following respects.

- *Complexity and flexibility of training.* The training time of statistical classifiers is linear with the number of classes, and it is easy to add a new class to an existing classifier since the parameters of the new class are estimated from the new samples only. Also, adapting the density parameters of a class to new samples is possible. In contrast, the training time of discriminative classifiers is proportional to square of the number of classes, and to guarantee the stability of parameters, adding new classes or new samples need re-training with all samples.
- *Classification accuracy.* When training with enough samples, discriminative classifiers give higher generalization accuracies than statistical classifiers. This is because discriminative classifiers are trained to separate the samples of different classes in the feature space, while the pre-assumed density form of statistical classifiers limits its capability to accommodate large variability of samples.
- *Dependence on training sample size.* The generalization accuracy of regularized statistical classifiers (like MQDF and RDA) are more stable against the training sample size than discriminative classifiers (see [68]). On small sample size, statistical classifiers can generalize better than discriminative ones.
- *Storage and execution complexity.* At same level of classification accuracy, discriminative classifiers tend to have less parameters than statistical classifiers. Hence, discriminative classifiers are more economical in storage and execution.
- *Confidence of decision.* The discriminant functions of parametric statistical classifiers are connected to the class conditional probability, and can be easily converted to a posteriori probabilities by the Bayes formula. On the other hand, the outputs of discriminative classifiers are directly connected to a posteriori probabilities.
- *Rejection capability.* Classifiers of higher classification accuracies tend to reject ambiguous patterns better, but not necessarily reject well outliers (patterns out of defined classes) [68]. Parametric statistical classifiers are resistant to outliers because of the assumption of compact density functions, whereas discriminative classifiers are susceptible to outliers because of open decision regions [71]. Outlier rejection is important to integrated segmentation and recognition of character strings [72]. The rejection capability of discriminative classifiers can be enhanced by training with outlier samples.

4.3 Neural Networks vs. SVMs

In addition to the common properties of discriminative classifiers as above, neural classifiers and SVMs show different properties in the following respects.

- *Complexity of training.* The parameters of neural classifiers are generally adjusted by gradient descent with the aim of optimizing an objective function on training samples. By feeding the training samples a fixed number of epochs, the training time is linear with the number of samples. SVMs are trained by quadratic programming (QP), and the training time is generally proportional to the square of number of samples. Some fast SVM training algorithms with nearly linear complexity are available, however.
- *Flexibility of training.* The parameters of neural classifiers (for character classification) can be adjusted in string-level or layout-level training by gradient descent with the aim of optimizing the global recognition performance [2, 73]. SVMs can only be trained at the level of holistic patterns.
- *Model selection.* The generalization performance of neural classifiers is sensitive to the size of the network structure, and the selection of an appropriate structure relies on cross-validation. The performance of SVMs also depends on the selection of kernel type and kernel parameters, but this dependence is not so influential as the structure selection of neural networks.
- *Classification accuracy.* SVMs have been demonstrated superior classification accuracies to neural classifiers in many experiments.
- *Storage and execution complexity.* SVM learning by QP often results in a large number of SVs, which should be stored and computed in classification. Neural classifiers have much less parameters, and the number of parameters are easy to control. For reducing the execution complexity of SVMs, SV reduction techniques are effective, but may sacrifice the classification accuracy to some degree.

5 Remaining Problems and Future Works

Though tremendous advances have been achieved in applying classification and learning methods to character recognition, there is still a gap between the needs of applications and the actual performance, and some problems encountered in practice have not been considered seriously. We list these problems and discuss the future research directions of classification and learning that can potentially solve or alleviate them.

5.1 Improvements of Accuracy

Recognition rates over 99% have been reported to handwritten numeral recognition and handprinted Chinese character recognition, but accuracies lower than 90% are often reported to some difficult cases like English letters, cursive words, unconstrained Chinese characters, etc. The recognition rate, even as high as 99.9%, is never sufficient. Any improvement to accuracy will make the recognition system more welcome by users. Improved accuracy can be achieved by carefully tuning every processing task: pre-processing, feature extraction, sample generation, classifier design, multiple classifier combination, etc. We hereof only discuss some issues related to classification and learning.

- *Feature transformation.* Feature transformation methods, including PCA and FDA, have been proven effective in pattern classification, but no method claims to find the best feature subspace. Generalized transformation methods based on relaxed density assumptions and those based on discriminative learning are expected to find better feature spaces.
- *Feature selection.* Character classification has been mostly performed on a limited number of features, which are usually artificially selected. Increasing the number of features complicates the design of classifier and may deteriorate the generalization performance. It is now possible to automatically select a good feature set from huge number of candidate features. With the aim of optimizing separability or description, the selected features may lead to better classification than artificially selected ones.
- *Sample generation and selection.* Training with distorted samples has resulted in improved generalization performance, but better methods of distorted sample generation are yet to be found. Since very large number of distorted samples can be generated and some of them may be misleading, the selection of samples then becomes important to guarantee the efficiency and quality of training.
- *Joint feature selection and classifier design.* To select features and design classifier jointly may lead to better classification performance. The Bayesian network belongs to such kind of classifiers and is now being studies intensively.
- *Hybrid statistical/discriminative learning.* A hybrid statistical/discriminative classifier may yielder high accuracy than both the pure statistical and the pure discriminative classifier [74]. A way to design such classifiers is to adjust the parameters of parametric statistical classifiers discriminatively on training samples [75, 32], to improve both generalization accuracy and resistance to outliers. Also, combining the decisions of statistical and discriminative classifiers is preferred to combining similar classifiers.
- *Ensemble learning.* The performance of combining multiple classifiers primarily relies on the complementariness of classifiers. Maximizing the diversity of classifiers is now receiving increasing attention. A heuristic is to combine classifiers with different properties: training data, pre-processing, feature extraction, classifier structure, learning algorithm, etc. Among the methods that explore the diversity of data, the Boosting is considered as the best ensemble classifier. It has not been widely tested in character recognition yet.

5.2 Reliable Confidence and Rejection

Since we cannot achieve 100% correct recognition in practice, it is desirable to reject or delay the decision for those patterns with low confidence. There maybe two kinds of confidence measures: class conditional probability-like (conditional confidence) and posterior probability-like (posterior confidence). Rejecting ambiguous patterns (those confused between different classes) is generally based on posterior confidence, and rejecting outliers (those out of defined classes) is generally based on conditional confidence. If we can estimate the conditional confidence reliably, it would help reject ambiguous patterns as well. Both confidence measures can be unified into the posterior probabilities of open world: normalization to unity for defined classes and an outside world. Transforming classifier outputs to probability measures facilitates contextual processing which integrates information from multiple sources. The following ways may help improve the rejection capability of current character recognition methods.

- *Elaborate density estimation.* Probability density estimation is a traditional problem in statistical pattern recognition, but is not well-solved yet. Good density models for character classes can yield both high classification accuracy and rejection capability, especially outlier rejection. The Gaussian mixture model is being studied intensively, and many efforts are given to automatically estimating the number of components. For density estimation in high-dimensional spaces, combining feature transformation or selection may result in good classification performance. Density estimation in kernel space would be a choice to explore nonlinear subspace.
- *One-class classification.* One-class classifiers separate one class from the remaining world with parameters estimated from the samples of the target class only. Using one-class classifiers as class verifiers added to a multi-class classifier can improve rejection. The distribution of a class can be described by a good density model (as discussed above) or support vectors in kernel space [76]. Structural analysis, though do not compete with statistical and discriminative classifiers in classification accuracy, may serve as good verifiers.
- *Hybrid statistical/discriminative learning.* Hybrid statistical/discriminative classifiers, as discussed in 5.1, may yield both high classification accuracy and resistance to outliers. This principle of learning is to be extended to more statistical models than Gaussian discriminant function and may be combined with feature transformation.
- *Multiple classifier combination.* Different classifiers tend to disagree on ambiguous patterns, so the combination of multiple classifiers can better identify and reject ambiguous patterns [77]. Generally, combining complementary classifiers can improve the classification accuracy and the tradeoff between error rate and reject rate.

5.3 Improvements to Large Category Set

Discriminative learning methods have not been extensively applied to the recognition of large character set. Using quadratic classifiers with sophisticated normalization and feature extraction, high accuracies have been reported to handprinted sample databases like the ETL9B. However many misrecognized samples are easily recognized by humans and are potentially solvable by discriminative learning. The probable reasons that neural networks could not perform competitively (e.g. [55]) are: (1) There are few training samples per class (less than 200 in ETL9B); (2) The class-modular network only takes the samples of confusing classes as negative samples, so the resulting network is not resistant to the samples of un-trained classes. The application of SVMs to Chinese character recognition [37] is successful though its accuracy cannot be compared directly to other works because of different pre-processing and feature extraction procedures. To better utilize discriminative classifiers for discriminating similar characters, several issues should be considered seriously.

- *Training sample size.* To demonstrate the benefit of discriminative classifiers, we should use a large number of samples per class for training. If using the discriminative classifier for discriminating a subset of classes only, these selected classes should have more samples than other classes. A new public database released in Japan, called JEITA-HP, contain more than 500 samples per class. The Chinese database HCL2000 contains over 1,000 samples for each of 3,755 classes.
- *Confusing set selection.* The subset of confusing classes is usually selected heuristically according to the classification results of a statistical classifier on training samples. This procedure need to be considered more rigorously, say, from probabilistic view.
- *Type of discriminative classifier.* When the discriminative classifier is used to discriminate a subset of classes, the resistance to outliers is preferable because the patterns of un-trained classes are often presented to the discriminative classifier in execution. In this respect, the hybrid statistical/discriminative classifier is a good choice.
- *Samples for training discriminative classifier.* If a neural or SVM classifier is used to discriminate a subset of classes, it should be trained with samples of other classes too, in order to enhance the resistance to outliers.
- *Fusion of cascaded classifiers.* Using either a multi-class classifier or pairwise classifier for the second-stage classification, fusing the decisions of first-stage and second-stage classifiers probabilistically will benefit the global recognition accuracy.

5.4 Incremental Learning

As discussed in Section 4.2, when adding new classes or new samples to defined classes, discriminative classifiers need to be re-trained with all the accumulated samples. Incremental learning for adapting existing classifiers to new classes and new samples has rarely been considered in character recognition. Some published works of incremental learning in the neural networks community can be referred for our application. Statistical models can be adapted on new samples without forgetting past data distribution in the framework of Bayesian learning, and hybrid statistical/discriminative models can also be stabilized to past distribution while adapting to new data. For all classifier models, an ensemble classifier can be generated by combining new classifiers trained with new samples with existing ones [78].

The samples for classifier training and adaptation are ever increasing. In addition to adaptation to new labeled data, training with unlabeled data is another topic that is intensively studied in machine learning, called semi-supervised learning [79]. This learning scheme is applicable to character recognition since we cannot attach class labels to all samples artificially.

5.5 Benchmarking of Methods

Fair comparison of classifiers is difficult because many classifiers, especially neural networks, are flexible in implementation and their performance are affected by human factors [80]. In the character recognition field, the comparison of methods is more difficult because many processing steps (pre-processing, feature extraction, classification) are involved. Even on experiments using the same training and test sets, researchers often compare the performance at system level: the final recognition rate obtained by integrating all the processing steps. In these circumstances, it is hard to decide what method at which step is the most influential to the final result.

To conduct a fair comparison of methods instead of overall systems, we recommend to use standard techniques for all steps except the step under comparison. For example, to compare classifiers, standard pre-processing and feature extraction techniques should be applied to all the classifiers to compare. Many techniques, e.g. nonlinear normalization and direction feature extraction, are variable in implementation details. It is hoped that open source codes of standard techniques for every processing step of character recognition are released, such that other researchers can fairly compare the methods of a special step. For comparing classifiers, to release common feature data instead of sample images is meaningful.

Acknowledgements

This work was supported by the Hundred Talents Program of CAS, and the Natural Science Foundation of China under grants No.60543004 and No.60121302.

References

1. Mori, S., Suen, C.Y., Yamamoto, K.: Historical review of OCR research and development. Proc. IEEE 80(7) (1992) 1029–1058
2. LeCun, Y., Bottou, L., Bengio, Y., Haffner, P.: Gradient-based learning applied to document recognition. Proc. IEEE 86(11) (1998) 2278–2324
3. Suen, C.Y., Liu, K., Strathy, N.W.: Sorting and recognizing cheques and financial documents. In: Lee SW, Nakano Y (eds) Document Analysis Systems: Theory and Practice. Springer, LNCS 1655 (1999) 173–187
4. Liu, C.L., Nakashima, K., Sako, H., Fujisawa, H.: Handwritten digit recognition: Benchmarking of state-of-the-art techniques. Pattern Recognition 36(10) (2003) 2271–2285
5. Marinai, S., Gori, M., Soda, G.: Artificial neural networks for document analysis and recognition. IEEE Trans. Pattern Anal. Mach. Intell. 27(1) (2005) 23–35
6. Jain, A.K., Duin, R.P.W., Mao, J.: Statistical pattern recognition: A review. IEEE Trans. Pattern Anal. Mach. Intell. 22(1) (2000) 4–37
7. Fukunaga, K.: Introduction to Statistical Pattern Recognition. Academic Press, 2nd edition (1990)
8. Bishop, C.M.: Neural Networks for Pattern Recognition. Oxford University Press (1995)
9. Duda, R.O., Hart, P.E., Stork, D.G.: Pattern Classification. Wiley Interscience, 2nd edition (2001)
10. Friedman, J.H.: Regularized discriminant analysis. J. Am. Statist. Ass. 84(405) (1989) 165–175
11. Kimura, F., Takashina, K., Tsuruoka, S., Miyake, Y.: Modified quadratic discriminant functions and the application to Chinese character recognition. IEEE Trans. Pattern Anal. Mach. Intell. 9(1) (1987) 149–153
12. Kimura, F., Wakabayashi, T., Tsuruoka, S., Miyake, Y.: Improvement of handwritten Japanese character recognition using weighted direction code histogram. Pattern Recognition 30(8) (1997) 1329–1337
13. Ikeda, M., Tanaka, H., Motooka, T.: Projection distance method of recognition of handwritten characters. Trans. IPS Japan 24(1) (1983) 106–112
14. Nakajima, T., Wakabayashi, T., Kimura, F., Miyake, Y.: Accuracy improvement by compound discriminant functions for resembling character recognition. Trans. IEICE Japan J83-D-II(2) (2000) 623–633
15. Hinton, G.E., Dayan, P., Revow, M.: Modeling the manifolds of images of handwritten digits. IEEE Trans. Neural Networks 8(1) (1997) 65–74
16. Kim, H.C., Kim, D., Bang, S.Y.: A numeral character recognition using the PCA mixture model. Pattern Recognition Letters 23 (2002) 103–111
17. Tsay, M.K., Shyu, K.H., Chang, P.C.: Feature transformation with generalized LVQ for handwritten Chinese character recognition. IEICE Trans. Information and Systems E82-D(3) (1999) 687–92
18. Zhang, P., Bui, T., Suen, C.Y.: Hybrid feature extraction and feature selection for improving recognition accuracy of handwritten numerals. In: Proc. 8th ICDAR, Seoul, Korea, 1 (2005) 136–140
19. Kawatani, T., Shimizu, H.: Handwritten Kanji recognition with the LDA method. In: Proc. 14th ICPR, Brisbane, 2 (1998) 1031–1035
20. Wakabayashi, T., Shi, M., Ohyama, W., Kimura, F.: Accuracy improvement of handwritten numeral recognition by mirror image learning. In: Proc. 6th ICDAR, Seattle (2001) 338–343

21. Simard, P.Y., Steinkraus, D., Platt, J.C.: Best practices for convolutional neural networks applied to visual document analysis. In: Proc. 7th ICDAR, Edinburgh, UK, 2 (2003) 958–962
22. Oh, I.S., Suen, C.Y.: A class-modular feedforward neural network for handwriting recognition. Pattern Recognition 35(1) (2002) 229–244
23. Pao, Y.H.: Adaptive Pattern Recognition and Neural Networks. Addison-Wesley, MA (1989)
24. Schürmann, J.: Pattern Classification: A Unified View of Statistical and Neural Approaches. Wiley Interscience (1996)
25. Kreßel, U., Schürmann, J.: Pattern classification techniques based on function approximation. In: Bunke H, Wang PSP (eds) Handbook of Character Recognition and Document Image Analysis, World Scientific (1997) 49–78
26. Franke, J.: Isolated handprinted digit recognition. In: Bunke H, Wang PSP (eds) Handbook of Character Recognition and Document Image Analysis, World Scientific (1997) 103–121
27. Liu, C.L., Sako, H.: Class-specific feature polynomial classifier for pattern classification and its application to handwritten numeral recognition. Pattern Recognition 39(4) (2006) 669–681
28. Kimura, F., Inoue, S., Wakabayashi, T., Tsuruoka, S., Miyake, Y.: Handwritten numeral recognition using autoassociative neural networks. In: Proc. 14th ICPR, Brisbane, 1 (1998) 166–171
29. Zhang, B., Fu, M., Yang, H.: A nonlinear neural network model of mixture of local principal component analysis: Application to handwritten digits recognition. Pattern Recognition 34(2) (2001) 203–214
30. Kohonen, T.: The self-organizing map. Proc. IEEE 78(9) (1990) 1464–1480
31. Liu, C.L., Nakagawa, M.: Evaluation of prototype learning algorithms for nearest neighbor classifier in application to handwritten character recognition. Pattern Recognition 34(3) (2001) 601–615
32. Liu, C.L., Sako, H., Fujisawa, H.: Discriminative learning quadratic discriminant function for handwriting recognition. IEEE Trans. Neural Networks 15(2) (2004) 430–444
33. Vapnik, V.: The Nature of Statistical Learning Theory. Springer-Verlag, New Work (1995)
34. Burges, C.J.C.: A tutorial on support vector machines for pattern recognition. Knowledge Discovery and Data Mining 2(2) (1998) 1–43
35. Kressel, U.: Pairwise classification and support vector machines. In: Schölkopf B, Burges CJC, Smola AJ (eds) Advances in Kernel Methods: Support Vector Learning, MIT Press (1999) 255–268
36. Bellili, A., Gilloux, M., Gallinari, P.: An MLP-SVM combination architecture for offline handwritten digit recognition: Reduction of recognition errors by support vector machines rejection mechanisms. Int. J. Document Analysis and Recognition 5(4) (2003) 244–252
37. Dong, J.X., Krzyzak, A., Suen, C.Y.: An improved handwritten Chinese character recognition system using support vector machine. Pattern Recognition Letters 26(12) (2005) 1849–1856
38. Xu, L., Krzyzak, A., Suen, C.Y.: Methods of combining multiple classifiers and their applications to handwriting recognition. IEEE Trans. System Man Cybernet. 22(3) (1992) 418–435
39. Ho, T.K., Hull, J., Srihari, S.N.: Decision combination in multiple classifier systems. IEEE Trans. Pattern Anal. Mach. Intell. 16(1) (1994) 66–75

40. Rahman, A.F.R., Fairhurst, M.C.: Multiple classifier decision combination strategies for character recognition: A review. Int. J. Document Analysis and Recognition 5(4) (2003) 166–194
41. Kittler, J., Hatef, M., Duin, R.P.W., Matas, J.: On combining classifiers. IEEE Trans. Pattern Anal. Mach. Intell. 20(3) (1998) 226–239
42. Duin, R.P.W.: The combining classifiers: To train or not to train. In: Proc. 16th ICPR, Quebec, Canada, 2 (2002) 765–770
43. Liu, C.L.: Classifier combination based on confidence transformation. Pattern Recognition, 38(1) (2005) 11–28
44. Suen, C.Y., Lam, L.: Multiple classifier combination methodologies for different output levels. In: Kittler J, Roli F (eds) Multiple Classifier Systems, Springer, LNCS 1857 (2000) 52–66
45. Breiman, L.: Bagging predictors. Machine Learning 24(2) (1996) 123–140
46. Freund, Y., Schapire, R.E.: A decision-theoretic generalization of on-line learning and an application to boosting. J. Computer and System Sciences 55(1) (1997) 119–139
47. Ha, T., Bunke, H.: Off-line handwritten numeral recognition by perturbation method. IEEE Trans. Pattern Anal. Mach. Intell. 19(5) (1997) 535–539
48. Dahmen, J., Keysers, D., Ney, H.: Combined classification of handwritten digits using the virtual test sample method. In : Kittler J, Roli F (eds) Multiple Classifier Systems, Springer, LNCS 2096 (2001) 99–108
49. Tang, Y.Y., et al.: Offline recognition of Chinese handwriting by multifeature and multilevel classification. IEEE Trans. Pattern Anal. Mach. Intell. 20(5) (1998) 556–561
50. Wang, Q.R., Suen, C.Y.: Analysis and design of a decision tree based on entropy reduction and its application to large character set recognition. IEEE Trans. Pattern Anal. Mach. Intell. 6(4) (1984) 406–417
51. Suzuki, M., Omachi, S., Kato, N., Aso, H., Nemoto, Y.: A discrimination method of similar characters using compound Mahalanobis function. Trans. IEICE Japan J80-D-II(10) (1997) 2752–2760
52. Kato, N., Suzuki, M., Omachi, S., Aso, H., Nemoto, Y.: A handwritten character recognition system using directional element feature and asymmetric Mahalanobis distance. IEEE Trans. Pattern Anal. Mach. Intell. 21(3) (1999) 258–262
53. Fu, H.C., Xu, Y.Y.: Multilinguistic handwritten character recognition by Bayesian decision-based neural networks. IEEE Trans. Signal Processing 46(10) (1998) 2781–2789
54. Kimura, Y., Wakahara, T., Tomono, A.: Combination of statistical and neural classifiers for a high-accuracy recognition of large character sets. Trans. IEICE Japan J83-D-II(10) (2000) 1986–1994
55. Saruta, K., Kato, N., Abe, M., Nemoto, Y.: High accuracy recognition of ETL9B using exclusive learning neural network-II (ELNET-II). IEICE Trans. Information and Systems 79-D(5) (1996) 516–521
56. Fukumoto, T., Wakabayashi, T., Kumura, F., Miyake, Y.: Accuracy improvement of handwritten character recognition by GLVQ. In: Proc. 7th IWFHR, Amsterdam (2000) 271–280
57. Liu, C.L.: High accuracy handwritten Chinese character recognition using quadratic classifiers with discriminative feature extraction. In: Proc. 18th ICPR, Hong Kong, 2 (2006) 942–945

58. Liu, H., Ding, X.: Handwritten character recognition using gradient feature and quadratic classifier with multiple discrimination schemes. In: Proc. 8th ICDAR, Seoul, Korea, 1 (2005) 19–23
59. Liu, H., Ding, X.: Handwritten Chinese character recognition based on mirror image learning and the compound Mahalanobis distance. J. Tsinghua Univ. (Sci & Tech) 46(7) (2006) 1239–1242 (in Chinese)
60. Liu, H., Ding, X.: Improve handwritten character recognition performance by heteroscedastic linear discriminant analysis, In: Proc. 18th ICPR, Hong Kong, 1 (2006) 880–883
61. Biem, A., Katagiri, S., Juang, B.H.: Pattern recognition using discriminative feature extraction. IEEE Trans. Signal Processing 45(2) (1997) 500–504
62. Huo, Q., Ge, Y., Feng, Z.D.: High performance Chinese OCR based on Gabor features, discriminative feature extraction and model training. In: Proc. ICASSP'01, Salt Lake City, Utah, 3 (2001) 1517–1520
63. Liu, C.L., Mine, R., Koga, M.: Building compact classifier for large character set recognition using discriminative feature extraction. In: Proc. 8th ICDAR, Seoul, Korea (2005) 846–850
64. Suen, C.Y., Nadal, C., Legault, R., Mai, T.A., Lam, L.: Computer recognition of unconstrained handwritten numerals. Proc. IEEE 80(7) (1982) 1162–1180
65. Liu, C.L., Nakashima, K., Sako, H., Fujisawa, H.: Handwritten digit recognition: investigation of normalization and feature extraction techniques. Pattern Recognition 37(2) (2004) 265–279
66. Teow, L.N., Loe, K.F.: Robust vision-based features and classification schemes for off-line handwritten digit recognition. Pattern Recognition 35(11) (2002) 2355–2364
67. Holmström, L., Koistinen, P., Laaksonen, J., Oja, E.: Neural and statistical classifiers—taxonomy and two case studies. IEEE Trans. Neural Networks 8(1) (1997) 5–17
68. Liu, C.L., Sako, H., Fujisawa, H.: Performance evaluation of pattern classifiers for handwritten character recognition. Int. J. Document Analysis and Recognition 4(3) (2002) 191–204
69. Tsukumo, J., Tanaka, H.: Classification of handprinted Chinese characters using non-linear normalization and correlation methods. In: Proc. 9th ICPR, Rome (1988) 168–171
70. Yamada, H., Yamamoto, K., Saito, T.: A nonlinear normalization method for hanprinted Kanji character recognition—line density equalization. Pattern Recognition 23(9) (1990) 1023–1029
71. Gori, M., Scarselli, F.: Are multilayer perceptrons adequate for pattern recognition and verification? IEEE Trans. Pattern Anal. Mach. Intell. 20(11) (1998) 1121–1132
72. Liu, C.L., Sako, H., Fujisawa, H.: Effects of classifier structures and training regimes on integrated segmentation and recognition of handwritten numeral strings. IEEE Trans. Pattern Anal. Mach. Intell. 26(11) (2004) 1395–1407
73. Liu, C.L., Marukawa, K.: Handwritten numeral string recognition: Character-level training vs. string-level training. In: Proc. 17th ICPR, Cambridge, UK, 1 (2004) 405–408
74. Raina, R., Shen, Y., Ng, A.Y., McCallum, A.: Classification with hybrid generative/discriminative models. In: Advances in Neural Information Processing System 16 (2003)

75. Dahmen, J., Schluter, R., Ney, H.: Discriminative training of Gaussian mixtures for image object recognition. In: Proc. 21st Symposium of German Association for Pattern Recognition, Bonn, Germany (1999) 205–212
76. Tax, D.M.J., Duin, R.P.W.: Support vector data description. Machine Learning 54(1) (2004) 45–66
77. Suen, C.Y., Tan, J.: Analysis of error of handwritten digits made by a multitude of classifiers. Pattern Recognition Letters 26(3) (2005) 369–379
78. Polikar, R., Udpa, L., Udpa, A.S., Honavar, V.: Learn++: An incremental learning algorithm for supervised neural networks. IEEE Trans. System Man Cybernet. Part C 31(4) (2001) 497–508
79. Chawla, N.V., Karakoulas, G.: Learning from labeled and unlabeled data: An empirical study across techniques and domains. J. Artificial Intelligence Research 23 (2005) 331–366
80. Duin, R.P.W.: A note on comparing classifiers. Pattern Recognition Letters, 17 (1996) 529–536

Combining Classifiers with Informational Confidence

Stefan Jaeger, Huanfeng Ma, and David Doermann

Institute for Advanced Computer Studies, University of Maryland, College Park, MD 20742, USA {jaeger,hfma,doermann}@umiacs.umd.edu

Summary. We propose a new statistical method for learning normalized confidence values in multiple classifier systems. Our main idea is to adjust confidence values so that their nominal values equal the information actually conveyed. In order to do so, we assume that information depends on the actual performance of each confidence value on an evaluation set. As information measure, we use Shannon's well-known logarithmic notion of information. With the confidence values matching their informational content, the classifier combination scheme reduces to the simple sum-rule, theoretically justifying this elementary combination scheme. In experimental evaluations for script identification, and both handwritten and printed character recognition, we achieve a consistent improvement on the best single recognition rate. We cherish the hope that our information-theoretical framework helps fill the theoretical gap we still experience in classifier combination, putting the excellent practical performance of multiple classifier systems on a more solid basis.

1 Introduction

Multiple classifier systems are a relatively new field of research, which is currently under intensive investigation. We have seen a major boost of publications during the last ten years. A considerable number of publications is older though. This chapter is in good company with many other papers applying multiple classifier systems to handwritten or printed character recognition [1, 2, 3, 4]. In fact, character recognition seems to be one of the most popular application targets for multiple classifier systems so far. Research on multiple classifier systems has developed into two main branches: One major research field addresses the problem of generating classifiers and compiling classifier ensembles. Typical examples are Bagging [5] and Boosting [6, 7, 8]. Several other techniques, which are out of the scope of this chapter, have been proposed during the recent years and have gained some importance [9, 10]. The second major research field investigated in the recent literature deals with combination schemes for integrating the diverse outputs of classifier ensembles

S. Jaeger et al.: *Combining Classifiers with Informational Confidence*, Studies in Computational Intelligence (SCI) **90**, 163–191 (2008)
www.springerlink.com

into a single classification result. Our work falls into this second category, though we will address Boosting shortly at the end of this chapter.

Multiple classifier systems offer an appealing alternative to the conventional monolithic classifier system. Instead of pressing all information into a single classifier, the idea of multiple classifier systems is to distribute information among more than one classifier. Distributing information over several classifiers reduces the average complexity of each classifier, making them easier to train and optimize. Another intriguing feature is that a multiple classifier system can outperform each of its individual classifiers. In fact, a multiple classifier system can achieve outstanding recognition rates even if all its constituent classifiers provide only moderate performance. Classifiers highlighting different aspects of the same classification problem are especially suitable for combination purposes since they can complement each other. For complementary classifiers, the weakness of one classifier is the strength of at least one other classifier, which compensates for false judgments made by its co-classifier. We can generate complementary classifiers by using, for example, multiple feature sets, training methods, and classification architectures.

However, the combination of multiple classifiers requires integration of individual classifier outputs into a single answer. In virtually all practical cases, classifier output is only a rough approximation of the correct a-posteriori class probabilities. These approximations do not meet usually the mathematical requirements for probabilities. While this is not a major problem for single classifier systems, it can lead to severe performance degradation in multiple classifier systems, where the individual outputs need to be combined in a mathematically reasonable manner. In this chapter, we presume that each classifier returns a confidence value for its recognition result and each potential candidate in an n-best list. These confidence values denote the confidence of the classifier in its output. In order to compare confidences from different classifiers, confidence values need to be compatible with each other. Identical ranges and scales are necessary, but not sufficient, prerequisites of compatible confidence values. Each classifier in a multiple classifier system should output confidence values that are related to its actual performance. We must avoid combining classifiers that are "too optimistic" with classifiers that are "too pessimistic." While the confidence values of the former are higher than their actual performance suggests, the latter provide confidence values that are too low and do not reflect their good performance properly [11]. In this sense, we need techniques for making confidence values of different classifiers compatible with each other.

We propose an information-theoretical approach to this problem. Our method normalizes confidence values of different classifiers, acting as a repair mechanism for imprecise confidence values. The idea is to combine the information actually conveyed by each confidence value, instead of directly combining the values themselves. In order to do so, we first compute a performance estimate for each confidence value. Based on this performance estimate, we compute the informational content for each confidence value according to the well-known logarithmic notion introduced by Shannon [12]. The newly computed informational confidence values will then replace the old values, serving as a standard representation enabling "fair" comparisons.

The following sections present the theoretical background of our approach, which implies a clear statement regarding the optimal classifier combination scheme: Since the natural way of combining information from different sources is to simply add the amount of information provided by each source, the sum-rule seems to be the logical combination scheme. In fact, being additive is part of the definition of information. Accordingly, we strongly advocate the use of the sum-rule for classifier combination, i.e. adding the confidence values of all classifiers, because it is optimal in the information-theoretical sense. We do not see our approach in competition with more complex combination schemes, rather with elementary schemes such as max-rule or product-rule. In fact, our approach can be used in combination with almost any other scheme, though some schemes (e.g. product-rule) would theoretically make less sense. We therefore concentrate on four elementary combination schemes in this chapter: sum-rule, max-rule, product-rule, and majority vote.

Applying the plain sum-rule does not automatically guarantee performance improvements. On the contrary, it can degrade performance. One of the main goals of this chapter is therefore to show that sum-rule in combination with informational confidence values can provide more consistent performance improvements. We cherish the hope that our findings will contribute to overcome the theoretical gap we have been experiencing in classifier combination for quite some time now. In addition, the proposed framework can prove useful in all kinds of sensor fusion tasks and learning in general.

We structured the chapter as follows: Section 2 explains the idea and motivation of the proposed approach, followed by the theoretical background in Section 3. Section 4 explains how we can learn informational confidence values in practical applications and Section 5 then shows practical examples for script identification and character recognition. Finally, a summary of the main results concludes the chapter.

2 Motivation

Many combination schemes have been proposed in the literature. They range from simple voting schemes [13, 14, 15] to relatively complex combination strategies, such as the approach by Dempster/Shafer [16] or the Behavior-Knowledge Space (BKS) method introduced in [17]. The large number of proposed techniques shows the uncertainty researchers still have in this field. Up till now, researchers have not been able to show the general superiority of a particular combination scheme, neither theoretically nor empirically. Though several researchers have come up with theoretical explanations supporting one or more of the proposed schemes, a general commonly accepted theoretical framework for classifier combination is still missing.

Despite the high complexity of several proposed techniques, many researchers still continue to experiment with much simpler combination schemes. In fact, simple combination schemes have resulted in high recognition rates, and it is by no means clear that more complex methods are superior to simpler ones, such as sum-rule or product-rule. The sum-rule (product-rule), for instance, just adds (multiplies) the score for every class provided by each classifier of a classifier ensemble, and assigns the class with the maximum score to the input pattern. Interesting theoretical results have been derived for those simple combination schemes. For instance, Kittler et al. showed that the sum-rule is less sensitive to noise than other rules [18]. As said above, we also advocate the use of sum-rule, but for different reasons.

The main motivation of our chapter lies in an idea introduced by Oberlaender more than ten years ago [11]. Though the approach presented here is more elaborate, and common ground between both works may not be easy to detect anymore, it was his approach that gave the main impetus for the technique described. Oberlaender worked on a method for improving the recognition rate of a single classifier system by adjusting the classifier's confidence values. He measured the classification rate of a confidence value by computing the recognition rate on all test patterns classified with this particular confidence value. For each classification, he then replaced each output confidence value with its corresponding recognition rate stored in a look-up table. Figure 1 illustrates this idea for two confidence values: On the left-hand side, confidence

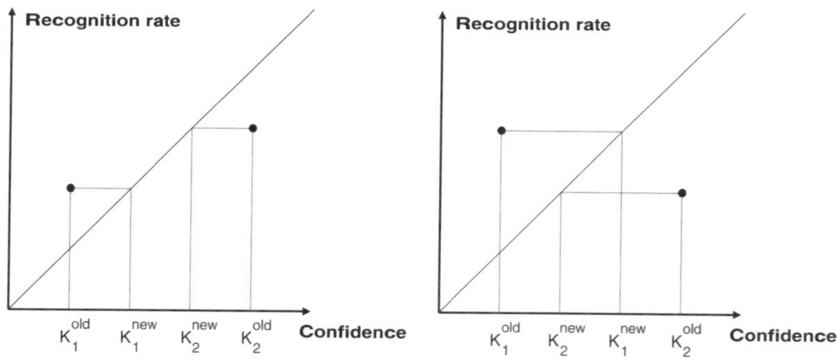

Fig. 1. Alignment of confidence values with recognition rates

value K_1^{old} performs better than its nominal value suggests. Consequently, this confidence value is mapped to the larger value K_1^{new}, which better reflects the true classification performance. According to Oberlaender, the classifier is "too pessimistic" when he outputs K_1^{old} since the confidence value is lower than its actual recognition rate. The left-hand side of Figure 1 also shows an example of a confidence value that is too optimistic: confidence value K_2^{old}

has a nominal value that is higher than its corresponding recognition rate. In this case, the lower value K_2^{new} reflects better the actual performance.

A potential problem with Oberlaender's approach is that the order of confidence values may be reversed after this process. The right-hand side of Figure 1 shows an example where this is the case: While, in the previous example, K_1^{old} is smaller than K_2^{old} and K_1^{new} is smaller than K_2^{new}, i.e. monotony has not been violated by the replacement process, the example on the right-hand side of Figure 1 shows a reversed order after replacing each confidence value with its new value. The new value K_2^{new} is now smaller than K_1^{new}.

Generally speaking, a classifier should not provide confidence values violating monotony, i.e. larger confidence values should always entail better performance in the application domain. In practice, however, this goal is usually hard to achieve due to the limited number of training samples for instance. One can argue that reversing the order of confidence values is in general a bad idea. After all, the classifier has been trained with typical examples from the application domain so that its confidence values, and in particular their relative order, should be meaningful. This apparent drawback motivated another approach based on an idea from Velek et al., who proposed a method in which confidence values still depend on their respective recognition rates, but their relative order is never reversed. This approach has been successfully applied to Japanese character recognition [19, 20, 21]. The key idea is to let confidence values progress according to the partial sum of correctly recognized patterns. This process, which warps the old confidence values, can be stated mathematically as follows:

$$K_i^{new} = R_i = \frac{\sum_{k=0}^{i} n_{correct}(K_k^{old})}{N}, \tag{1}$$

where K_i is the ith confidence value, assuming discrete values, and N is the number of all patterns tested. The term R_i denotes the relative number of patterns correctly classified with a confidence smaller than or equal to K_i^{old}. Equation 1 expresses this number using the help function $n_{correct}(K_k^{old})$, which returns the number of patterns correctly classified with confidence K_k^{old}. Since R_i is a monotonously increasing function, the new confidence values K_i^{new} will also increase monotonously and will thus not change the relative order of the old confidence values.

The following section introduces a new method that picks up ideas presented above, but realizes them in a general information-theoretical framework.

3 Informational Confidence

Our approach is based on a simple observation: If the confidence values of a classifier convey more information than provided by the classifier itself, the additional information lacks foundation and is thus redundant. On the other hand, if the confidence values convey less information than suggested by the classifier, essential information is lost. Consequently, confidence values that contain exactly the same amount of information as provided by their classifier are optimal, or fair, because they allow a fair comparison between different classifiers. Throughout our chapter, we will refer to these optimal values as "optimal confidence" or "informational confidence." In practical classifier systems, confidence values are usually just approximations of their optimal values. While this does not necessarily hamper the operation of a single classifier, which usually only depends on the relative proportion of confidence values, it causes problems in multiple classifier systems, where we need the correct amount of information conveyed for combination purposes.

We are going to propose an information-theoretical learning mechanism for informational confidence. We assume that the amount of information conveyed by a confidence value depends directly on its performance, i.e. its positive feedback within the application domain. The higher the performance of a confidence value, the higher its informational content and thus the informational confidence. For the time being, we do not further specify what "performance" actually means in this context. In fact, we will see that the precise mathematical definition of "performance" follows from the observations and considerations made above, as soon as we assume a linear relationship between information and confidence, and use the negative logarithm as the definition of information.

We begin the formalization with notations for confidence and performance: Let K_C be a set of confidence values for a classifier C:

$$K_C = \{K_0, K_1, \ldots, K_i, \ldots, K_N\} \tag{2}$$

Furthermore, let $p(K_i)$ denote the performance of the i-th confidence value K_i. We assume that the set of confidence values is either finite or that we can map it to a finite set by applying an appropriate quantization. Let us agree on identifying the lowest confidence with K_0 and the highest confidence with K_N. Note that this restriction of confidence values to a finite set of discrete values is not required per se, but it will later allow us to estimate performance values on a training set. Discrete confidence values are implicitly introduced whenever we process continuous values on a digital computer.

Claude Shannon's theory of communication serves as the starting point for the idea of informational confidence [12]. Shannon's definition of information has numerous applications in many fields, and can be considered a major milestone of research in general. Readers interested in Shannon's theory will find a good introduction to information theory in [22, 23, 24]. According to Shannon, the information of a probabilistic event is the negative logarithm of its probability. Figure 2 shows his definition of information for both a probabilistic event with probability x and its complement with probability $1 - x$. Shannon's logarithmic definition of information is motivated by two main characteristics, which in combination are unique features of the negative logarithm: First, a likely event conveys less information than an unlikely one.

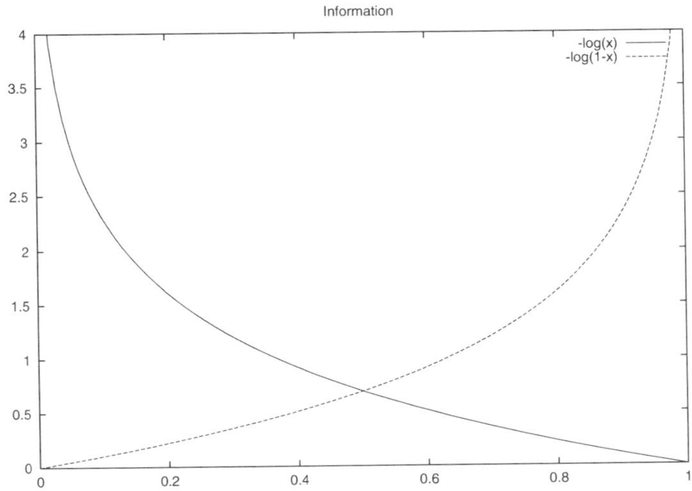

Fig. 2. Information $I(x) = -\log(x)$ for probability x, and $I(1-x) = -\log(1-x)$ for $\overline{x}$

Or, in other words, a very likely event does not cause much "surprise" since it happens very often and we thus know it partly beforehand. Second, the information of two independent events occurring simultaneously equals the sum of information independently provided by each event:

$$I(x * y) = I(x) + I(y) \tag{3}$$

The only degree of freedom Shannon's definition allows is the base of the logarithm, which is the unit that information is measured in. Base 2 is very often used by researchers and programmers alike. However, we choose the natural logarithm ln with base e as information bit for reasons becoming clear below. In Figure 2, we already used the natural logarithm to show the information conveyed by a probabilistic event.

The last requisite needed for defining the information of a confidence value K_i is the performance function $p(K_i)$, which measures the performance, or reliability, of K_i under C. As mentioned above, we impose no further constraints on this function here. The reader should think of $p(K_i)$ as a function returning values between 0 and 1, which are related to the probability that a pattern is correctly classified given the confidence value K_i. We assume that the information conveyed by a confidence value depends directly on its reliability. In particular, we assume that a highly reliable confidence value K_i (a confidence value with high performance $p(K_i)$) provides more information than a less reliable confidence value. Mathematically, we achieve this desired behavior by using the complement $\overline{p}(K_i)$ as the argument for the logarithm, instead of $p(K_i)$, when computing the information of a confidence value. The complement $\overline{p}(K_i)$, which is computed simply as $1 - p(K_i)$, denotes the uncertainty one has to take into account when a pattern is classified with confidence K_i.

Having made these reasonable assumptions, we now establish the following linear relationship between old and new informational confidence values:

$$K_i^{new} = E * I\left(\overline{p}(K_i^{old})\right) + C \tag{4}$$

In (4), the variable E is a multiplying factor just influencing the scale. However, we will later see that E represents also an expectation value in the statistical sense. The term C is merely a constant specifying an offset. It will play no further role in this chapter and will thus be equal to zero in the following. The expression $I\left(\overline{p}(K_i^{old})\right)$ in (4) denotes the information conveyed by the complement $\overline{p}(K_i^{old})$ of $p(K_i^{old})$, and implements the desired feature of providing more information for better performances of K_i. Inserting the definitions of information and performance complement into (4), and setting C to 0, leads to the following central relationship between confidence and information:

$$\begin{aligned} K_i^{new} &= E * I\left(1 - p(K_i^{old})\right) \\ &= -E * \ln\left(1 - p(K_i^{old})\right) \end{aligned} \tag{5}$$

We require confidence values to satisfy this equation in order to qualify as informational confidence. The functional relationship defined by Equation (5) ensures that old, raw confidence values with no performance, i.e. $p(K_i) = 0$, will be mapped to an informational confidence value equal to zero. On the other hand, confidence values showing perfect performance, i.e. $p(K_i) = 1$, will be assigned infinite confidence. Generally speaking, we can regard Equation (5) as a function mapping raw confidence values to their informational counterparts, depending on their performance. Each performance $p(K_i)$ determines a corresponding informational confidence value K_i, and vice versa, each confidence value K_i requires a specific performance $p(K_i)$. Hence, Equation (5) defines an equilibrium in which informational confidence values are fixed points, i.e. $K_i^{new} = K_i^{old}$. In the following, we will therefore not further distinguish between K_i^{old} and K_i^{new}, but just write K_i for a fixed point of Equation (5). Information and confidence become basically the same in the state of equilibrium.

Using the mathematical definition of informational confidence in Equation (5), we are now able to specify also the performance function $p(K_i)$, which has not been further investigated so far. The next subsection is going to derive the still missing mathematical specification of $p(K_i)$, which is essential for a practical computation of informational confidence.

3.1 Performance Function

The performance function is a direct consequence of the fixed point equation in (5). By resolving (5) for $p(K_i)$, the following straightforward transformation produces its mathematical specification:

$$\begin{aligned} K_i &= E * I\,(1 - p(K_i)) \\ \Longleftrightarrow \frac{K_i}{E} &= -\ln\,(1 - p(K_i)) \\ \Longleftrightarrow e^{-\frac{K_i}{E}} &= 1 - p(K_i) \\ \Longleftrightarrow p(K_i) &= 1 - e^{-\frac{K_i}{E}} \end{aligned} \tag{6}$$

This result shows that the performance function $p(K_i)$ describes the distribution of exponentially distributed confidence values. We can therefore consider confidence as a random variable with exponential density and parameter $\lambda = \frac{1}{E}$. Let us repeat some of the basics of statistics in order to clarify this crucial result.

The general definition of an exponential density function $e_\lambda(x)$ with parameter λ is:

$$e_\lambda(x) = \begin{cases} \lambda * e^{-\lambda x} & : \quad x \geq 0 \\ 0 & : \quad x < 0 \end{cases} \qquad \lambda > 0 \tag{7}$$

The mathematical key requirement for a function to be a density function is that its enclosed area equals 1. Accordingly, exponential densities with parameter λ meet the following criterion:

$$\int_{-\infty}^{\infty} e_\lambda(x)\,dx = \int_{0}^{\infty} \lambda * e^{-\lambda x}\,dx = 1 \quad \forall\,\lambda > 0 \tag{8}$$

Figure 3 shows three different exponential densities differing in their parameter λ, with $\lambda = 100$, $\lambda = 20$, and $\lambda = 10$, respectively. We see that parameter

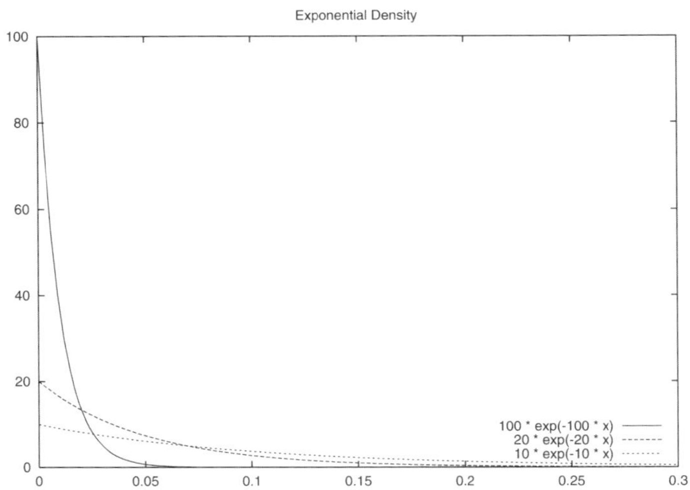

Fig. 3. Exponential density for $\lambda = 100$, $\lambda = 20$, and $\lambda = 10$

λ has a direct influence on the steepness of the exponential density: The higher λ, the steeper the corresponding exponential density function.

The mathematical companion piece of a density function is its distribution. Based on a random variable's density, a distribution describes the probability that the random variable assumes values lower than or equal to a given value k. For a random variable with exponential density $e_\lambda(x)$, we can compute the corresponding distribution $E_\lambda(k)$ as follows:

$$\begin{aligned} E_\lambda(k) &= \int_{-\infty}^{k} e_\lambda(x)\, dx \\ &= \int_{0}^{k} \lambda * e^{-\lambda x}\, dx \\ &= \left[-e^{-\lambda x}\right]_0^k \\ &= 1 - e^{-\lambda k} \end{aligned} \tag{9}$$

Consequently, a random variable with exponential density is also called "exponentially distributed random variable." Figure 4 shows the corresponding distributions for the three different densities depicted in Figure 3, with $\lambda = 100$, $\lambda = 20$, and $\lambda = 10$. The distribution converges on 1 with increasing confi-

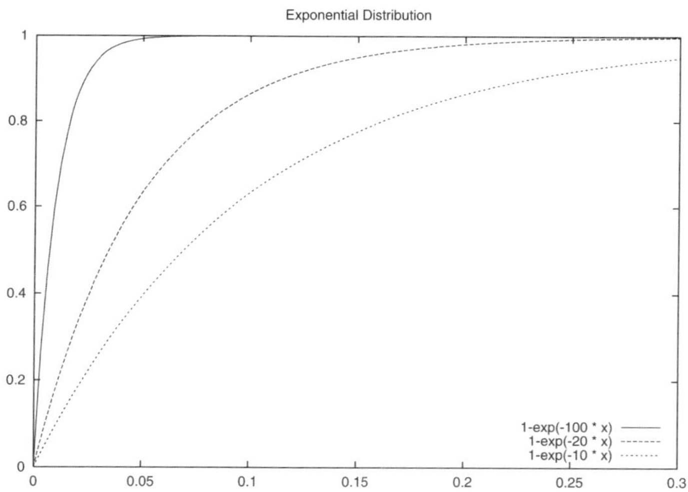

Fig. 4. Exponential distribution for $\lambda = 100$, $\lambda = 20$, and $\lambda = 10$

dence. Again, we see the influence of the parameter λ on the steepness of the distribution function. The higher λ, the steeper the corresponding exponential distribution. Moreover, there is a direct relation between λ and the expectation value E of the exponentially distributed random variable. Both are in inverse proportion to each other, i.e. $E = \frac{1}{\lambda}$. Accordingly, the expectation values belonging to the exponential densities in Figure 3, or distributions in Figure 4, are $E = \frac{1}{100}$, $E = \frac{1}{20}$, and $E = \frac{1}{10}$, respectively.

3.2 Performance Theorem

When we compare the performance specification in (6) with the exponential distribution in (9), we see that the only difference lies in the exponent of the exponential function. In fact, we can make performance and exponential distribution the same by simply setting λ to $\frac{1}{E}$. This relationship between performance and distribution now also sheds light on the parameter E. As mentioned above, the expectation value of an exponentially distributed random variable with parameter λ is $\frac{1}{\lambda}$. The parameter E therefore denotes the specific expectation value for classifier C. We summarize this important result in the Performance Theorem:

Performance Theorem:
A classifier C with performance $p(K)$ provides informational confidence $K = -E * \ln(1 - p(K))$ if, and only if, $p(K)$ is an exponential distribution with expectation E.

Proof. The theorem follows from (5), (6), and (9).

3.3 Expected Information

We have already found out that parameter E is an expectation value in the statistical sense. The performance theorem, however, reveals even more information about E. Mapping raw confidence values with identical performance to the same informational confidence value, the performance theorem describes a warping process over the set of raw confidence values for monotonously increasing performances [20]. The average information $I_{avg}(C)$ provided by a classifier C satisfying the performance theorem can be written in the form of a definite integral:

$$\begin{aligned} I_{avg}(C) &= E \int_0^1 -\ln(1 - p(K))\, dp(K) \\ &= E \int_0^1 -\ln(p(K))\, dp(K) \\ &= E \int_0^1 -\ln(K)\, dK \\ &= E \left[K - \ln(K) * K\right]_0^1 \\ &= E \end{aligned} \tag{10}$$

We see that the average information provided by C is exactly E bits, or more precisely E Euler-bits. This is in accordance with the fact that E is an expectation value. In fact, parameter E stands for both, expected information and expected confidence, because the performance theorem is based on the assumption that information and confidence are the same.

The mathematical results presented up to this point explain the meaning and implications of the parameters $p(K)$ and E. They say nothing about how to actually compute them. For classifiers that do not provide informational confidence and thus violate the performance theorem, we need to know the specific values of E and $p(K)$ in order to compute their correct informational confidence values. The next section is going to explain how we can actually use the performance theorem to first estimate these values and then compute informational confidence values for a given classifier.

4 Learning

In most practical cases, classifiers do not provide informational confidence values. Their confidence values typically violate the fixed point equation in the performance theorem, indicating a distorted equilibrium between information and confidence. In order to combine those classifiers, we need a second training process in addition to the classifier-specific training methods, which teach classifiers the decision boundaries for each class. Technically, there are two different ways of restoring the equilibrium so that confidence values satisfy the performance theorem: We can adjust the expectation E and/or the confidence K. In both cases, we need to estimate the expectation and the performance for each confidence value on a training set, or rather evaluation set, which should be different from the set that classifier C was originally trained with. For discrete confidence values, the fixed point equation of the performance theorem then formulates as follows:

$$\hat{K}_i = -\hat{E} * \ln\left(1 - \hat{p}(\hat{K}_i)\right), \tag{11}$$

where $\hat{K}_i$ is the estimated fixed point confidence for the expectation estimate $\hat{E}$ and performance estimate $\hat{p}(\hat{K}_i)$. Accordingly, learning informational confidence can be considered a 3-step process: In the first step we train classifier C with its specific training method and training set. In the second step, we estimate the expectation and performance for each confidence value on an evaluation set. Finally, we compute new informational confidence values according to (11) and store them in a look-up table. In all future classifications, confidence values provided by classifier C will then always be replaced with their informational counterparts stored in the look-up table.

The following two subsections explain how we can compute estimates of expectation and performance on the evaluation set.

4.1 Expectation Estimation

In the following, expectation E will be an invariable constant for each classifier, depending solely on the classifier's global recognition rate R measured on the evaluation set. In the practical experiments that are going to follow later in this chapter, we will use the expectation per Euler-bit as an estimate of E, instead of directly using R as E [25, 26]. In analogy to the computation of information for confidence values in (4) and (5), we are computing an estimate $\hat{I}(C)$ of the information $I(C)$ conveyed by classifier C as follows:

$$\hat{I}(C) = I(1 - R) = -\ln(1 - R), \tag{12}$$

where R denotes the overall recognition rate of C on the evaluation set. Based on the estimate $\hat{I}(C)$, $\hat{E}$ now computes as $\sqrt[\hat{I}(C)]{R}$, which maps the global recognition rate R to its corresponding rate for a one-bit classifier. The fixed point equation in the performance theorem then formulates as follows:

$$\hat{K}_i = -\sqrt[\hat{I}(C)]{R} * \ln\left(1 - \hat{p}(\hat{K}_i)\right), \tag{13}$$

This leaves us with the performance estimates as the only missing parameters we still need to compute the final informational confidence values.

4.2 Performance Estimation

Motivated by the performance theorem, which states that the performance function follows an exponential distribution, we propose an estimate that expresses performance as a percentage of the maximum performance possible. Accordingly, our relative performance estimate describes the different areas delimited by the confidence values under their common density function. Mathematically, it is based on accumulated partial frequencies and can be described by the following formula:

$$\hat{p}(\hat{K}_i) = \frac{\sum_{k=0}^{i} n_{correct}(K_k)}{N} \tag{14}$$

This is the same formula as in (1), which the reader has already encountered in Section 2. The use of accumulated partial frequencies guarantees that the newly computed estimates of the informational confidence values will not affect the order of the original confidence values. Based on this performance estimate, the mapping of old confidence values to their informational counterparts becomes a monotonous function satisfying the following relationship:

$$K_i \leq K_j \Longrightarrow \hat{K}_i \leq \hat{K}_j \tag{15}$$

Moreover, our performance estimate in (14) ensures that informational confidence values have no affect on the recognition rate of a classifier C, except for ties introduced by mapping two different confidence values to the same informational confidence value. Ties can happen when two neighboring confidence values show the same performance and become indistinguishable due to insufficient evaluation data. In most applications, this should be no problem, though. Typically, the effect of informational confidence values shows only when we combine C with other classifiers in a multiple classifier system.

Accumulated partial frequencies act like a filter in that they do not consider single confidence values but a whole range of values. They average the estimation error over all confidence values in a confidence interval. This diminishes the negative effect of inaccurate measurements of the estimate $\hat{p}(\hat{K}_i)$ in application domains with insufficient or erroneous evaluation data.

Before inserting the relative performance estimate $\hat{p}(\hat{K}_i)$ into (13), we still normalize again to the one-bit classifier using $\hat{I}(C)$, as we already did for the expectation value. For expectation and performance estimates, the final version of the fixed point equation in the performance theorem then reads as follows:

$$\hat{K}_i = - \sqrt[\hat{I}(C)]{R} * \ln\left(1 - \sqrt[\hat{I}(C)]{\hat{p}(\hat{K}_i)}\right) \tag{16}$$

The next section presents practical experiments using the proposed theoretical framework and performance estimate.

5 Experimental Results

Using informational confidence values, we will show that combined recognition rates outperform the single best recognition rates in several multiple classifier systems trained for handwritten or printed character recognition, and script identification. Let us begin with handwritten character recognition.

5.1 Japanese On-Line/Off-Line Handwritten Character Recognition

Handwriting recognition is a very promising application field for classifier combination, as it is both challenging and practically important. Handwriting, together with speech, is one of the most natural interfaces for interacting with computers. Multiple classifier systems therefore have a long tradition in handwriting recognition [1, 4]. In particular, the duality of handwriting recognition, with its two branches off-line recognition and on-line recognition, makes it suitable for multiple classifier systems. While off-line classifiers process static images of handwritten words, on-line classifiers operate on the dynamic data and expect point sequences over time as input signals. On-line systems require special hardware that is able to recover the dynamic information during the writing process. Compared to the time-independent pictorial representations used by off-line classifiers, on-line classifiers suffer from the many stroke-order and stroke-number variations inherent in human handwriting and thus in on-line data. On the other hand, dynamic information provides valuable information that on-line classifiers can exploit to better discriminate between classes. In fact, off-line and on-line information complement each other. This relationship between off-line and on-line classifiers suggests the combination of both types of classifiers to overcome the problem of stroke-order and stroke-number variations [27, 28]. This is especially important in Japanese and Chinese character recognition because the average number of strokes per character, and thus the number of variations, is much higher than in the Latin alphabet [29, 30].

Classifiers

We tested a multiple classifier system comprising two classifiers for on-line handwritten Japanese characters. One of these two classifiers, however, transforms the captured on-line data into a pictorial representation by connecting neighboring on-line points using a sophisticated painting method [19, 31]. This transformation happens in a pre-processing step before feature computation and actual classification. We can therefore consider this classifier to be an off-line classifier. Both on-line and off-line classifiers are nearest neighbor classifiers. Each classifier was trained with more than one million handwritten Japanese characters. The test and evaluation set contains 54,775 handwritten characters. From this set, we took about two thirds of the samples to estimate, and about one third to test the informational confidence values. For more information on the classifiers and data sets used, we refer readers to the references [31, 32, 33].

Table 1 lists the individual recognition rates for the off-line and on-line classifier. It shows the probabilities that the correct class label is among the n-best class labels in which the off-line or on-line classifier has the most confidence in, with n ranging from 1 to 3. We see that the off-line recognition rates

Japanese	offline	online	AND	OR
1-best	**89.94**	81.04	75.41	95.56
2-best	94.54	85.64	82.62	97.55
3-best	95.75	87.30	84.99	98.06

Table 1. Single n-best rates for handwritten Japanese character recognition

are much higher than the corresponding on-line rates. Clearly, stroke-order and stroke-number variations are largely responsible for this performance difference. They complicate considerably the classification task for the on-line classifier. The last two columns of Table 1 show the percentage of test patterns for which the correct class label occurs either twice (AND) or at least once (OR) in the n-best lists of both classifiers. The relatively large gap between the off-line recognition rates and the numbers in the OR-column suggests that on-line information is indeed complementary and useful for classifier combination purposes.

Combination Experiments

Table 2 shows the recognition rates for combined off-line/on-line recognition, using sum-rule, max-rule, and product-rule as combination schemes. As their

Japanese (**89.94**)	Raw Conf.	Perf. Conf.	Inf. Conf.	Norm. Inf. Conf.
Sum-rule	93.25	93.31	93.16	**93.78**
Max-rule	91.30	91.35	91.30	91.14
Product-rule	92.98	49.04	57.69	65.16

Table 2. Combined recognition rates for handwritten Japanese character recognition

names already suggest, sum-rule adds the confidence values provided by each classifier for the same class, while product-rule multiplies the confidence values and max-rule simply takes the maximum confidence without any further operation. The class with the maximum overall confidence will then be chosen as the most likely class for the given test pattern, followed by the n-best alternatives ranked according to their confidence. Note that the sum-rule is the mathematically appropriate combination scheme for integrating information from different sources [12]. Also, the sum-rule is very robust against noise, as was already mentioned in Section 2, at least theoretically [18]. The upper left cell of Table 2 lists again the best single recognition rate from Table 1, achieved by the off-line recognizer. We tested each combination scheme on four different confidence types:

1) raw confidence K_i as provided directly by the respective classifier
2) performative confidence $\hat{p}(\hat{K}_i)$ as defined by the performance estimate in (14)
3) informational confidence $\hat{K}_i$ as defined in (11) with $\hat{E} = R$
4) normalized informational confidence $\hat{K}_i$ as defined in (16)

Compared to the individual rates, the combined recognition rates in Table 2 are clear improvements. The sum-rule on raw confidence values already accounts for an improvement of almost 3.5%. The best combined recognition rate achieved with normalized informational confidence is 93.78%. It outperforms the off-line classifier, which is the best individual classifier, by almost 4.0%. Sum-rule performs better than max-rule and product-rule, a fact in accordance with the results in [18]. Mathematically, informational confidence values only make sense when used in combination with sum-rule. For the sake of completeness, however, we list the combined recognition rates for the other combination schemes as well. Generally speaking, the recognition rates in Table 2 compare favorably with the state-of-the art, especially when we consider the difficult real-world test data used for training and testing. Also, sum-rule in combination with informational confidence exploits the candidate alternatives in the n-best lists to a fairly large extent, as indicated by the small difference between the "OR"-column and the practical recognition rates actually achieved. This should also benefit syntactical post-processing.

Figure 5 depicts the performance estimate for the off-line classifier (left-hand side) and on-line classifier (right-hand side). In both cases, the perfor-

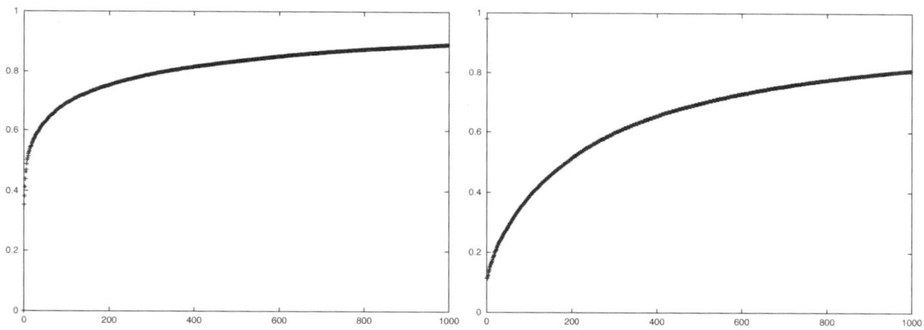

Fig. 5. Performance estimate for off-line (left) and on-line (right) recognition

mance estimates describe a monotonously increasing function. The off-line function is steeper and reaches a higher level though. Figure 6 shows the corresponding informational confidence values computed according to (16). On

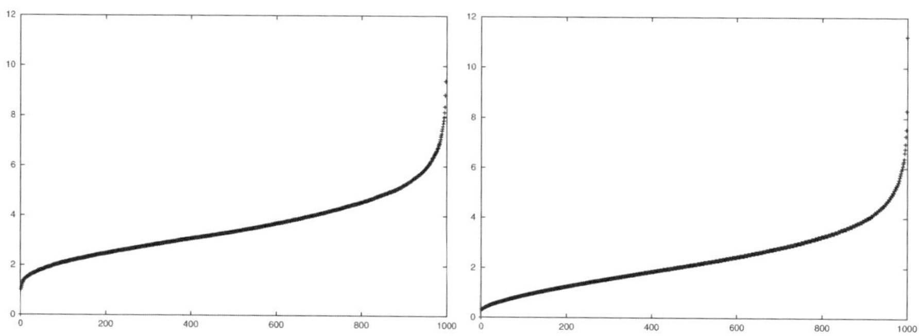

Fig. 6. Informational confidence for off-line (left) and on-line (right) recognition

average, due to the better performance of the off-line classifier, the off-line confidence values are higher than the on-line confidence values.

5.2 Printed Character Recognition

The second application we experimented with is optical character recognition (OCR) for a transliterated Arabic-English bilingual dictionary scanned with a resolution of 300dpi. Our character set comprises all Latin characters and a large number of dictionary-specific special characters, which let the overall number of categories grow to 129. Figure 7 shows a sample page of the dictionary.

ʔ-x-r

ʔaxxar **1. to delay, make late, hold up.** ***ð̣all yiḥči w-ʔaxxarni nuṣṣ saaʕa.*** **He kept talking and delayed me a half hour.** ***l-muṭar ʔaxxar taqaddumna.*** **The rain slowed our advance.** ***ʔariid il-ʔiijaar waaḥid biš-šahar. la-ykuun itʔaxxra.*** **I want the rent the first of the month. You're not to hold it up. 2. to postpone, put off.** ***ʔaxxraw il-ijtimaaʕ ʕala muudi.*** **They postponed the meeting on my account.** ***dizz il-maktuub il-yoom; la-tʔaxxra ʔila baačir.*** **Send the letter today; don't put it off till tomorrow.**

tʔaxxar **1. to be delayed.** ***wuṣuul iṭ-ṭiyyaara tʔaxxar nuṣṣ saaʕa.*** **The plane's arrival was delayed a half hour. 2. to become late, get late, be late.** ***l-wakt itʔaxxar. xalli nruuḥ.*** **It's gotten late.**

Fig. 7. Arabic-English dictionary

Classifiers

We used two different nearest neighbor classifiers for recognition: a classifier based on a weighted Hamming distance operating on the pictorial data and a classifier using Zernike moments. The Hamming distance of the first classifier describes the distance between the size-normalized character image and a binary template map computed for each character class from the training set. On the other hand, moment descriptors have been studied for image recognition and computer vision since 1960s [34]. Teague [35] first introduced the use of Zernike moments to overcome the shortcomings of information redundancy present in the popular geometric moments. Zernike moments are a class of orthogonal moments which are rotation invariant and can be easily constructed to an arbitrary order. They are projections of the image function onto a set of complex, orthogonal polynomials defined over the interior of a unit circle $x^2 + y^2 = 1$. It has already been shown in [36, 37] that Zernike moments can be useful for OCR. In our work, the order of Zernike moments was chosen to be 12, which provides a 48-dimensional moment vector. Due to space constraints, we refer readers again to the references for more information on Zernike moments, e.g. [36, 37].

Table 3 lists the n-best single recognition rates and the corresponding AND and OR rates. In our experiments, the weighted Hamming distance

Latin	Hamming	Zernike	AND	OR
1-best	**96.35**	93.66	91.01	99.01
2-best	96.97	97.75	95.46	99.26
3-best	97.00	98.64	96.34	99.30

Table 3. Single n-best rates for printed Latin character recognition

is superior to the Zernike moments. It achieves a recognition rate of 96.35 versus 93.66 achieved by the Zernike moments. The OR-column of Table 3 is again an indication that a combination of both classifiers has the potential to improve the single best recognition rate, i.e. the recognition rate provided by the Hamming distance.

Combination Experiments

Table 4 is an overview of the combined recognition rates we computed. It is

Latin (**96.35**)	Raw Conf.	Perf. Conf.	Inf. Conf.	Norm. Inf. Conf.
Sum-rule	98.12	98.16	97.94	**98.50**
Max-rule	97.06	96.69	96.72	96.97
Product-rule	97.78	79.87	84.98	93.83

Table 4. Combined recognition rates for printed Latin character recognition

organized similar to Table 2, with the single-best recognition rate listed in the upper left corner of the table. Each column contains the recognition rate for a particular confidence type and each row contains the rates for a specific combination scheme. The recognition rates in Table 3 and 4 represent the average over twenty randomly selected pages with 64570 test patterns from which we took every second pattern to compute the performance estimate. As we have already observed for Japanese character recognition, the sum-rule yields higher improvements than max-rule and product-rule. On raw confidence values, it achieves a considerable improvement on the best single recognition rate. Normalized informational confidence values increase the recognition rate further to 98.50, which is more than two percent higher than the best individual rate.

5.3 Script Identification

In our third application, we investigate the effect of informational confidence values on a multiple classifier system integrating different classification architectures. The application domain is bi-lingual script identification.

A significant number of today's documents can only be accessed in printed form. A large portion of them are multilingual documents, such as patents or bilingual dictionaries. In general, automatic processing of these documents requires that the scripts must be identified before they can be fed into an appropriate OCR system for automatic processing. Earlier work on script identification includes template-based approaches [38], approaches exploiting character-specific characteristics [39, 40, 41], as well as text line projections [42] and font recognition based on global texture analysis [43]. For our classifier combination experiments, we have developed a different approach that operates on word-level [44, 45]: A modified Docstrum algorithm first segments the document into words, then we use Gabor filters to compute features for each script class [46]. Our experiments focus on bilingual documents with one script being English and the second script being either Arabic, Chinese, Hindi, or Korean. Accordingly, script identification then becomes a 2-class classification problem for each word. Figure 8 shows a page from a bilingual Arabic-English word dictionary, with the Arabic words successfully identified.

Classifiers

This subsection is a brief description of the different classifiers used in our multiple classifier system for script identification, including their recognition rates. More information on the features used and pre-processing steps is given in [44].

Our multiple classifier system comprises four classifiers based on different classification principles: nearest neighbors, weighted Euclidean distances, support vectors, and Gaussian mixtures.

- **Nearest Neighbor (KNN):** First introduced by Cover and Hart [47] in 1967, the Nearest Neighbor (NN) classifier was proven to be a very effective and simple classifier. It found applications in many research fields. Using the NN classifier, a test sample $\mathbf{x}$ is classified by assigning it the

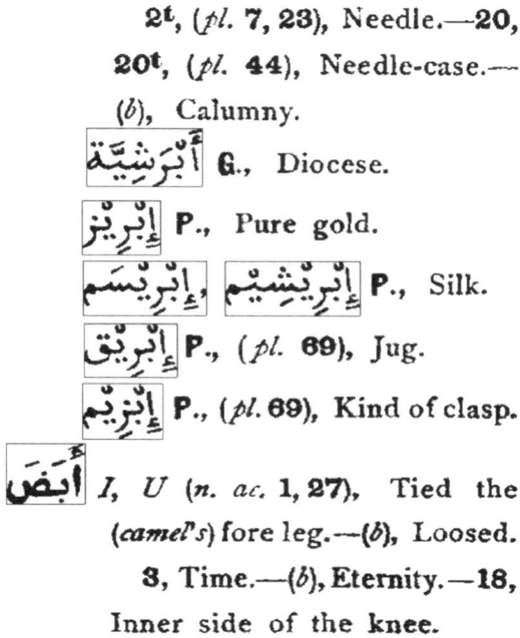

Fig. 8. Arabic word segmentation

label that its nearest sample represents in the training samples. The distance between two feature vectors is usually measured by computing the Euclidean distance. In the experiments reported here, the classifier's confidence in a class label λ_i assigned to a testing sample $\mathbf{x}$ computes as follows:

$$conf(\mathbf{x}|\lambda_i) = \min_{\mathbf{x}_t \in \mathbf{S}_{\lambda_i}} (dis(\mathbf{x}, \mathbf{x}_t)),$$

where $dis(\mathbf{x}, \mathbf{x}_t)$ is the Euclidean distance between two vectors $\mathbf{x}$ and $\mathbf{x}_t$, and $\mathbf{S}_{\lambda_i}$ is the set containing all training samples with label λ_i. In our experiments, $\mathbf{x}$ represents a 32-dimensional feature vector.

- **Weighted Euclidean Distance (WED):** Based on the mean $\boldsymbol{\mu}^{(i)}$ and standard deviation $\boldsymbol{\alpha}^{(i)}$ of the training samples for each class λ_i, the distance between a test sample $\mathbf{x}$ and λ_i is computed as follows:

$$dis(\mathbf{x}, \lambda_i) = \sum_{k=1}^{d} \left| \frac{x_k - \mu_k^{(i)}}{\alpha_k^{(i)}} \right| \quad i = 1...M$$

where d is the feature dimension and M is the number of classes. Testing samples are assigned the class label with the minimum distance, and the classifier's confidence in a class label λ_i is computed as:

$$conf(\mathbf{x}|\lambda_i) = \frac{\sum_{j=1, j\neq i}^{M} dis(\mathbf{x}, \lambda_j)}{\sum_{j=1}^{M} dis(\mathbf{x}, \lambda_j)}$$

- **Support Vector Machine (SVM):** SVMs were first introduced in the late seventies, but are now receiving increased attention. The SVM classifier constructs a 'best' separating hyperplane (the maximal margin plane) in a high-dimensional feature space which is defined by nonlinear transformations from the original feature variables. The distance between a test sample and the hyperplane reflects the confidence in the final classification. The larger the distance, the more confidence the classifier has in its classification result. Hence, we can directly use this distance as confidence. A detailed description of the computation of the separating hyperplanes is given, for instance, in [48]. The experiments reported here are based on the SVM implementation *SVM-light*, which uses a polynomial kernel function [49].
- **Gaussian Mixture Model (GMM):** The Gaussian Mixture Model (GMM) classifier is used to model the probability density function of a feature vector, $\mathbf{x}$, by the weighted combination of M multi-variate Gaussian densities (Λ):

$$p(\mathbf{x}|\Lambda) = \sum_{i=1}^{M} p_i g_i(\mathbf{x}),$$

where the weight (mixing parameter) p_i corresponds to the prior probability that feature $\mathbf{x}$ was generated by component i, and satisfies $\sum_{i=1}^{M} p_i = 1$. Each component λ_i is represented by a Gaussian mixture model $\lambda_i = N(p_i, \boldsymbol{\mu}_i, \Sigma_i)$ whose probability density can be described as:

$$g_i(\mathbf{x}) = \frac{1}{\sqrt{(2\pi)^d|\Sigma_i|}} exp(-\frac{1}{2}(\mathbf{x} - \boldsymbol{\mu}_i)^T \Sigma_i^{-1}(\mathbf{x} - \boldsymbol{\mu}_i)),$$

where $\boldsymbol{\mu}_i$ and Σ_i are the mean vector and covariance matrix of Gaussian mixture component i, respectively. Details about the estimation of these parameters can be found in [44]. In the following experiments, the likelihood value $ln(p(\lambda_i|\mathbf{x}))$ describes the classifier's confidence in labeling test pattern $\mathbf{x}$ as λ_i.

Table 5 lists the individual recognition rates of each classifier for all scripts, i.e. the percentage of successful discriminations between English and respective non-English words. The KNN and SVM classifiers provide the best over-

Script/Classifier	KNN	WED	GMM	SVM	AND	OR
Arabic	90.90	80.07	88.14	**90.93**	68.55	98.04
Chinese	92.19	84.89	90.59	**93.34**	76.04	98.66
Korean	**94.04**	84.68	90.72	92.54	76.57	98.53
Hindi	**97.51**	91.97	93.11	97.27	86.48	99.51

Table 5. Single Recognition rates for script identification

all performance. Classification rates are lowest for Arabic script since Arabic words are more similar to English (Latin) words than they are to words of other scripts. The OR-column indicates that the combination of all four classifiers has the potential to improve recognition.

Combination Experiments

In addition to the elementary combination schemes sum-rule, max-rule, and product-rule, we also experimented with combinations based on majority vote. Majority vote implements a simple voting among all four classifiers. The class with the maximum number of votes will then be chosen as the most likely class for a given test pattern. In case of ties, we take the label provided by the classifier with the highest recognition rate as ultimate decision. We roughly labeled between 6000 and 10000 evaluation patterns for each script, and used again 50% of these sets to estimate the performance of each confidence value. Tables 6, 7, 8, and 9 show the combined recognition rates for each script and each combination scheme.

Arabic (**90.93**)	Raw Conf.	Perf. Conf.	Inf. Conf.	Norm. Inf. Conf.
Sum-rule	91.03	92.39	92.39	**92.73**
Max-rule	88.18	90.67	90.87	90.80
Product-rule	91.03	91.13	91.17	91.23
Majority vote	91.46	—	—	—

Table 6. Combined recognition rates for Arabic script identification

Chinese (**93.34**)	Raw Conf.	Perf. Conf.	Inf. Conf.	Norm. Inf. Conf.
Sum-rule	94.24	94.26	94.24	**94.50**
Max-rule	90.83	93.24	93.24	93.30
Product-rule	93.45	93.45	93.45	93.45
Majority vote	93.79	—	—	—

Table 7. Combined recognition rates for Chinese script identification

The results show that the combined recognition rates based on normalized informational confidence consistently outperform the best single recognition rate, which is again displayed in the upper left cell of each table. We achieve the biggest improvement of 1.8% on the most difficult script: Arabic. In particular, normalized informational confidence values perform better than other confidence types. The unnormalized informational confidence values also lead to improvements in all four cases. A very important point to notice is that the

Korean (**94.04**)	Raw Conf.	Perf. Conf.	Inf. Conf.	Norm. Inf. Conf.
Sum-rule	93.73	93.75	94.16	**94.35**
Max-rule	90.99	93.30	93.49	94.06
Product-rule	92.52	92.38	92.38	92.38
Majority vote	94.06	—	—	—

Table 8. Combined recognition rates for Korean script identification

Hindi (**97.51**)	Raw Conf.	Perf. Conf.	Inf. Conf.	Norm. Inf. Conf.
Sum-rule	98.05	97.98	97.83	**98.08**
Max-rule	93.13	95.64	95.81	96.65
Product-rule	97.32	97.19	97.19	97.19
Majority vote	97.93	—	—	—

Table 9. Combined recognition rates for Hindi script identification

plain sum-rule is not able to improve the single best recognition rate for Korean, whereas normalized informational confidence leads to an improvement. This is another confirmation of the fact that sum-rule does not necessarily improve recognition rates, as has already been observed by many researchers before. It performs better than max-rule or product-rule in our experiments, though. The simple performative confidence estimate also performs well except for Korean. Majority voting outperforms sum-rule on Arabic and Korean. On the other hand, it is outperformed by sum-rule on Chinese and Hindi. This somewhat inconsistent behavior also reflects very well the practical experiences of other authors.

AdaBoost

This subsection reports recognition experiments with Boosting, i.e. AdaBoost, in order to have a comparison with other, more complex state-of-the-art combination schemes. AdaBoost (Adaptive Boosting) was introduced by Freund and Schapire in 1995 to expand the boosting approach originally introduced by Schapire. We focus on the AdaBoost approach called AdaBoost.M1 [7]. The AdaBoost algorithm generates a set of classifiers and combines them. It changes the weights of the training samples based on classifiers previously built (trials). The goal is to force the final classifiers to minimize the expected error over different input distributions. The final classifier is formed using a weighted voting scheme. Details of AdaBoost.M1 can be found in [7]. Table 10 is a representative example of our Boosting experiments, with the number of trials being 20. In most cases, Boosting provides slight improvements of the single recognition rates in Table 5. However, the improvements are smaller than those achieved with informational confidence and do not outperform the single best recognition rate for Hindi in our experiments.

Classifier	KNN	WED	GMM	SVM
Arabic (**90.93**)	**90.95**	82.24	88.03	90.48
Chinese (**93.34**)	92.22	85.17	91.02	**93.49**
Korean (**94.04**)	**94.12**	85.18	91.14	93.03
Hindi (**97.51**)	97.39	92.13	92.85	97.38

Table 10. Recognition rates with Boosting

We must also note that the nearest neighbor classifier does not lend itself to boosting because it never misclassifies on the training set. For this reason and the fact that boosting can still be applied on top of informational confidence, we do not report more results for boosting in this chapter.

6 Summary

The classifiers of a multiple classifier system typically provide confidence values with quite different characteristics, especially classifiers based on different classification architectures. While this poses no problem for the individual classifiers themselves, it considerably complicates the integration of different classification results into a single outcome. We propose an information-theoretical framework to overcome this problem and make confidence values from different sources comparable. Our approach is based on two main postulates. First, we assume that confidence is basically information as introduced by Shannon's logarithmic definition. Second, we require that the confidence of a classifier depends on its performance function in a given application domain. These two simple assumptions lead to a fixed point equation for what we call informational confidence. Our main result, which we summarize in the Performance Theorem, states that classifiers providing informational confidence feature an exponential distribution as a performance function.

The performance theorem motivates a learning method for informational confidence: We can estimate the performance of each confidence value on an evaluation set and insert the estimate directly into the fixed point equation given in the Performance Theorem. This provides us directly with the informational confidence values, which are fixed points. As performance estimate, we use a monotonously increasing function whose main purpose is to estimate the values of an exponential distribution for each confidence value.

In our experiments for character recognition and script identification, combined recognition rates based on informational confidence consistently outperform the best single recognition rate. In particular, sum-rule in combination with informational confidence outperforms other elementary combination schemes, namely max-rule, product-rule, majority vote, and sum-rule itself on raw confidence values.

For monotonously increasing performance estimates, informational confidence values do not affect the recognition rates of individual classifiers. They take effect only when used in combination with informational confidence values from other classifiers. Nevertheless, a single recognizer can of course take advantage of informational confidence values in the post-processing steps following classification, such as syntactic context analysis of n-best lists, etc.

While sum-rule is the most natural combination scheme for informational confidence values, integration of informational confidence is by no means a contradiction to other, more complex integration techniques. On the contrary, informational confidence can be considered a general standard representation for confidence on which other post-processing techniques can rest.

References

1. Xu, L., Krzyzak, A., Suen, C.: Methods of Combining Multiple Classifiers and Their Applications to Handwriting Recognition. IEEE Trans. on Systems, Man, and Cybernetics **22** (1992) 418–435
2. Gader, P., Mohamed, M., Keller, J.: Fusion of Handwritten Word Classifiers. Pattern Recognition Letters **17** (1996) 577–584
3. Sirlantzis, K., Hoque, S., Fairhurst, M.C.: Trainable Multiple Classifier Schemes for Handwritten Character Recognition. In: 3rd International Workshop on Multiple Classifier Systems (MCS), Cagliari, Italy, Lecture Notes in Computer Science, Springer-Verlag (2002) 169–178
4. Wang, W., Brakensiek, A., Rigoll, G.: Combination of Multiple Classifiers for Handwritten Word Recognition. In: Proc. of the 8th International Workshop on Frontiers in Handwriting Recognition (IWFHR-8), Niagara-on-the-Lake, Canada (2002) 117–122
5. Breiman, L.: Bagging Predictors. Machine Learning **2** (1996) 123–140
6. Freund, Y., Schapire, R.: A Short Introduction to Boosting. Journal of Japanese Society for Artificial Intelligence **14** (1999) 771–780
7. Freund, Y., Schapire, R.: Experiments with a New Boosting Algorithm. In: Proc. of 13th Int. Conf. on Machine Learning, Bari, Italy (1996) 148–156
8. Guenter, S., Bunke, H.: New Boosting Algorithms for Classification Problems with Large Number of Classes Applied to a Handwritten Word Recognition Task. In: 4th International Workshop on Multiple Classifier Systems (MCS), Guildford, UK, Lecture Notes in Computer Science, Springer-Verlag (2003) 326–335
9. Ho, T.K.: The Random Subspace Method for Constructing Decision Forests. IEEE Trans. on Pattern Analysis and Machine Intelligence (TPAMI) **20** (1998) 832–844
10. Ianakiev, K., Govindaraju, V.: Architecture for Classifier Combination Using Entropy Measures. In: 1st International Workshop on Multiple Classifier Systems (MCS), Cagliari, Italy, Lecture Notes in Computer Science, Springer-Verlag (2000) 340–350
11. Oberlaender, M.: Mustererkennungsverfahren (1995) German Patent DE 4436408 C1 (in German).

12. Shannon, C.E.: A Mathematical Theory of Communication. Bell System Tech. J. **27** (1948) 379–423
13. Ho, T., Hull, J., Srihari, S.: Decision Combination in Multiple Classifier Systems. IEEE Trans. on Pattern Analysis and Machine Intelligence (TPAMI) **16** (1994) 66–75
14. Erp, M.V., Vuurpijl, L.G., Schomaker, L.: An Overview and Comparison of Voting Methods for Pattern Recognition. In: Proc. of the 8th International Workshop on Frontiers in Handwriting Recognition (IWFHR-8), Niagara-on-the-Lake, Canada (2002) 195–200
15. Kang, H.J., Kim, J.: A Probabilistic Framework for Combining Multiple Classifiers at Abstract Level. In: Fourth International Conference on Document Analysis and Recognition (ICDAR), Ulm, Germany (1997) 870–874
16. Mandler, E., Schuermann, J.: Combining the Classification Results of Independent Classifiers Based on the Dempster/Shafer Theory of Evidence. In E.S. Gelsema, L.K., ed.: Pattern Recognition and Artificial Intelligence. (1988) 381–393
17. Huang, Y., Suen, C.: A Method of Combining Multiple Experts for Recognition of Unconstrained Handwritten Numerals. IEEE Trans. on Pattern Analysis and Machine Intelligence (TPAMI) **17** (1995) 90–94
18. Kittler, J., Hatef, M., Duin, R., Matas, J.: On Combining Classifiers. IEEE Transactions on Pattern Analysis and Machine Intelligence **20** (1998) 226–239
19. Velek, O., Liu, C.L., Jaeger, S., Nakagawa, M.: An Improved Approach to Generating Realistic Kanji Character Images from On-Line Characters and its Benefit to Off-Line Recognition Performance. In: 16th International Conference on Pattern Recognition (ICPR). Volume 1., Quebec (2002) 588–591
20. Velek, O., Jaeger, S., Nakagawa, M.: A New Warping Technique for Normalizing Likelihood of Multiple Classifiers and its Effectiveness in Combined On-Line/Off-Line Japanese Character Recognition. In: 8th International Workshop on Frontiers in Handwriting Recognition (IWFHR), Niagara-on-the-Lake, Canada (2002) 177–182
21. Velek, O., Jaeger, S., Nakagawa, M.: Accumulated-Recognition-Rate Normalization for Combining Multiple On/Off-line Japanese Character Classifiers Tested on a Large Database. In: 4th International Workshop on Multiple Classifier Systems (MCS), Guildford, UK, Lecture Notes in Computer Science, Springer-Verlag (2003) 196–205
22. Pierce, J.R.: An Introduction to Information Theory: Symbols, Signals, and Noise. Dover Publications, Inc., New York (1980)
23. Sacco, W., Copes, W., Sloyer, C., Stark, R.: Information Theory: Saving Bits. Janson Publications, Inc., Dedham, MA (1988)
24. Sloane, N.J.A., Wyner, A.D.: Claude Elwood Shannon: Collected Papers. IEEE Press, Piscataway, NJ (1993)
25. Jaeger, S.: Informational Classifier Fusion. In: Proc. of the 17th Int. Conf. on Pattern Recognition, Cambridge, UK (2004) 216–219
26. Jaeger, S.: Using Informational Confidence Values for Classifier Combination: An Experiment with Combined On-Line/Off-Line Japanese Character Recognition. In: Proc. of the 9th Int. Workshop on Frontiers in Handwriting Recognition, Tokyo, Japan (2004) 87–92
27. Jaeger, S., Manke, S., Reichert, J., Waibel, A.: Online Handwriting Recognition: The Npen++ Recognizer. International Journal on Document Analysis and Recognition **3** (2001) 169–180

28. Jaeger, S.: Recovering Dynamic Information from Static, Handwritten Word Images. PhD thesis, University of Freiburg (1998) Foelbach Verlag
29. Jaeger, S., Liu, C.L., Nakagawa, M.: The State of the Art in Japanese Online Handwriting Recognition Compared to Techniques in Western Handwriting Recognition. International Journal on Document Analysis and Recognition **6** (2003) 75–88
30. Liu, C.L., Jaeger, S., Nakagawa, M.: Online Recognition of Chinese Characters: The State-of-the-Art. IEEE Trans. on Pattern Analysis and Machine Intelligence (TPAMI) **26** (2004) 198–213
31. Jaeger, S., Nakagawa, M.: Two On-Line Japanese Character Databases in Unipen Format. In: 6th International Conference on Document Analysis and Recognition (ICDAR), Seattle (2001) 566–570
32. Nakagawa, M., Akiyama, K., Tu, L., Homma, A., Higashiyama, T.: Robust and Highly Customizable Recognition of On-Line Handwritten Japanese Characters. In: Proc. of the 13th International Conference on Pattern Recognition. Volume III., Vienna, Austria (1996) 269–273
33. Nakagawa, M., Higashiyama, T., Yamanaka, Y., Sawada, S., Higashigawa, L., Akiyama, K.: On-Line Handwritten Character Pattern Database Sampled in a Sequence of Sentences without Any Writing Instructions. In: Fourth International Conference on Document Analysis and Recognition (ICDAR), Ulm, Germany (1997) 376–381
34. Teh, C.H., Chin, R.T.: On image anlaysis by the methods of moments. IEEE Transactions on Pattern Analysis and Machine Intelligence **10** (1988) 496–513
35. Teague, M.: Image analysis via the general theory of moments. Journal of the Optical Society of America **70** (1979) 920–930
36. Khotanzad, A., Hong, Y.H.: Rotation invariant image recognition using feature selected via a systematic method. Pattern Recognition **23** (1990) 1089–1101
37. Khotanzad, A., Hong, Y.H.: Invariant image recognition by zernike moments. IEEE Transactions on Pattern Analysis and Machine Intelligence **12** (1990) 489–497
38. Hochberg, J., Kelly, P., Thomas, T., Kerns, L.: Automatic script identification from document images using cluster-based templates. IEEE Trans. Pattern Analysis and Machine Intelligence **19** (1997) 176–181
39. Sibun, P., Spitz, A.L.: Language determination: Natural language processing from scanned document images. In: Proc. 4th Conference on Applied Natural Language Processing, Stuttgart (1994) 115–121
40. Spitz, A.L.: Determination of the script and language content of document images. IEEE Trans. Pattern Analysis and Machine Intelligence **19** (1997) 235–245
41. Tan, C., Leong, T., He, S.: Language Identification in Multilingual Documents. In: Int. Symposium on Intelligent Multimedia and Distance Education (ISI-MADE'99), Baden-Baden, Germany (1999) 59–64
42. Waked, B., Bergler, S., Suen, C.Y.: Skew detection, page segmentation, and script classification of printed document images. In: IEEE International Conference on Systems, Man, and Cybernetics (SMC'98), San Diego, CA (1998) 4470–4475
43. Zhu, Y., Tan, T., Wang, Y.: Font recognition based on global texture analysis. IEEE Trans. Pattern Analysis and Machine Intelligence **23** (2001) 1192–1200

44. Ma, H., Doermann, D.: Word level script identification for scanned document images. Proc. of Int. Conf. on Document Recognition and Retrieval (SPIE) (2004) 178–191
45. Jaeger, S., Ma, H., Doermann, D.: Identifying Script on Word-Level with Informational Confidence. In: Int. Conf. on Document Analysis and Recognition (ICDAR), Seoul, Korea (2005) 416–420
46. O'Gorman, L.: The document spectrum for page layout analysis. IEEE Trans. Pattern Analysis and Machine Intelligence **15** (1993) 1162–1173
47. Cover, T.M., Hart, P.E.: Nearest neighbor pattern classification. IEEE Trans. Information Theory **IT-13** (1967) 21–27
48. Burges, C.J.C.: A tutorial on support vector machines for pattern recognition. Data Mining and Knowledge Discovery **2** (1998) 121–167
49. Joachims, T.: Making Large-Scale SVM Learning Practical. In: Advances in Kernel Methods-Support Vector Learning. B. Schölkopf, C. Burges, and A. Smola, MIT-Press (1999) 41–56

Self-Organizing Maps for Clustering in Document Image Analysis

Simone Marinai, Emanuele Marino, and Giovanni Soda

University of Florence
Dipartimento di Sistemi e Informatica (DSI)
Via S. Marta, 3, I-50139, Firenze, Italy
{marinai,marino,soda}@dsi.unifi.it

Summary. In this chapter, we discuss the use of Self Organizing Maps (SOM) to deal with various tasks in Document Image Analysis. The SOM is a particular type of artificial neural network that computes, during the learning, an unsupervised clustering of the input data arranging the cluster centers in a lattice. After an overview of the previous applications of unsupervised learning in document image analysis, we present our recent work in the field. We describe the use of the SOM at three processing levels: the character clustering, the word clustering, and the layout clustering, with applications to word retrieval, document retrieval and page classification. In order to improve the clustering effectiveness, when dealing with small training sets, we propose an extension of the SOM training algorithm that considers the tangent distance so as to increase the SOM robustness with respect to small transformations of the patterns. Experiments on the use of this extended training algorithm are reported for both character and page layout clustering.

1 Introduction

Supervised classifiers are important components of most Document Image Analysis (DIA) systems. Few systems rely on unsupervised learning, or clustering. In unsupervised learning there is no explicit teacher and the training algorithm takes into account unlabeled samples. Systems that include clustering algorithms discover "natural" groupings, or clusters, of the input patterns that can be used in subsequent processing steps. Some specific knowledge about the nature of the patterns to be processed is considered in the design of the clustering algorithms, for instance to fix in advance the desired number of clusters to be found in the training data. An appropriate choice of this parameter is essential for maximizing the performance of clustering algorithms. Another important feature is the type of distance function that is embedded in the clustering algorithm.

In this chapter, we focus our attention on a particular kind of clustering algorithm, the Self Organizing Map (SOM) [1], that is well suited to

S. Marinai et al.: *Self-Organizing Maps for Clustering in Document Image Analysis*, Studies in Computational Intelligence (SCI) **90**, 193–219 (2008)
www.springerlink.com

dimensionality reduction and exploratory data analysis. We analyze various applications of SOM-based clustering in document image analysis sub-tasks and the incorporation of the tangent distance in the SOM training.

The chapter is organized as follows. In Section 2 we summarize the main features of SOM, as well as its standard training algorithm. In Section 3 we survey some recent applications of clustering algorithms in DIA with a particular emphasis on the use of the SOM. In Section 4 and in Section 5 we discuss our work related to SOM clustering at the character, word, and page levels. Lastly, we present our concluding remarks in Section 6.

2 Self Organizing Maps

Self Organizing Maps are a particular kind of unsupervised Artificial Neural Network (ANN). In this section we summarize some aspects of ANNs that can be relevant to appreciate the SOM peculiarities. We describe also some strategies that have been proposed in the literature to obtain invariant training systems. Additional information on ANNs can be found in other chapters of this book or in specific survey papers (see e.g. [2] for an overview of ANNs applications in the field of Document Image Analysis).

2.1 Artificial Neural Networks

Artificial Neural Networks are biologically inspired processing systems composed of a set of units, referred to as neurons, and a set of weighted connections between neurons where the signals are propagated. One of the first models for artificial neurons was the perceptron, which operates on continuous inputs and returns a Boolean output usually regarded as a classification of the input pattern. Perceptrons were dismissed primarily because of their limited capacity for function approximation. In contrast, Multi-Layer Perceptrons (MLPs) exhibit a universal interpolation capacity (see e.g. [3, 4]). In Multi-Layer Perceptrons the neurons are arranged into layers, and the connections link one layer to the next one. The input is regarded as a special layer which is propagated forward to hidden layers and, lastly, to the output. The term MLP usually denotes networks with sigmoidal neurons. Also RBF networks present the same layered architecture, but are often organized in two layers composed of RBF units and sigmoidal units, respectively.

The neural network's learning process can be either supervised or unsupervised. In the former case an expected target value is assigned to each example. In the latter case there is no external teacher that pre-defines the desired behavior of the network for the training samples. The supervised training of feed-forward architectures is performed by searching in the weight space for a set of parameters minimizing the mismatch between the target values and the network outputs. This search is typically made with the Back-propagation algorithm [5]. Competitive learning is another class of training algorithms where the output neurons compete among themselves. Both unsupervised (e.g. *Vector Quantization* and *Self Organizing Maps* [6]) and supervised learning (e.g. *Learning Vector Quantization*) can be considered. In the Self-Organizing Map the neurons are usually arranged in a two dimensional lattice and each neuron receives inputs from the input layer and from the other neurons in the lattice. During the learning process the network performs clustering. The mapping of the neurons to class membership can be carried out upon completion of the learning process. For instance, each neuron can be labeled with the most frequent class among the training patterns belonging to its cluster.

2.2 Invariant Recognition

One of the most important goals in pattern recognition is the ability to achieve invariant recognition with respect to irrelevant distortions of the input patterns. For this purpose, prior knowledge on the pattern recognition problem to be solved can be considered in the training process adopting three main strategies: invariance by feature extraction, invariance by training, and invariance by structure [7].

In **invariance by feature extraction**, the prior knowledge is considered in the design of the feature extraction algorithm. A pattern can be represented either with a flat feature vector or with higher level structural representations (e.g. lists, trees, or graphs). In pattern recognition applications, including DIA, this is the prevailing strategy and large emphasis is placed on the design of appropriate feature extraction algorithms adopting standard paradigms for the classification.

In **invariance by training**, the prior knowledge is used when building the training set that is designed to contain samples depicting different aspects of the patterns to be processed. For instance, in the case of OCR, invariance to translations can be obtained by "building" training samples that are slightly shifted with respect to the original position (in [8] this type of processing is referred to as dithering). A similar approach has been proposed, for instance, for the generation of touching characters required to train neural segmentation algorithms [9]. A related application for the generation of synthetic handwritten data is described in the chapter by Varga and Bunke in this book.

In **invariance by structure** the classifier is designed to produce the same output when transformed versions of one pattern are presented to it. Convolutional networks [10] are examples of architectures based on feedforward neural networks which are invariant with respect to translations (for instance in handwriting recognition). Convolutional neural networks are based on local *receptive fields* where each hidden neuron (the receptive field) is only connected to a set of units in the previous layer. The first layers are organized as local receptive fields to extract some elementary visual features such as orientated edges, end-points and corners, independently of horizontal and vertical translation. Since the receptive fields of neighboring units overlap, a large number of connections should be trained, thus giving raise to overfitting. To reduce the number of free parameters and, consequently, the risk of overfitting, the units in a layer are organized in planes within which all the units share the same set of weights.

One extension of this approach relies on the use of the tangent distance in the training algorithm so as to incorporate into the learning process some tolerance with respect to small known transformations [11]. For example, in [12] the tangent distance was used for computing the input-output distance of autoassociators, obtaining a recognition of handwritten characters that is invariant with respect to a set of eight transformations (x- and y-translation, rotation, scaling, axis-deformation, diagonal-deformation, x- and y-thickness). An autoassociator ([13], pp. 55, 161) is an MLP with the same number of input and output units and less neurons in the hidden layer. During the training the network, that is forced to reproduce the input to the output, is fed only with samples of one class. A modular classifier can be built by feeding in parallel one autoassociator for each class, and including one decision module which interprets the distances between the output vectors and the input pattern. The lower the distance, the higher the similarity between the pattern and the corresponding autoassociator class.

In this chapter the integration of the tangent distance in the context of character clustering will be discussed in Section 4.1, whereas the layout clustering based on tangent distance will be addressed in Section 5.2.

2.3 Self-Organizing Map

The Self Organizing Map (SOM [1]) is an artificial neural network that performs clustering by means of unsupervised competitive learning. In the SOM the neurons are usually arranged in a two dimensional lattice (the feature map). Each neuron gets information from the input layer and from the other neurons in the map. The training samples are usually described by real vectors $x(p) \in R^n$, where p is the index of the sample ($p \in [1, N_P]$), and N_P is the number of training patterns. Each node in the SOM contains a model vector $m_i \in R^n$ that can be regarded as a prototype of the patterns in the cluster. During the learning, the network performs clustering and the model vectors are modified so as to reflect cluster similarity. The goal of the mapping is to represent the patterns in the source space by corresponding points in a lower dimensional target space (e.g. a 2D space). In this mapping the distance and proximity relationships should be preserved as much as possible.

The initial values of the model vectors, $m_i(0)$, can be selected at random or can be initialized in some orderly fashion, for instance arranging the vectors along a two-dimensional subspace spanned by the two principal eigenvectors of the input data. The two most widely used learning algorithms are the on-line and the batch. We report in Algorithm 1 a summary of the on-line version that is the basis of the modified algorithm that is described later in this chapter. N_N is the number of neurons and N_{Train} is the number of training cycles.

Algorithm 1: On-line training

for all $i = 1$ to N_N **do**
  Initialize $m_i(0)$
end for
$p \leftarrow 0$;
for all $j = 0$ to N_{Train} **do**
  $p \leftarrow p + 1$;
  Find the BMU ($b(x)$)

$$||x(p) - m_{b(x)}|| = min_i\{||x(p) - m_i(j)||\} \tag{1}$$

  for all $i \in$ neighborhood of $b(x)$ **do**
    Adapt model vectors

$$m_i(j+1) = m_i(j) + h_{b(x),i}(j)(x(p) - m_i(j)), \tag{2}$$

  where:

$$h_{b(x),i}(j) = \alpha(j)exp(-\frac{||r_i - r_{b(x)}||^2}{2\sigma^2(j)}), \tag{3}$$

  end for
  for all $i \notin$ neighborhood of $b(x)$ **do**
    $m_i(j+1) = m_i(j)$
  end for
  decrease $\alpha(j)$
  decrease $\sigma(j)$
  if $p = N_P$ **then**
    $p \leftarrow 0$
  end if
end for

The algorithm is based on two main stages.

1. The training vector, $x(p)$, is compared with all the model vectors $m_i(j)$ and the *best matching unit* (*BMU*) on the map is identified. The *BMU* is the node having lowest distance with respect to $x(p)$. The final topological organization of the map is heavily influenced by the distance function considered in this step. Usually, the Euclidean distance is considered and the best matching unit $b(x)$ is identified by Eq. (1).
2. The model vector of the *BMU* as well as those of its neighboring nodes are changed so as to "move" towards the current input pattern $x(p)$ according to Eq. (2), where $h_{b(x),i}$ is the neighborhood function implemented with a smoothing kernel that is time-variable and is defined over the lattice points. The neighborhood function is a decreasing function of the distance between the i-th and the $b(x)$-th models on the map grid. The extension

of the kernel is also decreasing monotonically during the iterations. A common neighborhood function is based on the Gaussian function, Eq. (3), where $0 < \alpha(j) < 1$ is the learning-rate factor that decreases with the iterations, $r_i \in \Re^2$ and $r_{b(x)} \in \Re^2$ are the locations of the neurons in the lattice, and $\sigma(j)$ defines the width of the neighborhood function that also decreases monotonically.

One advantage of the use of the SOM with respect to other clustering algorithms is the spatial organization of the feature map that is achieved after the learning process. Basically, more similar clusters are closer than more different ones. Consequently, the distance among prototypes in the output layer of the map can be considered as a measure of similarity between patterns in the clusters, and this feature can be considered in several application domains.

3 Unsupervised Learning in Document Analysis

The aim of unsupervised learning, or clustering, is to find some structure in a set of patterns without the interaction with an explicit teacher. In particular, the goal of clustering is to identify a finite and discrete set of groupings in the patterns. There is no universally agreed definition of clusters, but in general the similarity between objects in a group is required to be larger than the similarity between objects belonging to different clusters.

When using clustering algorithms two important issues should be addressed: the choice of an appropriate similarity measure (or distance function) and the identification of a rule to select the number of clusters to be found. It is important to remark that the use of different distance measures with a given clustering algorithm can give rise to different groupings with significant differences in the final results.

The applications of clustering in pattern recognition follow three main lines ([14], page 517) that are briefly analyzed in the following.

Clustering algorithms are designed to deal with *unlabeled data*, this feature is therefore particularly helpful in applications where the human validation of the pattern membership needed to train a supervised classifier is difficult. A suitable combination of supervised and unsupervised approaches relies on the preliminary identification of the clusters on the basis of unlabeled patterns. A class is attached to the clusters relying on a reduced number of labeled patterns. We will discuss a page layout classification method based on this idea in Section 5.2.

A second approach relies on the application of unsupervised learning to *extract features* to be used, for instance, as input to a discriminant classifier. Features computed by means of unsupervised clustering can be considered also for retrieval systems. An application of this approach in DIA is character clustering that is discussed in Section 3.3.

Exploratory data analysis is an application of clustering techniques that allows to discover natural orderings of the patterns and can be suitably integrated in complex pattern recognition systems. Several techniques described in this chapter can be used in this context.

In the rest of this section we analyze some recent applications of clustering, with a special emphasis on SOM-based approaches, in the field of document image analysis.

3.1 Symbol Thinning

Thinning algorithms are used, in pre-processing, to extract features based on the symbol skeleton. These features, when used in handwritten character recognition systems, allow a recognition independent from the stroke thickness. Ahmed proposed in [15] a clustering-based skeletonization algorithm (CBSA) implemented with SOM. The CBSA is composed by two main steps: in the first step some clusters, corresponding to adjacent pixels, are located in the input image; in the second step the skeleton is built connecting together the neighboring cluster centers. The clustering step is implemented in [15] by means of a particular SOM (the self-organizing graph) where the adjacency of neurons can change during learning. More recently, a topology-adaptive self-organizing neural network has been proposed for skeletonization [16]. The map grows in size over time and improves the performance with respect to a SOM having a fixed dimension. The system can handle rotated patterns and works with binary and gray level images. A similar approach is described in [17], whereas a multi-scale skeletonization method based on SOM is described in [18].

3.2 Layout Analysis

Layout analysis is executed after the pre-processing with the aim of extracting homogeneous regions from the document image assigning a semantic meaning to each region. When dealing with color documents, the layout analysis consists in the identification of regions with uniform color. To this purpose, different colors are first identified and then pixels having the same color are grouped together. The color identification is frequently addressed with clustering in the color space. Global color clustering methods are described in [19] and in [20]. In [19] the clustering is obtained by using the Euclidean Minimum Spanning Tree (EMST) in color space, that is claimed to provide better results than the k-means algorithm. In [20] three-dimensional morphological operators are adopted to erode the regions in the color space. Iterating this erosion the cluster centers are identified. In [21], a color grouping algorithm, the *LOCUSi* method, is proposed. This algorithm extracts the clusters in the RGB space and is based on the analysis of the expected shape of clusters in this space.

Clustering at the spatial level is adopted in [22] for document image segmentation. In this case, the text lines are represented by means of the proposed *interval encoding* and subsequently clustered with the k-means algorithm. A hierarchical clustering algorithm is adopted in [23] for grouping closest connected components in an OCR system that is aimed at processing mathematical equations.

3.3 Character Clustering

Character clustering is the basis of some *character-like coding* techniques where similar objects, usually corresponding to characters, are clustered on the basis of their shape. Each word is then represented by concatenating the codes assigned to the individual objects. In character-like coding, in contrast with OCR, no alphabetical class is assigned to symbols. The query is encoded with the same algorithm used during the indexing and is compared with the indexed words taking into account various matching approaches. For instance, in [24] the words are represented with strings and compared by means of an inexact string matching technique. By adopting this symbolic representation indexed words can be sorted allowing users to retrieve words printed with different fonts as well as to satisfy partial-match queries.

Character clustering is applied also in some document image compression algorithms that first group together similar symbols roughly corresponding to characters. The characters and the background image are afterwards compressed with specific algorithms [25, 26]. In [26] the characters are compressed by extracting the *marks* (connected components) in each page, and a library of marks is built. Each mark is then replaced with a pointer to the closest item in the library. The library is obtained with a simple template matching clustering algorithm that provides good results when processing documents with a small variability of fonts and a low level of noise.

A related method is addressed in [27], where a hierarchical clustering algorithm is used to enhance degraded document images. Bitmaps of the symbols belonging to each cluster are identified and an average symbol for each class is computed. By replacing the original symbols with the average bitmap it is possible to render the document at arbitrary resolutions and enhance degraded document images.

3.4 Handwriting

Clustering algorithms, and in particular the SOM, have been widely used for the recognition of isolated handwritten digits. A three-stage recognition system dealing with handwritten numerals is described in [28]. The first stage is based on a SOM whose aim is to create prototypes representing parts of the characters that capture the similarities between digits. The gradual shape variations are represented in the feature map, since closer prototypes generally represent similar patterns. In the second stage a suitable function is used to convert the distances between the input (unknown) pattern and the prototypes into membership values so as to obtain a fuzzy membership. The third stage performs the final classification by means of a fully connected MLP.

A three-dimensional SOM for unconstrained handwritten numeral recognition is described in [29]. The third dimension is defined by taking into account 11 layers of 9x9 SOMs. The neighborhood consists of the units that are within a cube centered on the BMU.

A hybrid handwritten word recognition system using SOM, discrete HMM, and evolutionary programming has been proposed in [30]. In this system the purpose of the SOM clustering is to partition the feature space into a set of codeword vectors to limit the number of observation symbols in discrete HMM training. The weight vectors of the trained map are used as codewords to describe the word frames, that are represented by a sequence of 2-dimensional codeword positions in the map. The neighborhood information preserved by the SOM is used for smoothing the trained HMM parameters.

In [31] handprinted character recognition is addressed by representing the characters with a fixed number of ellipses that capture the local structure of the strokes. Each ellipse is described with the center, the length and the orientation of the mayor axis. A modified SOM is used to accomplish one elastic matching and find the correspondence between the feature points. When the network converges, a mapping between the input feature patterns and the neuron support is obtained.

3.5 Structure Adaptive Classifiers

Structure adaptive classifiers automatically adjust their structure to the uneven distribution of classes in the pattern space. These classifiers are particularly useful in applications where a large number of classes is addressed, such as oriental character recognition. In this context, clustering algorithms are frequently used for grouping together most similar patterns in modular classifiers, that are mostly based on parallel and serial combinations [32].

In serial combinations [2] the classifiers are arranged in a list. For each unknown pattern the first classifier decides if a further refinement of the decision is required by one or more subsequent classifiers. The individual classifiers are usually applied in increasing order of complexity and the "simplest" symbols are recognized first, whereas the more difficult ones are processed by next classifiers. Some methods analyze the recognition results of the first classifier in order to identify the most confused classes. Examples of this approach are proposed in [33] and in [34]. In [33] maxima in the confusion table are automatically identified. In [34] the confusion table of the first stage, a feature-based OCR, is manually analyzed to find nine sets of classes corresponding to most confused groups of characters (e.g. {S,5,6}; {B,D,O,8}), which are recognized by appropriate MLPs. Alternatively, clustering algorithms can be explicitly applied to the characters belonging to the learning set. For instance, Su *et al.* [35] cluster together similar characters and organize the classifiers so that most indistinguishable classes (e.g. “4” and “A”) are recognized by the last networks. Each network is subsequently trained to recognize the patterns belonging to its classes, and to reject the patterns that should be recognized by the other classifiers.

In hierarchical combinations, the classifiers are arranged in a tree. A structure adaptation method for the recognition of Korean characters using a self organizing neural tree has been proposed in [36]. The basic idea is to automatically find a network structure and size suitable for the classification of large-set and complex patterns. The tree-structured network is based on subnetworks that are logically connected to nodes in the previous level. As a matter of fact, subnetworks define with higher resolution regions of the pattern space containing more patterns. Another hierarchical structure adaptation SOM for the recognition of handwritten digits is proposed in [37].

3.6 Text

The WEBSOM [38] is a SOM-based system that is able to organize large document collections according to textual similarities. The feature vectors describing the documents are statistical representations of their vocabularies. One peculiarity of the system described in [38] is the scaling up of the SOM algorithm in order to process large collections of high-dimensional data. In the experiments 6,840,568 patent abstracts have been mapped onto a 1,002,240-node SOM. To reduce the feature vector size some random projections of weighted word histograms are computed, obtaining 500-dimensional vectors. A similar application has been described in [39].

In [40] the documents are represented with feature vectors containing the occurrences of 489 terms in each document so as to reveal the document similarity. One hierarchical feature map is then built considering this document representation. This hierarchical representation is claimed to be well suited for text archive organization.

The SOM clustering is used in [41] to build a lexical analyzer designed to focus on a very limited sub-set of the whole dictionary. Each string S is represented by a vector $X = [X_0, X_1, \ldots, X_{25}]$ where X_i corresponds to the number of characters of class C_i ($C_0 =$ ‘A’ $, \ldots, C_{25} =$ ‘Z’) in the string S (the anagrams of S share the same representation). The map is a two dimensional array, organized as a tore to avoid singularity effects on the sides. The neighborhood relations in the projected space of the map are used to define a short list of hypotheses considered for spell checking.

3.7 Discussion

In this section, we analyzed some applications of unsupervised clustering in the domain of DIA. The main advantage of SOM clustering, with respect to other clustering algorithms like k-means, is the spatial organization of the neurons that reflects cluster similarity into prototype proximity in the 2D space. The distance among prototypes in the SOM map can therefore be considered as an estimate of the similarity between objects belonging to the clusters. It is important to remark that in general the use of SOM for multivariate data projection on large data sets is not advisable due to the high computational cost [42]. However, usually a reduced number of objects (obtained from few documents) are used to compute the mapping that is subsequently adopted to label all the patterns to be processed.

In the next sections we describe the use of the SOM at three processing levels: the character clustering, the word clustering, and the layout clustering, with applications to word retrieval, to document retrieval and to page classification.

4 Word Indexing

Word indexing, that aims at a fast retrieval of words in a document collection, can either process the output of OCR engines or directly work on the document image. Several strategies have been proposed to deal with OCR errors [43, 44]. In most approaches the uncorrected OCR output is used for text indexing and the words are compared with the query by means of *string edit distance* algorithms. This strategy has been improved by modifying the edit costs for the most common OCR errors (e.g. [45]).

When the use of OCR is not advisable, either due to the low quality of images or to the presence of non-standard fonts, then image-based word retrieval is a viable alternative. Two main strategies have been considered in this framework: character-like coding and holistic word representation.

In methods based on character-like coding some objects, that potentially correspond to characters, are extracted from each word. The word is then represented by concatenating the codes assigned to the objects. In so doing similar words share most of the codes.

In the holistic approach each word image is encoded by means of some global features (e.g. the number of characters or the number of ascenders/descenders) [46]. A particular case of holistic word representation is zoning (e.g. [47]) that consists of overlapping the word image with a fixed-size grid. Suitable features, such as the density of black pixels, are computed in each grid region. Most keyword spotting methods are based on this kind of representation. For instance, signal processing techniques are used in [48] to allow scale and translation invariance. Holistic shape features for handwritten word image retrieval are described also in [49] where a training set is used to learn a joint probability distribution between word features and their transcriptions.

In this section, we discuss the use of SOM clustering to perform word indexing with two approaches: a character-based and a holistic one. In both cases, during the indexing each document image is first processed with a layout analysis tool that identifies the text regions and extracts the words using an RLSA-based algorithm. To perform the word retrieval a query word is inserted into one text field in the user interface. A word image is computed by processing the query word with the LaTeX software and this image is used to compute a suitable word representation either based on character clustering or on holistic features. Lastly, the indexed words are sorted estimating their similarity to the query by appropriate processing steps that are described in detail in [50] for the character-based encoding and in [51] for the holistic-based approach. In this section we focus our attention on the peculiarities of character and word clusterings.

4.1 Character Clustering

In word indexing based on character coding each word is first split into Character Objects (*CO*) that in most cases correspond to isolated characters (sometimes one *CO* can comprise two touching characters). The *CO*s extracted from a few random pages are used to compute appropriate collection-specific character prototypes by means of SOM clustering. To improve the clustering we include the tangent distance into the on-line training algorithm as detailed in the following.

Tangent Distance

The SOM training algorithm generally relies on the Euclidean distance to compare the training patterns and the model vectors. Unfortunately, the Euclidean distance between two patterns is very sensitive to small transformations and a limited displacement of a character can give rise to a large value of the distance, since many pixels in the two patterns are no longer aligned. To address this problem, we propose the use of the tangent distance in the SOM training algorithm.

The invariance of the training algorithm with respect to transformations of the patterns (Section 2.2) can be achieved with techniques based on invariance by structure, invariance by feature extraction, and invariance by training. The use of the tangent distance into a training algorithm can be considered an invariance by structure technique whose goal is the incorporation in the distance function of the tolerance with respect to small transformations in the pattern space.

We shortly outline in the following the tangent distance principle. Readers interested in more details can refer, for instance, to [11]. Let us suppose to transform a pattern P with a non linear transformation t that is controlled by one parameter β (for instance t can be the rotation of the pattern with the angle β). In the pattern space the set of all the transformed patterns $S_P = \{x \mid \exists \; \beta \; x = t(P, \beta)\}$ can be regarded as a one-dimensional curve parametrized by β. In the general case several transformations are combined together. The possible transformations can be characterized by a vector of n parameters ($\overrightarrow{\beta}$), that is expected to describe all the possible deformations that can be applied to the patterns. The patterns in S_P are now arranged in a manifold and two objects P and Q in the same class are expected to generate identical manifolds $S_P \equiv S_Q$. Two problems should be addressed to compute the distance between patterns taking into account these manifolds.

The first problem is related to the identification of appropriate transformations that should generate realistic patterns. In the case of handwritten characters, some standard transformations have been proposed, such as rotation, translation, and line thickening. However, real patterns are usually subjected to transformations that are difficult to model and actual patterns usually stay close to the manifold but are not perfectly described by it.

The second problem is related to the computational cost required to evaluate the distance between two manifolds. The solution addressed by the tangent distance approach is based on a local approximation of one manifold by means of the hyperplane tangent to it in the point P. The tangent plane is defined with the linear combination of the n vectors computed by applying small independent transformations to the original pattern.

The tangent distance between patterns P and Q can be computed by first defining the tangent planes to P and Q:

$$\begin{array}{c} T_P(\overrightarrow{\beta_P}) = P + L_P\overrightarrow{\beta_P} \\ T_Q(\overrightarrow{\beta_Q}) = Q + L_Q\overrightarrow{\beta_Q} \end{array} \quad (4)$$

where L_P and L_Q are the matrices containing the tangent vectors which are usually pre-computed. The *double-sided* tangent distance is defined as the minimum Euclidean distance between the tangent planes:

$$TD(P,Q) = min_{x \in T_P, y \in T_Q} ||x - y||^2. \quad (5)$$

Computing the tangent distance amounts to solve a linear least squares problem (e.g. see [11]). In some cases it is simpler to use the *one-sided* tangent distance where the minimum distance between one pattern and the plane tangent to the other pattern is computed:

$$TD1(P,Q) = min_{x \in T_P} ||x - Q||^2. \quad (6)$$

Character Clustering with Tangent Distance

Character object labeling is made on the basis of the prototypes that are obtained by clustering the *CO*s contained in some pages of the collection to be indexed. Each *CO* image is scaled to fit a fixed size grid obtaining an 80-dimensional feature vector that is used as input to the clustering algorithm. The SOM-based character clustering computed using the Euclidean distance is very sensitive to small local transformations of the character. To reduce these problems in the proposed SOM_TD model we use the tangent distance instead of the Euclidean distance during the SOM training. To speed-up the

Fig. 1. Two SOMs built on the Gothic data-set. Left: map computed with the SOM_TD. Right: map computed with the standard SOM training

training we pre-compute the tangent vectors for each pattern (CO) in the training set. Even if the estimation of the tangent vectors is computationally expensive, it is important to remark that the vectors are computed only once for each pattern. During the training, we use the one-sided tangent distance, Eq. (6), to identify the BMU for each pattern, replacing Eq. (1) with:

$$||x(p) - m_{b(x)}|| = min_i\{TD1(x(p), m_i(j))\}. \tag{7}$$

In the next section we compare the results obtained performing the character clustering with the standard SOM training algorithm and with the SOM_TD model.

Experimental Results

To evaluate the effectiveness of the proposed SOM training model we used two data-sets having complementary features described in [50]: the Gothic and the French one. The Gothic data-set contains 14 pages printed with a Gothic font that is not recognized by current off-the-shelf OCR packages. The second data-set contains more than 600 pages printed with a standard font.

The tangent distance is a technique appropriate for data-sets having a small number of training examples. It is therefore not surprising that the experimental results on the French data-set do not show significant differences between the standard and the tangent SOM. Therefore, we will not discuss results for this collection in the chapter.

On the other hand, when dealing with the Gothic data we have some interesting results that it is worth to analyze. A pictorial comparison of the maps computed with the two methods is shown in Figure 1. The quantitative comparison of the two approaches is obtained by testing the retrieval system described in [50] without the use of the word alignment method that is described in that paper. The precision-recall plots shown in Figure 2 are

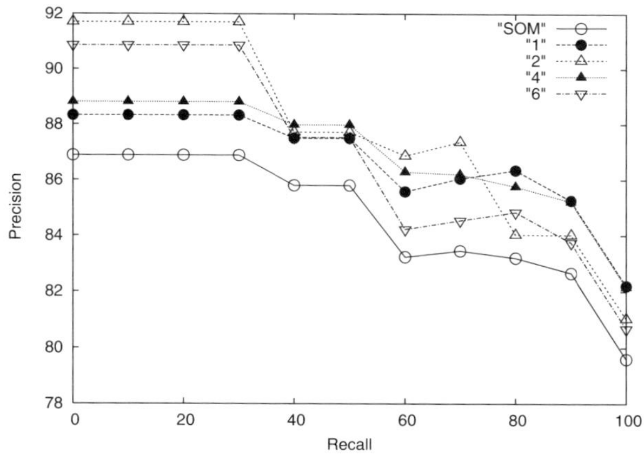

Fig. 2. Comparison of precision-recall plots with and without the use of the tangent distance. The transformations 1,2, 4, and 6 correspond to vertical shift, hyperbolic, scale and line thickness, respectively

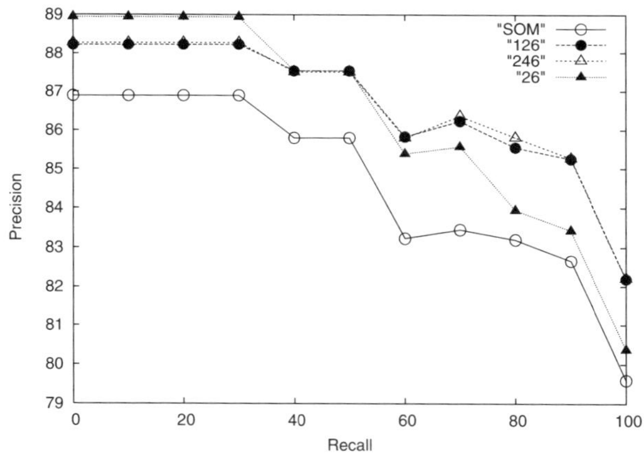

Fig. 3. Comparison of different combinations of various tangent vectors. For instance, the black circle corresponds to the combination of vertical shift, hyperbolic, and line thickness

computed by running the word retrieval system with 26 query words representative of both frequent and rare words and averaging the single plots. The five plots correspond to the standard map ("SOM") and to the results obtained with the SOM_TD model using some of the most important transformations. The tangent vectors computed with these transformations allow us to obtain better results with respect to the standard model, since the precision-recall plots are always better than the SOM one. Figure 3 shows the results that can be obtained with few combinations of the transformations analyzed in Figure 2.

Fig. 4. Graphical representation of a WordSOM

4.2 Holistic Word Clustering

In holistic word indexing the SOM is used to cluster together most similar words (from a graphic point of view). The word images are linearly scaled to appropriate normalized dimensions, obtaining a vectorial representation where the value in each vector item is the average gray level of the pixels belonging to the corresponding grid cell. To allow a uniform representation for variable size words we compute six maps. Each map contains the words having aspect-ratio in a pre-defined interval. The main drawback of this approach is the large vector size (hundreds of items) that is reflected into a long training time. However, it should be remarked that the training is performed during the indexing, that can be considered an off-line stage.

In Figure 4 we show a *WordSOM* computed by processing the pages in a book belonging to the French data-set previously mentioned. A deeper analysis of the contents of four neurons is depicted in Figure 5 where the words in the clusters are ordered from top to bottom on the basis of the distance with respect to the model vector of the neuron. As we can expect, the farthest words are generally loosely related with those closer to the model vector.

The large vector size affects the retrieval performance for problems related to the curse of dimensionality. To speed-up the search in high dimensional spaces we proposed in [51] a method based on the combination of SOM clustering with the search in a lower dimensional space obtained by the PCA projection. The main steps performed in the word retrieval are as follows. We first identify the three clusters closer to the query. In the second step we search the most similar words sorting the PCA-projected vectors. Lastly, to merge the three lists and refine the final ranking, we compute the distance in the original space between the query word and the most similar words in the three lists. Some detailed experiments on the use of the SOM for holistic word indexing are described in [51].

Fig. 5. Contents of four neurons of the *WordSOM* shown in Figure 4. The words are sorted on the basis of the distance from the cluster centroid

5 Page Indexing

In this section, we discuss two approaches aimed at indexing documents at the page level. The first application (Section 5.1) targets the page retrieval considering text similarities computed combining word image clustering and the *tf-idf* weighting. In Section 5.2 we explore the SOM clustering of the page layout for page classification applications. In this approach, we analyze also the use of the tangent distance between vectorial representations of the page layout in order to improve the recognition rate for data-sets containing few labeled pages.

5.1 Document Retrieval

The proposed document retrieval based on word image clustering originates from one classical approach in text-based Information Retrieval: the *vector model* (see [52], Chapter 2). This approach is based on a vectorial description of the document contents where the vector items are related to the occurrences of index terms, usually corresponding to words, in the document. Vector values are weighted to give more importance to most discriminant terms. To this purpose one common approach relies on the well known *tf-idf* weighting scheme. The basic idea is that index terms that are present in many documents of the collection should have a low weight since their presence is not discriminant. With the *tf-idf* approach the weight assigned to the k-th word in the document D_i is computed by:

$$w_{i,k} = f_{i,k} \cdot log\left(\frac{N}{n_k}\right), \tag{8}$$

where $f_{i,k}$ is the frequency of the k-th word in D_i normalized with respect to the maximum word frequency in D_i, N is the total number of documents, and n_k is the number of documents containing the k-th word.

The vector model has been designed to process textual documents where the word identification and clustering (with possible stemming and stopword removal) is quite straightforward. The document retrieval discussed in this section relies on the use of the WordSOM clusters in lieu of ASCII words in the $tf-idf$ weighting. Basically, a document is represented by a vector whose items correspond to the neurons of the six maps computed (Section 4.2). Each vector item contains the number of words in the document that are assigned to the corresponding neuron (i.e. the words having that neuron as BMU). In analogy with the vector model, we apply the $tf - idf$ weighting to this representation. It is worth to remark that the size of the overall vector (on the average 1,800 items) is smaller than the typical size of the dictionaries considered with the vector model (for instance for the well known *Reuters 21578* corpus the dictionary contains around 19,000 terms). After the indexing, a weighted vector is associated to each document.

To perform document retrieval a vector is computed from the query document taking into account the neurons assigned to each word in the page. The similarity is evaluated by comparing the query (q) with each indexed document (d) by using the *cosine of the angle* between the vectors:

$$sim(q,d) = \frac{\sum_{i=0}^{n-1}(q_i \cdot d_i)}{\sqrt{\sum_{i=0}^{n-1} q_i^2} \cdot \sqrt{\sum_{i=0}^{n-1} d_i^2}} \tag{9}$$

To obtain a global ranking of the indexed documents we compute the similarity of all the documents with respect to the query and sort the documents on the basis of the measure computed in Eq. (9).

Rank	Page	Sim.	Most frequent words
1	448	0.1624	*carbonate, carbon, sulfate* (*carbonate, coal, sulphate*)
2	446	0.1492	*four* (*oven*)
3	1007	0.1368	*mouvement, cylindre, ouvrier, roues* (*movement, cylinder, worker, wheels*)
4	822	0.1348	–
5	1164	0.1320	similar to 1007
6	1254	0.1313	–
7	455	0.1309	*ammoniaque, sulfate* (*ammonia, sulfate*)

Table 1. Most frequent words in the top ranked pages of the example query for document retrieval

Experimental Results

The experiments described in this section are made on two books (containing 1280 pages) that are part of an encyclopedia addressing machineries and techniques of the industry of the XIXth Century [1]. The images are quite clean and the OCR works well on these documents. The interest for this data-set lies on the homogeneity of the contents of the pages in each chapter, so that the evaluation is simplified.

An accurate evaluation of document retrieval systems requires a judgment of the relevance which is provided by human experts. In our application this information is not available and we evaluated the system in two ways. First, we made a global test by performing several queries for each chapter. Second, we carefully analyzed some queries by checking the contents of the neurons providing the highest contribution to the similarity measure (Eq. 9).

We describe in the following the results obtained with one query page by highlighting the most important neurons for each selected page. The query page (number 452) belongs to chapter 4 *'sodium and potassium'* and describes one specific machinery. Therefore, the page contains several technical terms (e.g. *cylinder*, *wheels*, *oven*, *movement*, *combustion*, *handle*, *heat*, *worker*, *rotation*). This machinery is used for the production of sodium and the following words are contained in the page: *sulfate*, *reaction*, *carbonate*, *chaux* (lime). Table 1 reports the first ranked pages together with the most frequent words contained in the clusters providing the higher contribution to Eq. (9). Pages 448 and 455 belong to the same chapter as the query page. Page 446 describes a machinery very similar to the one addressed in the query page. Also the machineries described at page 1007 and 1164 are very similar to the query, but are designed to work with other materials. Another interesting point is the presence among the top ranked pages of few pages belonging to chapter 4

[1] *Les Merveilles De l'Industrie*: downloaded from the web site of the *National Library of France*.

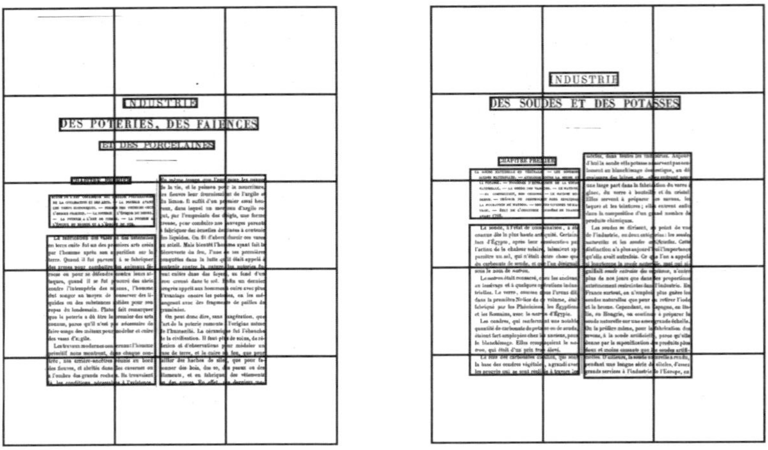

Fig. 6. Two pages of the Issue2 class. We show the regions identified by the layout analysis and the grid overlapped to the pages

(pages 419-480), since the query page contains few references to sodium and potassium.

5.2 Layout Clustering

Another application of SOM clustering is the page layout classification. The approach proposed in this section is particularly useful when dealing with data sets where only few pages have been manually annotated, and therefore the training of a discriminant classifier can be difficult.

Page classification methods generally represent the page layout either with fixed-size feature vectors [53, 54] or with graphical structures (e.g. graphs and trees [55]). In [55] we compared some approaches for SOM-based page clustering relying on the MXY-tree page representation. Graphical representations usually suffer important non-linearities and small transformations (e.g. translations) in the image space can give rise to significant changes in the subsequent representation.

In this chapter we propose the use of the tangent distance for layout clustering. This clustering is subsequently used to perform the page classification. We propose a vectorial page representation that smoothly depends on small transformations in the image space. The latter is a requirement to adopt the tangent distance approach.

A layout analysis tool is used to extract the homogeneous regions in the pages. Each page is then represented with a fixed-size feature vector obtained by computing appropriate features in the cells defined by a regular grid superimposed to the page. Two segmented pages of the same class are shown in Figure 6 together with the overlapping grid. From the figure we can notice

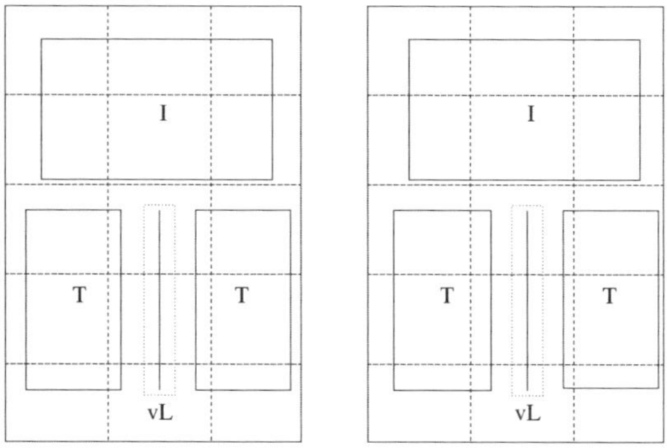

Fig. 7. Simulation of a horizontal displacement of a page. For each page we show the layout regions and the overlapping grid

that the right page has a similar layout but is vertically shifted. For each cell we compute the percentage of its area that is covered by text, image and line regions, respectively. For horizontal and vertical lines we assign a virtual area around each line (see the dotted area around vL in Figure 7) and evaluate the percentage of the cell covered by the union of all the virtual areas. In this way broken lines provide a contribution similar to continuous ones.

Two of the most important transformations that should be addressed in the layout clustering are horizontal and vertical displacements. To simulate these transformations in the tangent distance paradigm we compute, for each training page, the new representations that are obtained by slightly shifting the grid in the horizontal and vertical directions (an example of horizontal displacement is shown in Figure 7). By means of these transformed page representations we can compute a sort of tangent vectors that can be used to train the SOM_TD with the method described in Section 4.1.

Experimental Results

In this section we discuss the performance of the SOM_TD model for page layout clustering. The clustering is afterwards used to perform a layout-based page classification with particular interest on data-sets having a reduced number of labeled training patterns. Two experiments have been performed to analyze the approach under different conditions. In the first experiment we considered a data-set that is typical of digital library repositories that contains a digitized Encyclopedia. The second experiment deals with a collection of commercial invoices.

The Encyclopedia data-set is composed by 6 books containing 4035 pages that have been assigned to more than ten classes[2]. We made our experiments considering the seven most important classes (Figure 8). The SOM training

[2] This data-set is a superset of the one used in Section 4.1.

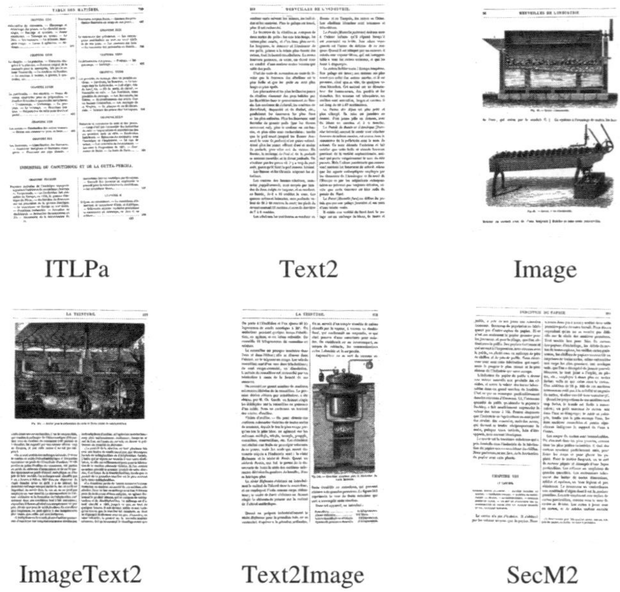

Fig. 8. Examples of pages of the main classes considered in the layout retrieval

10x8		12x10		15x10	
SOM	**SOM_TD**	**SOM**	**SOM_TD**	**SOM**	**SOM_TD**
55.59	67.51	59.91	60.97	70.74	71.41

Table 2. Encyclopedia Data Set: average recognition rate for various sizes of the SOM

has been made with the 617 pages of the first book. Figure 9 shows one SOM trained on this data-set. Some map regions containing similar pages have been identified in order to make easier the visual analysis of the map. After training the SOM map either with the standard or with the SOM_TD algorithm we label each neuron on the basis of the most frequent class among the pages in the training set firing the neuron.

During the test (that is made considering the remaining five books) we classify all the pages in the test set belonging to the nine classes of interest. The class assigned to each page is the label of the closest neuron. The comparison of the two approaches is made computing the recognition rate for each class when using the standard SOM and the SOM_TD. To reduce the variations in the training we used the same starting map for both learning methods. In Table 2 we compare the average recognition rate for the two approaches considering various map sizes. We can notice that in all the cases the tangent map provides better results. Moreover, larger maps allow us to obtain higher recognition rates. To further investigate these results we compare, in

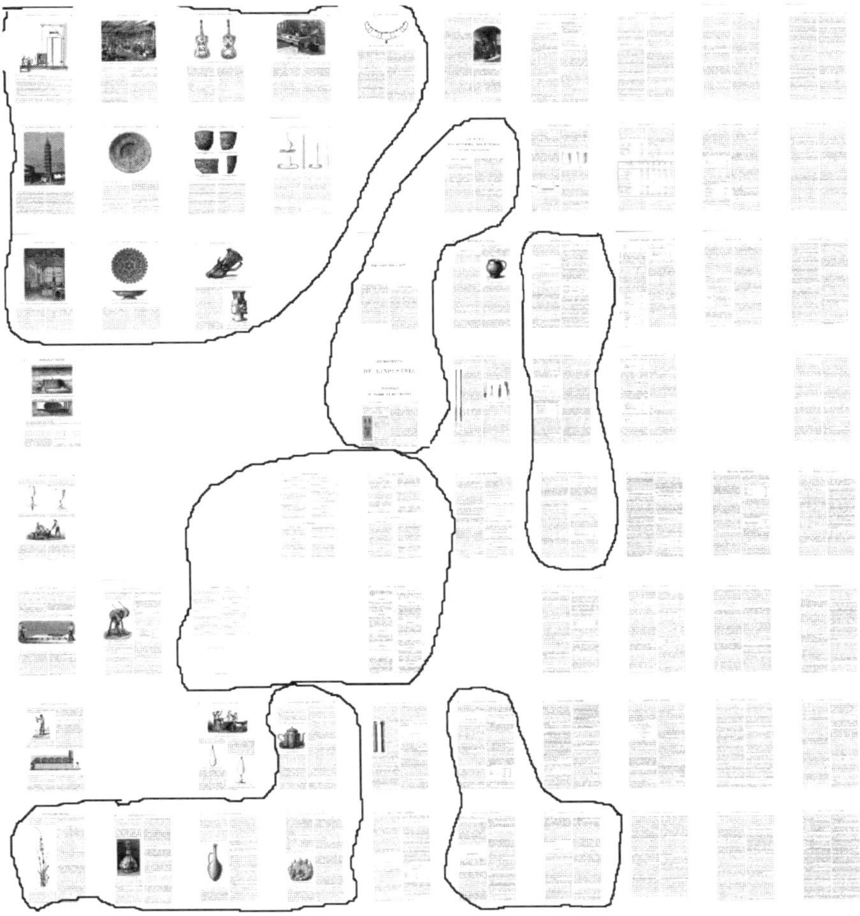

Fig. 9. A graphical representation of the layout SOM computed for the Encyclopedia data-set. Some regions containing similar pages are shown on the map

Table 3, the recognition rate for each class when using the largest map. We can observe that the tangent map provides better results for all the classes with the exception of the classes **Text2Image** and **Text2**. As a matter of fact pages of the two classes are sometimes confused. The reason for this behavior is probably due to the fact that the **Text2** class is the most populated one and has a very simple layout (text on two columns). In this case the use of tangent vectors is not useful since several training patterns already exist, and the addition of local distortions can be confusing for the clustering.

The experiments performed on the invoice data-set have been made with the aim of evaluating the effectiveness of the proposes SOM_TD model when dealing with a large collection of patterns that are labeled only in a small percentage.

	SOM	SOM_TD
ImageText2	81.33	82.14
Text2Image	69.08	67.05
SecM2	36.30	40.84
Image	96.72	96.72
Issue2	84.21	87.50
ITLPa	28.00	32.00
Text2	96.03	93.65

Table 3. Encyclopedia Data Set: average recognition rate of SOM and SOM_TD for the main classes

	12x10	15x10	18x10
SOM	59%	52%	59%
SOM_TD	58%	66%	61%

Table 4. Invoice Data Set: average recognition rate for various sizes of the SOM

The SOM training and labeling has been performed in two steps: first we trained a map with the SOM_TD algorithm considering all the patterns belonging to the training set. In the second step we labeled some neurons in the map on the basis of a majority voting computed with 64 pages belonging to 15 classes. The classification is obtained, as before, assigning each unknown page to the class of the closest neuron. Since a reduced number of labeled samples is available we made the experiments with a leave-one-out approach. From Table 4, containing the average recognition rate for the invoice data-set, we can notice that the best results are obtained with the 15x10 map when using the SOM_TD training algorithm. In this case we needed larger maps with respect to the Encyclopedia data-set since there are more classes anf the pages not belonging to the 15 labeled classes are more than a half of the whole data-set.

6 Conclusions

This chapter addresses the use of unsupervised learning in DIA applications with a special emphasis on SOM-based clustering. Clustering techniques are particularly appropriate in applications where a large number of labeled training samples is not available, or when there is no need to assign patterns into a pre-defined number of classes (for instance in document image retrieval applications). A well known technique for enhancing classifiers trained with a reduced number of training patterns is the tangent distance that we use, in this chapter, for SOM training.

The application of SOM clustering is discussed at three main processing levels: character, word, and layout clustering. In particular, the proposed

SOM_TD approach is evaluated on two applications: the word retrieval based on character clustering and the layout classification based on page clustering. In both cases the experimental results confirm the hint that the tangent distance is particularly suited when dealing with data sets having a small number of labeled training patterns.

References

1. Kohonen, T.: Self-organizing maps. Springer Series in Information Sciences (2001)
2. Marinai, S., Gori, M., Soda, G.: Artificial neural networks for document analysis and recognition. IEEE Transactions on PAMI **27**(1) (2005) 23–35
3. Poggio, T., Girosi, F.: Networks for approximation and learning. Proceedings of the IEEE **78**(9) (1990) 1481–1497
4. Hornik, K., Stinchcombe, M., White, H.: Multilayer feedforward networks are universal approximators. Neural Networks **2** (1989) 359–366
5. Rumelhart, D.E., Hinton, G.E., Williams, R.J.: Learning internal representations by error propagation. In Rumelhart, D.E., McClelland, J.L., eds.: Parallel Distributed Processing. Volume 1. MIT Press, Cambridge (1986) 318–362
6. Kohonen, T.: The self-organizing map. Proceedings of the IEEE **78**(9) (1990) 1464–1480
7. Bernard, E., Casasent, D.: Invariance and neural nets. IEEE Transactions on Neural Networks **2**(5) (1991) 498–508
8. Avi-Itzhak, H., Diep, T., Garland, H.: High accuracy optical character recognition using neural networks with centroid dithering. IEEE Transaction on PAMI **17**(2) (1995) 218–224
9. Oliveira, L., Britto, A.S., Sabourin, R.: A synthetic database to assess segmentation algorithms. In: Int'l Conference on Document Analysis and Recognition. (2005) 207–211
10. LeCun, Y., Bottou, L., Bengio, Y., Haffner, P.: Gradient-based learning applied to document recognition. Proceedings of the IEEE **86**(11) (1998) 2278–2324
11. Simard, P.Y., LeCun, Y., Denker, J.S.: Memory-based character recognition using a transformation invariant metric. In: Int'l Conference on Pattern Recognition. (1994) 262–267
12. Schwenk, H., Milgram, M.: Transformation invariant autoassociation with application to handwritten character recognition. In: Proc. NIPS. (1996) 991–998
13. Rumelhart, D.E., McClelland, J.L., the PDP Research Group: Parallel Distributed Processing: Explorations in the Microstructure of Cognition. Volume 1. MIT Press, Cambridge (1986)
14. Duda, R.O., Hart, P., Stork, D.G.: Pattern Classification. John Wiley & sons (2001)
15. Ahmed, P.: A neural network based dedicated thinning method. PRL **16**(6) (1995) 585 – 590
16. Datta, A., Parui, S.K., Chaudhuri, B.B.: Skeletonization by a topology-adaptive self-organizing neural network. Pattern Recognition **34** (2001) 617–629
17. Sasamura, H., Saito, T.: A simple learning algorithm for growing self-organizing maps and its application to skeletonization. In: Int'l Joint Conference on Neural Networks. Volume 1. (2003) 787–790

18. Palenichka, R.M., Zaremba, M.B.: Multi-scale model-based skeletonization of object shapes using self-organizing maps. In: Int'l Conference on Pattern Recognition. (2002) 143–146
19. Zhou, J., Lopresti, D.: Extracting text from WWW images. In: Int'l Conference on Document Analysis and Recognition. (1997) 248–252
20. Park, S., Yun, I., Lee, S.: Color image segmentation based on 3-D clustering: Morphological approach. Pattern Recognition **31**(8) (1998) 1061–1076
21. Worring, M., Todoran, L.: Segmentation of color documents by line oriented clustering using spatial information. In: Int'l Conference on Document Analysis and Recognition. (1999) 67–70
22. Hu, J., Kashi, R., Wilfong, G.: Document image layout comparison and classification. In: Int'l Conference on Document Analysis and Recognition. (1999) 285–288
23. Suzuki, M., Tamari, F., Fukuda, R., Uchida, S., Kanahori, T.: INFTY: An integrated OCR system for mathematical documents. In: Document Engineering. (2003) 95–104
24. Lu, Y., Tan, C.: Information retrieval in document image databases. IEEE Transactions on Knowledge and Data Discovery **16**(11) (2004) 1398–1410
25. Haffner, P., Bottou, L., Howard, P.G., LeCun, Y.: DjVu: analyzing and compressing scanned documents for Internet distribution. In: Int'l Conference on Document Analysis and Recognition. (1999) 625–628
26. Witten, I.H., Moffat, A., Bell, T.C.: Managing gigabytes: compressing and indexing documents and images. Academic Press (1999)
27. Hobby, J.D., Ho, T.K.: Enhancing degraded document images via bitmap clustering and averaging. In: Int'l Conference on Document Analysis and Recognition. (1997) 394–400
28. Chiang, J.H., Gader, P.: Recognition of handprinted numerals in VISA card application form. MVA **10**(3) (1997) 144–149
29. Reddy, N.S., Nagabhushan, P.: A three-dimensional neural network model for unconstrained handwritten numeral recognition: a new approach. Pattern Recognition **31**(5) (1998) 511–516
30. Dehghan, M., Faez, K., Ahmadi, M.: A hybrid handwritten word recognition using self-organizing feature map, discrete HMM, and evolutionary programming. In: Int'l Joint Conference on Neural Networks. (2000) 515–520
31. Liou, C.Y., Yang, H.C.: Handprinted character recognition based on spatial topology distance measurement. IEEE Transaction on PAMI **18**(9) (1996) 941–945
32. Rahman, A., Fairhurst, M.: A new hybrid approach in combining multiple experts to recognise handwritten numerals. PRL **18**(8) (1997) 781–790
33. Teo, R.Y.M., Shingal, R.: A hybrid classifier for recognizing handwritten numerals. In: Int'l Conference on Document Analysis and Recognition. (1997) 283–287
34. Wang, J., Jean, J.: Resolving multifont character confusion with neural networks. Pattern Recognition **26**(1) (1993) 175–188
35. Su, H., Wang, W., Li, X., Xia, S.: Hierarchical neural network for recognizing hand-written characters in engineering drawings. In: Int'l Conference on Document Analysis and Recognition. (1995) 46–49
36. Song, H.H., Lee, S.W.: A self-organizing neural tree for large-set pattern classification. In: Int'l Conference on Document Analysis and Recognition. (1995) 1111–1114

37. Cho, S.B.: Neural-network classifiers for recognizing totally unconstrained handwritten numerals. IEEE Transactions on Neural Networks **8**(1) (1997) 43–53
38. Kohonen, T., Kaski, S., Lagus, K., Salojarvi, J., Honkela, J., Paatero, V., Saarela, A.: Self organization of a massive document collection. IEEE Transactions on Neural Networks **11**(3) (2000) 574–585
39. O'Neil, P.: An incremental approach to text representation, categorization, and retrieval. In: Int'l Conference on Document Analysis and Recognition. (1997) 714–717
40. Merkl, D.: Text classification with self-organizing maps: some lessons learned. Neurocomputing **21**(1–3) (1998) 61–78
41. Ménier, G., Lorette, G.: Lexical analyzer based on a self-organizing feature map. In: Int'l Conference on Document Analysis and Recognition. (1997) 1067–1071
42. Konig, A.: Interactive visualization and analysis of hierarchical neural projections for data mining. IEEE Transactions on Neural Networks **11**(3) (2000) 615–624
43. Marukawa, K., Hu, T., Fujisawa, H., Shima, Y.: Document retrieval tolerating character recognition errors - evaluation and application. Pattern Recognition **30**(8) (1997) 1361–1371
44. Taghva, K., Borsack, J., Condit, A.: Evaluation of model-based retrieval effectiveness with OCR text. ACM TOIS **14**(1) (1996) 64–93
45. Lopresti, D.P.: Robust retrieval of noisy text. In: Proc. of ADL'96. (1996) 76–85
46. Madhvanath, S., Govindaraju, V.: The role of holistic paradigms in handwritten word recognition. IEEE Transactions on PAMI **23**(2) (2001) 149–164
47. Cesarini, F., Gori, M., Marinai, S., Soda, G.: INFORMys: A flexible invoice-like form reader system. IEEE Transactions on PAMI **20**(7) (1998) 730–745
48. Williams, W., Zalubas, E., Hero, A.: Word spotting in bitmapped fax documents. Information Retrieval **2**(2/3) (2000) 207–226
49. Rath, T.M., Manmatha, R., Lavrenko, V.: A search engine for historical manuscript images. In: ACM SIGIR 04. (2004) 369–376
50. Marinai, S., Marino, E., Soda, G.: Font adaptive word indexing of modern printed documents. IEEE Transactions on PAMI **28**(8) (2006) 1187–1199
51. Marinai, S., Faini, S., Marino, E., Soda, G.: Efficient word retrieval by means of SOM clustering and PCA. In: DAS 2006, Springer Verlag- LNCS 3872 (2006) 336–347
52. Baeza-Yates, R., Ribeiro-Neto, B.: Modern Information Retrieval. Addison Wesley (1999)
53. J.Hu, R.Kashi, G.Wilfong: Comparision and classification of documents based on layout similarity. Information Retrieval **2**(2/3) (2000) 227–243
54. Tzacheva, A., El-Sonbaty, Y., El-Kwae, E.A.: Document image matching using a maximal grid approach. In: Proceedings of the SPIE Document Recognition and Retrieval IX. (2002) 121–128
55. Marinai, S., Marino, E., Soda, G.: Tree clustering for layout-based document image retrieval. In: Proc. Second Int. Workshop on Document Image Analysis for Libraries. (2006) 243–251

Adaptive and Interactive Approaches to Document Analysis

George Nagy[1] and Sriharsha Veeramachaneni[2]

[1] RPI ECSE DocLab, Troy, NY 12180 USA, nagy@ecse.rpi.edu
[2] IRST Trento, Italy, sriharsha@itc.it

Summary. This chapter explores three aspects of learning in document analysis: (1) field classification, (2) interactive recognition, and (3) portable and networked applications. Context in document classification conventionally refers to language context, i.e., deterministic or statistical constraints on the sequence of letters in syllables or words, and on the sequence of words in phrases or sentences. We show how to exploit other types of statistical dependence, specifically the dependence between the shape features of several patterns due to the common source of the patterns within a field or a document. This type of dependence leads to field classification, where the features of some patterns may reveal useful information about the features of other patterns from the same source but not necessarily from the same class. We explore the relationship between field classification and the older concepts of unsupervised learning and adaptation. Human interaction is often more effective interspersed with algorithmic processes than only before or after the automated parts of the process. We develop a taxonomy for interaction during training and testing, and show how either human-initiated and machine-initiated interaction can lead to human and machine learning. In a section on new technologies, we discuss how new cameras and displays, web-wide access, interoperability, and essentially unlimited storage provide fertile new approaches to document analysis.

1 Introduction

The classical models for character recognition and document image analysis must be extended to accommodate the classification of multiple common-source patterns. We show how field classification exploits statistical dependence due to the common source of a field of patterns and also leads to a simple and operational definition of classifier adaptation. We explore diverse contextual constraints beyond those imposed by language models. Instead of ignoring the ever-present human-computer interaction, we propose more effective ways of exploiting it. We also examine the impact of recent technological developments on OCR and DIA and raise some research questions.

G. Nagy and S. Veeramachaneni: *Adaptive and Interactive Approaches to Document Analysis*, Studies in Computational Intelligence (SCI) **90**, 221–257 (2008)
www.springerlink.com

In this introductory section we list some of the limitations of conventional models for classification and propose extensions to field classification, adaptation and unsupervised learning. In the second section, *Field Classification*, we examine the contextual constraints that favor field classification and present some situations for which field classification algorithms have already been developed. We also attempt to clarify some distinctions between *trainable, supervised, semi-supervised, unsupervised, adaptive and self-organizing* algorithms for *training, teaching*, and *learning*.

The third section, *Interaction in training and testing*, considers paradigms where some interaction helps either or both the operator or the algorithmic parts of the system. Our premise is that interaction will remain necessary, for the foreseeable future, in most operational recognition systems. The fourth section, *Technology and Applications*, explores how advances in technology foster new applications in OCR and DIA. While the rest of this survey is mainly a retrospective and attempts to rationalize existing results, here we attempt to look ahead. In the *Conclusions* we list some trends and open research problems.

1.1 The Classical Paradigm for Pattern Recognition

Until the last decade, the customary framework for statistical pattern recognition in Optical Character Recognition (OCR), Hand-printed Character Recognition (HCR), and Document Image Analysis (DIA) was based on three key constraints:

1. *Representative training set.* The data was divided into two mutually exclusive sets of patterns for training and testing, each consisting of samples from a fixed number of classes with given or estimated prior probabilities. It was generally assumed that the patterns in both sets were independent samples produced by selecting a class label according to the prior class probabilities, then generating an observation (feature, attribute) vector from the corresponding class-conditional probability distributions. The labels of the test set were used only to determine classification accuracy. (Some researchers partitioned the training set further to provide a validation set for tuning parameters.)
2. *Singlet classification.* The patterns were classified one at a time. Each pattern was assigned a label on its own merits, independently of every other pattern. In OCR and HCR, each pattern was a single glyph (i.e., a letter, numeral, or ideograph). In DIA, it could be an entire word, a drawing or a photograph, or even an entire document. The only important exception to this constraint was the application of *linguistic context* in character recognition. Other entities, like forms, tables and documents, were also usually processed as though they occurred in isolation.
3. *No interaction.* Only algorithmic processes were considered of interest. It was understood that in real applications the labels in both the training

set and the test set would have to be provided by key entry, that human help was necessary to produce segmented character or word patterns for experimentation, and that intervention would be necessary at the operational level to deal with unclassified and misclassified patterns. However, these interactive components of the system were considered extraneous to the pattern recognition system, and in research settings and research publications little attention was devoted to optimizing them.

This architecture is typically represented by a data flow diagram similar to that shown in Fig. 1. No special provisions are made to indicate either the relevant data sets or the class labels.

Transducer → Feature Extraction → Classifier → Decision

Fig. 1. Generic first generation pattern recognition system

Over the last decade or two, many systems were proposed that did not fit neatly into the above paradigm. Some of the new approaches were the result of theoretical advances, while others arose from the realization that some important applications grossly violated the stated constraints. Many simply exploited technological advances: faster CPUs, larger amounts of storage, miniaturization, portability and connectivity, and better displays.

Our objective in this chapter is to construct a more general framework for pattern classification that encompasses recent research and may even leave room for new ideas. The notions at the core of the new paradigm are *learning and adaptation*, *styles*, *multi-pattern classification*, and *human-machine interaction*. We will give examples of methods and applications that fit the new paradigm, and discuss the technological advances that made them possible. We will also show how some widely used techniques, like clustering, expectation maximization, and active learning, fit naturally into the proposed framework.

We propose to define the new paradigm at a level of detail sufficient for probabilistic simulation of alternative classification algorithms. In this kind of simulation, the labels and patterns are generated by pseudo-random-number generators such as are readily found in most programming language libraries and in Matlab or Excel. Languages and software packages designed for simulation, like Simula, Modsim, Ross, Simscript, and Matlab toolboxes, offer a variety of built-in univariate probability distributions. It is, however, more difficult to generate multivariate distributions (e.g., Multinomial, Dirichlet, or Uniform), other than Gaussian, with the desired degree of statistical dependence completely specified by an arbitrary covariance matrix.

Our architecture for simulation does not address a critical component of all pattern recognition systems, *feature extraction*. Feature extraction is the step that transforms the output of the transducer (scanner, camera, tablet, microphone) into an abstract high-dimensional vector space where classification boundaries are defined. The error rate achievable by an ideal classifier depends on the chosen set of features. We are not, however, aware of any general technique for designing a good feature set, and even methods for selecting a subset of good features from a larger set leave much to be desired. We will therefore blithely assume that for each application some expert has already provided software for generating feature vectors. The simulations will start with a probabilistic feature space where the simulation parameters can simply be set to "good" features or "bad" features.

Another important aspect of OCR and DIA that we cannot simulate (but will discuss) is *segmentation*. Much effort has been devoted to separating text from illustrations, locating paragraphs and lines of print, and to word and character segmentation. Although the relevant algorithms fall in the realm of image processing rather than pattern recognition, segmentation and classification are often combined. As for feature extraction, there are few statistical models and tools for segmentation.

In the next subsection we define the components of a more comprehensive classifier architecture.

1.2 Definitions for an Expanded Paradigm

The definitions here pertain primarily to the role of various data sets in a classification system, with particular regard to simulated data. Merely envisaging it lends precision to definitions.

Training set. The training set consists of a set of labeled pattern (feature) vectors. For the purpose of analysis, one can assume that the feature vectors have either continuous or discrete valued components, but simulators can generate only discrete valued features. The number of components in the feature vectors, called the *dimensionality* of the problem, is fixed. There are four types of labels: *class labels, source labels, style labels,* and *instance labels*, as described below. The training set must have at least class labels, but the presence or absence of source and style labels leads to different types of classifiers. Patterns with the same source label share the same style, while patterns with different source labels may or may not be of the same style.

Test set. The test set consists of a sequence of feature vectors with source and class labels. The class labels must be used only for error counts. Each test pattern has a source label. The *source length* is the number of test patterns from the same source. The source distributions may be the same as in the training set, or different. Even if they are the same, the correspondence between the source labels of the test set and the source labels of the training set is assumed to be unknown. The test set has no style labels.

Field. For purpose of classification, the patterns of the test set are divided into fields. A field consists of a fixed number of patterns (called *field length*) from the same source. The choice of field length depends on the available computing resources, while the source length (the number of patterns from the same source in the test set) defines the scope of statistical dependence or context. Even in the absence of linguistic context, a field length of only two (i.e., *pair classification*) may lead to a significant increase in accuracy over singlet classification.

Source, style, and class labels. The assignment of these labels is the most time-consuming part of preparing a real dataset for experimentation. Under the assumption that each document is generated by the same source, only one *source label* per document need be entered. *Style labels* in the training set may be assigned by inspection, by font recognition for printed characters, by clustering or expectation maximization for handprint, or not assigned at all. Initial *class labels* are usually assigned by some classifier, and then the errors are found by proofreading and corrected manually. Sometimes data for experiments on printed characters is automatically generated by a script that generates so many samples of each font, in which case source, style and class labels can be assigned automatically.

Instance labels. Although not necessary for describing a classification scheme, it is good experimental practice to attach a unique label (*accession code, serial number, identifier*) to every pattern. This allows tracking changes in class label assignments when classifier parameters are changed, and whether errors committed by different classifiers are correlated. It may also serve as a *time stamp* for scenarios where the order of the patterns within the field matters, as in the case of linguistic context.

Example: Some NIST data sets have samples of isolated digits (10 classes). Each pattern is represented by a 24x30 binary array, therefore the dimensionality of the feature vector is 720 [1]. There are 600 writers, and the serial number of the writer is attached to each digit. These writer labels are our source labels. Writer consistency in the shape of the numerals is one aspect of style. Several writers may have the same style. To ensure that patterns of the same writer do not occur on both training and test set, the data is partitioned *by writer.* There are about 100 digits from each writer, so the source length is approximately 100. The NIST data set does not include style labels. As we will see, the presence of styles may improve classification accuracy even without the presence of explicit style labels.

Simulated data. The source label, which identifies patterns guaranteed to have the same style, is generated first. Then a style label is selected with fixed prior probability over the styles. Next, a sequence of class labels is generated according to *class priors.* The source length may be fixed or subject to a probability distribution that governs the number of patterns per source (for example, the number of digits in the courtesy field of a bank check, or the number of letters, digits and punctuation in a business letter). Finally, the feature vectors are generated from *class-and-style conditional feature distributions.* The patterns from each source are restricted to a single style; in other words, isogenous or common-source patterns of the same class are independently drawn samples from the same distribution. Before classification, the test set is partitioned into same-source fields. To simulate some applications, we may allow all of these probability distributions to change gradually. This makes a difference only if the classifier has a *bounded horizon*, i.e., if the field is shorter than the test set.

The important distinction between the new framework and old framework is the presence of multiple feature distributions that are constant within a source but may change from source to source. In order to exploit within-source consistency, *field classification* rather than singlet classification is necessary. A further distinction is that the distribution of the patterns may *change with time*. The nature of the classifiers appropriate for different scenarios within the above overall framework is elaborated in the next section.

2 Field Classification

We are now ready to consider situations where it is advantageous to classify entire groups of objects instead of classifying each object in isolation. This is generally the case in DIA and OCR, where a message (substantiated as a document) consists of an ordered collection of visual objects (*glyphs*). We show that many common constraints on acceptable sequences of symbols, and on the visual appearance of the glyphs used to represent them, can be expressed in terms of statistical dependence between patterns. Because the estimation and exploitation of the underlying joint probability distributions requires examination of more than one pattern at a time, we discuss *field classification*. We relegate the relevant mathematical formalisms to the cited references, but we present some tools that facilitate the study of the inter-pattern feature dependences, and state the assumptions under which optimal or approximate field classification algorithms have already been developed. We conclude the section with a discussion of adaptive classification and unsupervised learning.

2.1 Context

Information relevant to classifying an object (digit, letter, word, illustration or document) is often extraneous to the object itself. It may either reside in other objects that are also to be classified, or it can be considered part of the environment in which the classifier operates. In the first case, recognition accuracy can be improved by taking into account the characteristics of an entire group of objects to classify each one, i.e., by field classification. In the second case, the recognition can be improved by providing means to specialize or tune the classifier for either singlet patterns or fields to its environment. The additional information is generally called *context*, regardless of whether it can be derived from the available samples [2, 3].

In character and speech recognition, the word "context" is often reserved for linguistic context. It has, however, a much wider scope in Artificial Intelligence, as exemplified by the topics discussed at the biennial ACM Context conferences, which draw on several centuries of studies in epistemology [4, 5, 6, 7].

We will examine situations other than linguistic context where field classification is useful, but neglect broader considerations that need to be taken into account in preprocessing and feature extraction rather than in the classifier itself. In other words, we will concentrate on the kinds of context where the patterns to be classified provide information about each other.

Since the use of linguistic context is well established in both character and speech recognition, we will first look at *language models*. Then we will examine some relations between the *shapes* of the patterns. We will distinguish between *order-independent* and *order-dependent* relations, and also between forms of statistical dependence that arise *between labels*, *between shape features*, and *between labels and shape features* –of all the patterns within a field.

Language Models

Language models are approximate descriptions of natural language at the morphological, lexical, syntactic, semantic, or pragmatic levels. While many of the earlier models were rule-based, the advent of large computer-readable corpora for estimating parameters has given rise to statistical models.

Morphological models typically consist of polygram frequencies [8]. These frequencies vary from language to language and are always highly skewed [9]. In English text, for example, the probability of "e" is 0.1241, while that of "z" is only 0.0007. The skew increases with polygram length: P[th]=0.04, while P[qh]=0. It is clear that an ambiguity between an `e` and a `c` after a `d` should be resolved in favor of `e`, but is `e` or `c` more likely after `u`? Elaborate methods have been proposed to estimate the probabilities of rare letter or phoneme sequences [10].

Lexical models are based on word frequencies and word transition frequencies. The simplest systems are based on dictionaries (strictly speaking, *lexicons*) that report only the existence or non-existence of a sequence of letters as a valid word of a particular language, without its frequency of occurrence. (*Agglutinative* languages with many case endings and verb forms, like Italian, typically require lexicons at least three times larger than English.) Commercial OCR systems routinely use not only large general lexicons but also specialized lexicons of biological, chemical or legal terminology, and lists of abbreviations, acronyms, and proper names. Most often dictionary-lookup is carried out only as a post-processing step, which is generally suboptimal. Some examples of over-reliance on lexicons are given in [11], which also describes many other sources of OCR errors.

The best statistical *syntactic models* surpass the power of rule-based systems [12, 13, 14]. Estimates of transition frequencies between syntactic categories (noun, verb, adjective, adverb ...) can be obtained from large annotated corpora. Syntactic models are of limited use in English because of the multiple categories carried by many words (e.g., *Yellow* soap / I wonder where the *yellow* went, or To *fit* a dress / A *fit* athlete / A good *fit*). Applying semantic and pragmatic models is even harder [15].

Formally, all linguistic context in character recognition can be expressed as statistical dependence between the *labels* of patterns. The random variables whose joint probabilities must be estimated are letter, word, or part-of-speech labels. Linguistic context is always *order-dependent*, and therefore often modeled with transition frequencies in Markov Chains, Hidden Markov Models, and Markov Random Fields [16, 17, 18, 19, 20, 21, 22, 23, 24, 25, 26, 27, 28]. Linguistic variables are usually assumed to be independent of character shape, even though titles and headings in large or bold type have a different language structure than plain text. Optimal field classification of printed matter is often approximated by a post-processor that simply attempts to integrate confidence measures based on shape and language.

Style

We term *style* any difference between the statistical characteristics of a group of patterns generated by a single source and the characteristics of a group of patterns generated by several sources [29, 30, 31]. A single-source group usually exhibits some shape consistency. For instance, we may be able to distinguish numerals written by Alice from numerals written by Bob. Alice's numerals seem similar to each other, and Bob's numerals are also similar to each other, but Alice's numerals are different from Bob's. The same notion can be applied also to text printed in different fonts. Forensic analysts can tell whether two sets of letters or numerals were written with the same pen, or printed on the same printer.

More formally, style context is defined as the presence of statistical dependence arising between patterns (represented as random vectors) because they are from the same source. Unlike language context, it is independent of the order of the patterns in the field. It takes two distinct forms, which we call intra-class style and inter-class style [32].

Intra-class style is the shape consistency of a single class from each source. It reveals how consistent a writer is in writing a glyph. Does Alice always cross her 7s, while Bob never does? It is, of course, even more marked in print, where words, paragraphs, and entire documents are often composed in a single typeface. Experts can recognize dozens of typefaces by inspection. More subtle than typeface consistency is the intra-class style within documents printed by the same printer or scanned by the same scanner. In OCR, where each glyph (a letter, numeral or ideograph) is usually represented by a feature vector, we say that a data set exhibits intra-class style if the feature-vectors of patterns of the *same* source and class, considered as random variables, are (class-and-style-conditionally) statistically dependent.

Inter-class style determines how much the shape of a given class reveals about the appearance of *other* classes from the same source. The way Alice writes `1` helps predict the way she will write `7`. If the `n` has no serifs, neither will `h`, `m`, or `r`. We say that the data set exhibits inter-class style if the feature-vectors of patterns of *different classes*, considered as random variables, are (class-and-style-conditionally) statistically dependent. Fig. 2. illustrates two pairs that cannot both be recognized correctly as 17 by a singlet classifier, and an instance where the label assigned by a singlet classifier is corrected by the field classifier.

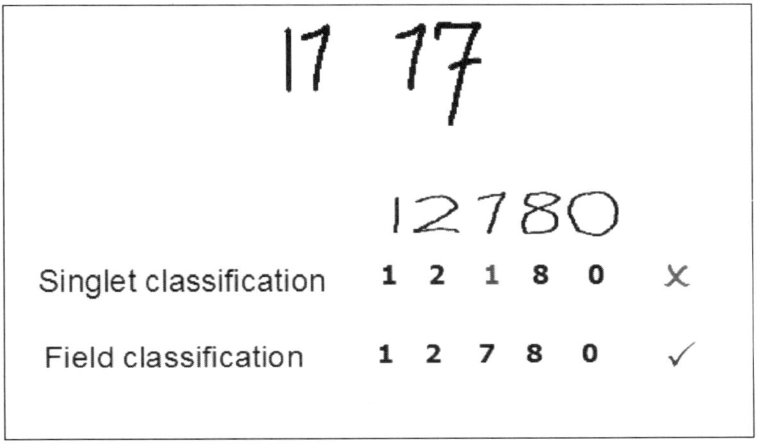

Fig. 2. Benefit of pair classification when there is inter-class style

We show in Fig. 3 a useful representation for visually comparing singlet and field classification boundaries. (This representation allows showing only a single feature for each pattern, hence only a 2-D *field feature.*) We plot some field features, which have bimodal Gaussian mixture distributions, for each field class (AA, AB, BA, BB) of a two-class problem (A,B) with a single feature x. We also show the 2-D decision boundaries of the singlet classifier and of the field classifier. The optimal field classification boundaries and the singlet boundaries are different, so we would expect some gain with a field classifier. Simulation of a field classifier shows a reduction in the error rate of single patterns from 15% to about 10%.

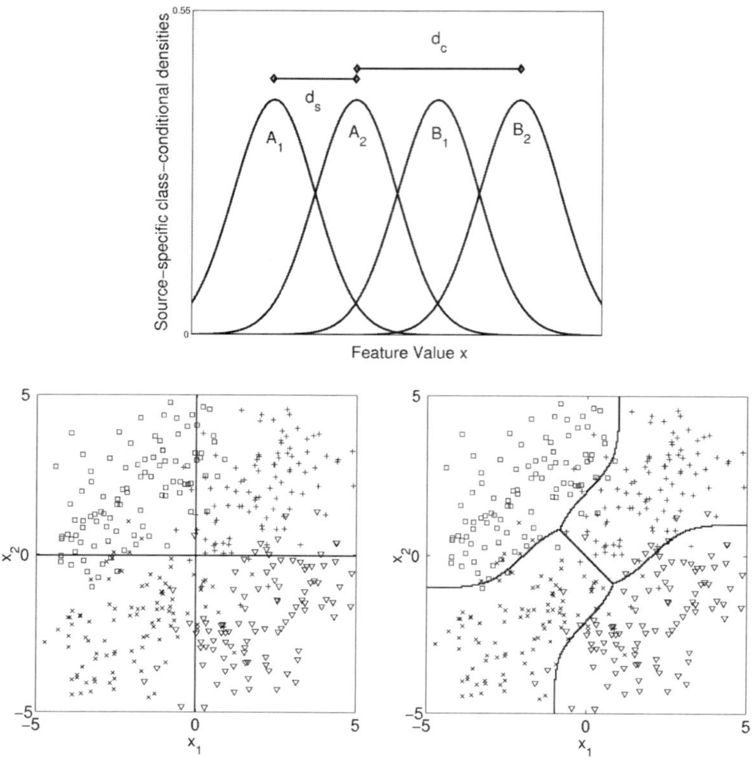

Fig. 3. Gaussian class-conditional distributions of a single feature x for classes A and B and styles 1 and 2. Below are the representations of singlet (left) and pair (right) feature (spaces x_1, x_2) and decision boundaries for a field of two patterns. The AA region is bottom left, AB is bottom right, BA top left, and BB top right

Order-Dependent Inter-Pattern Dependence

Inter-pattern class-feature dependence is fairly rare. It occurs when features of a pattern depend on the *class*, rather than on the rendering (features), of an adjacent pattern. For example, the vertical location of an apostrophe may depend on whether the previous letter had an ascender, but not on its font (Fig. 4). That is, the features of the apostrophe are independent of the preceding letter, *given its label.*

Feature dependence between adjacent patterns is common, as illustrated in Fig. 5 (from [3]). In cursive writing, the location of the last stroke of a pattern determines the nature of the ligature that joins it to the next pattern. It is different from style, because the ligature-sensitive features of the two patterns depend on the shape and *order* of the patterns. A similar phenomenon in speech is called co-articulation.

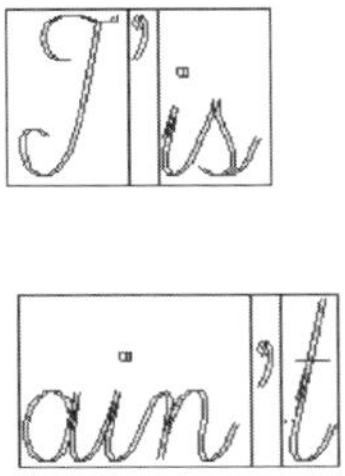

Fig. 4. Example of inter-pattern class-feature dependence

Fig. 5. Examples of order-dependent inter-pattern feature dependence: note the difference between the ligatures preceding the a's

2.2 Field Classifiers

Dependence between patterns suggests that a field classifier should assign a *field label*, consisting of a sequence of class labels, based on the feature vectors of *all the patterns* in a test field. The number of possible field labels rises exponentially with field length, thereby effectively limiting the maximum operational field length.

We discuss below several types of field classifiers that have been proposed under various assumptions. These field classifiers are generally based on the formulation of singlet classifiers: for instance, they may be Bayes classifiers or MAP classifiers, and either parametric or non-parametric classifiers. In addition to standard statistical classifiers, neural networks and support vector machines can also be exploited for field classification. To classify each pattern, all field classifiers combine information derived from the entire training set with information from the whole test field.

Field-Trained Classifiers

An obvious idea is to concatenate the features of singlet patterns to form field feature vectors, and train the classifier on every possible field class. All of the well-developed theory of singlet classification then applies. This method, however, requires training samples of every field class, and is therefore generally impractical with field lengths greater than two.

In text, not all combinations of letters occur. Word classifiers can therefore be trained on words, rather than on every possible sequence of characters. One version of this approach divides the letters of the alphabet into fewer and more easily classified categories based on character shape codes (ascenders and descenders) [33, 34]. Because character-level segmentation is error prone, most word classifiers are not based on concatenating singlet character features, but on features extracted from the entire word. This approach is particularly suitable for limited-vocabulary applications like postal addresses or legal amounts on bank checks [35, 36], and for correction of OCR errors [37]. Another example of holistic word classification is based on statistics extracted from each cell of a grid superimposed on the word [38]. For degraded documents with larger vocabularies, word level indexing (as opposed to keyword spotting) was proposed with a three-stage comparison based on word aspect ratios, vector features extracted with a grid superimposed on each word, and within-word connectivity. Experimental evidence for high precision and recall in retrieval was adduced from a multilingual collection of OCR-resistant documents spanning four centuries [39].

Font Classification

For printed matter, *font classifiers* and *font-specific character classifiers* can be trained on data sets of specific type faces or on broad groups (serif/sans-serif, italic, bold). The font classifier is then applied first to a test field, and its decision is used to select the appropriate character classifier [40, 41, 42, 43]. The same idea can be applied to writer identification [44]. Many words of text may be necessary to reliably identify the font. Furthermore, the resulting classifier is generally suboptimal, because the features in character classification are neglected in font classification, and those used in the font classifier are neglected in character classification. Style classifiers, discussed below, use the entire set of features.

Discrete-Style Classifier and Style-First Classifier

If the underlying feature distributions are Gaussian, and the training set has style labels that allow estimating the parameters of the class-and-style conditional feature distributions, then the joint posterior mixture-distributions of the field classes can be computed for fields of arbitrary length. The resulting optimal classifier is known as the *Discrete Style Classifier* [31]. The lengthy computation (exponential with field length) can be approximated by keeping track of frequently co-occurring (same-source) shapes [45] or, more consistently, by a *Style First Classifier* that computes the posterior probability based on the most likely style [46]. Non-parametric nearest-neighbor field classifiers are described in [47] and support vector machine field classifiers in [48].

Style-Conscious Quadratic Discriminant Field Classifier

In some applications, like handprint recognition with a multitude of writers, it is sensible to assume a continuous distribution of Gaussian styles instead of some predetermined fixed number. The posterior distribution for any field length can then be determined from only the cross-covariance matrix of *pairs* of same-source pattern feature vectors (*op. cit.* [46]). The resulting Style-Conscious Quadratic Discriminant Field classifier is optimal under the stated assumptions. The experiments described in the cited references indicate that all of these field classifiers achieve lower character error rate than the singlet classifiers on which they are based.

2.3 Adaptive Classifiers

The word *adaptive* (which surfaced in conjunction with stochastic approximation, potential functions, adelines, madelines, and perceptrons), is overloaded and has been used in many different ways since its appearance – first in automatic control then in pattern recognition – more than forty years ago [49]. Adaptation and learning were linked to stochastic approximation (Robbins-Monroe and Kiefer-Wolfovitz processes) by Aizerman, Tsypkin and Fu among others [50, 51, 52]. Nevertheless we need a word for a concept that fits with our definitions of training and test sets and of field classification, and that shares the connotation associated with adaptation. In our context, adaptation can be defined clearly and simply without introducing any additional notions. Our definition offers the advantage that it applies equally to structural adaptation, parameter adaptation, and to complex classification formulas that could be equivalent to either.

We define an *adaptive classifier* as a *field classifier with a field that encompasses the entire test set.*

Such a classifier can clearly use all of the information that is available in the patterns to be classified. Not only does the classification of the last pattern in the test set profit from information garnered from the first pattern, but the classification of the first pattern also benefits from the last pattern. In principle, the field-classification boundaries of an adaptive classifier can be determined entirely from the training set. This does not imply that the distribution of shorter subsequences of patterns in the field is stationary, but it does require all of the test patterns to be available at the same time.

For long test sets, the computation of the posterior field probabilities must be approximated. *Dynamic field classifiers* adjust their classification parameters after classifying a finite subset of the test field, thereby approximating an optimal adaptive classifier. The approximation may be necessary either because there are insufficient computational resources to classify the entire test set optimally, or because some of the test patterns must be classified before all of them are available.

Dynamic classifiers present the danger of wandering off course, perhaps because of a completely mislabeled subset of the test field, and never recover. It may therefore be prudent to test them periodically on some typical validation field and, if the error rate is too high, reset the parameters to those obtained from the original trusted training set. Adaptation in commercial OCR systems seldom exceeds page length (c. 2000 characters).

The style-constrained classifiers described below are not dynamic, because their decision depends only on the ensemble of patterns of the current field, which is generally only a subset of the test set. If a field reappears later, after many other fields from different sources and styles were classified, the result will be the same. (In contrast, a classifier that is dynamic according to our definition could well classify a subsequent but identical field differently.) Nevertheless, it may be appropriate to claim that these classifiers adapt to the style of each test field.

2.4 Supervised, Semi-supervised, and Unsupervised Learning

Algorithmic grouping or *clustering* of unlabeled patterns according to their distance to each other in feature space is often called *unsupervised* classification or learning [53]. Persistent attempts since the sixties to endow *self-organization*, *(self-)adaptivity*, *learning without a teacher*, *training without a trainer*, *self-produced pattern discrimination*, *self-correction*, and *unsupervised*, *semi-supervised* or *non-supervised classification* with a stable meaning have proved futile [54, 55, 56]. As mentioned above, we reserve the word *adaptive* for a more specific concept.

We take the position that pattern recognition in OCR and DIA cannot be entirely unsupervised, because documents, words, letters and numerals already have some prior meaning to human readers. At some point, this meaning must be communicated to the classifier so as to regain the correspondence between the arbitrary labels assigned by the machine and the labels of the user community (for instance ASCII character labels, or Reuters document categories). We attempt next to discover what is the "hidden" information used by various "unsupervised" pattern recognition methods.

The least information necessary to turn a mixture decomposition method into a classifier is admirably elucidated in [57, 58]. The unlabeled patterns are presented as a Gaussian mixture distribution with unknown mixing parameters. It is shown that with an increasing number of samples, the parameters of the constituent distributions can be estimated to arbitrary precision. However, in order to determine with better than chance accuracy which constituent corresponds to which class, we need at least *one* labeled pattern. Increasing the proportion of labeled to unlabeled samples brings such a classifier closer to the vanilla-flavored (supervised) classifier.

The idea of first partitioning unlabeled samples, and then assigning labels to each partition, was thoroughly explored in the sixties from the perspective of both signal-processing [59, 60, 61] and potential functions [62, 63, 64]. In 1966 Dorofeyuk presented several clustering algorithms, and then assigned labels to each cluster according to the known majority label in each cluster. He tested his algorithms on five classes of hand-printed digits. He called the procedure *teaching without a teacher*, because the labels were not used in the clustering process [65].

Examples of easy- and difficult-to-cluster pattern configurations are simple to visualize in two dimensions [66, 67]. The widely-used K-means clustering algorithm was popularized as a general method for "exploratory" multivariate analysis of unlabeled data [68, 69]. A variation that addressed some of the shortcomings of the elementary algorithm by splitting and merging classes was called Isodata [70]. In the communications community, iterative minimization of the sum-of-squared-error criterion became known as Vector Quantization [71, 72]. Among the first attempts at evaluating regiorously the effectiveness of clustering methods were Dubes and Jain [73]. Variations of the method with respect to initialization, cost function, splitting and merging clusters, and distance metrics, have been amply described [74, 75]. Current research focuses on combining multiple cluster configurations obtained by different algorithms, i.e., *clustering ensembles* [76].

Clustering with the K-means algorithm using *labeled seeds* (initial cluster centroids) circumvents the need for assigning labels after the clustering process. One of the simplest adaptive classifiers (called *decision-directed approximation* [77]) is a minimum-distance-to-class-centroid classifier that iteratively recomputes the class centroids according to the class labels assigned on the previous step (Fig 6). It is clear that the final class centroids depend, as do cluster centroids in K-means, only on the initial seeds (here the class means of the training set), and on the patterns in the test field. Several successful examples of decision-directed classification in OCR have been reported [78, 79, 80, 81, 82, 83]. Other applications include Morse Code transcription [84], adaptive equalization [85, 86], and thematic mapping in remote sensing [87, 88].

Crisp clustering algorithms assign each pattern to a single cluster. Fuzzy clustering algorithms assign membership functions. Agglomerative and divisive algorithms group or partition patterns according to a pre-calculated matrix of pairwise similarities (which need not have metric properties). Statistical methods based on the very general principle of Expectation Maximization [89, 90] attempt to decompose mixture distributions into their constituents. In each of these approaches, the problem is simplified if the number of classes is known. The number of classes may be replaced or augmented by other pertinent information, like constraints on the number of patterns per cluster, or on the cluster diameters.

We underline that any of the above methods for grouping patterns according to some predetermined measure of similarity may serve as the foundation for "unsupervised" classification. All of them exploit style consistency, albeit only intra-class style. Like other style classifiers, these methods reduce the need for labeled samples. The methods for addressing small-training-sample problems [2], [9] and 'wrong' (or non-representative) training-sample problems are essentially equivalent. However, classifiers based on only intra-class styles are obviously suboptimal when the data exhibits inter-class style as well.

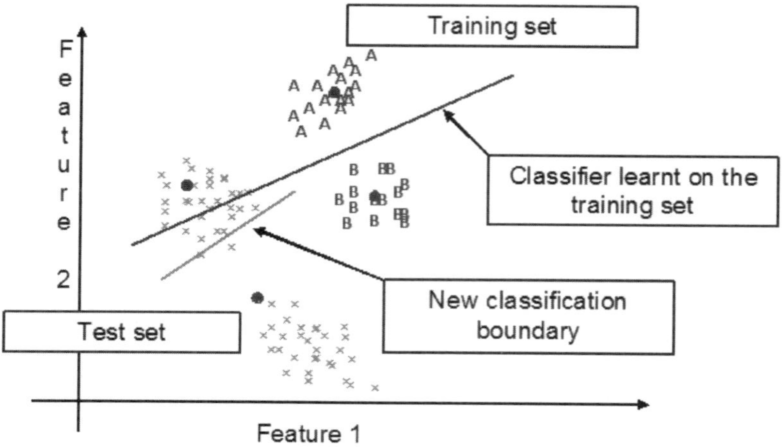

Fig. 6. Adjustment of the classification boundary in a decision-directed classifier. The new boundary is at equal distance from the class centroids of the patterns as classified by the original classifier learn on the training set

Patterns are often clustered for multi-stage classification of large- alphabets, like Chinese [92, 93]. As a demonstration of the power of language context, all the characters in a document can be clustered, and labels assigned to the cluster labels by solving a substitution cipher, without using any prior class-related shape information [94, 95, 96, 97, 98].

Clustering is not necessarily followed by assigning an object label to each cluster. Clustering the connected components (most of which correspond to individual characters) in an isogenous text image is the basis for efficient text-image compression, such as DjVu and JBIG2 [99, 100, 101]. Clustering of approximate representations of document *words* was used to reduce the number of comparisons of a query versus document words in the multilingual word indexing scheme mentioned above [102]. The method was called *font-adaptive* because the words in each source were clustered separately.

3 Interaction in Training and Classification

Research aimed at fully automating the processing of document images has received sustained attention over the past 40 years. Nevertheless, any of the dozens of surveys to date (one of our favorites is [103]) will reveal that progress in automatic recognition and interpretation has been slower than predicted. Further improvement on cursive handwriting and degraded print may be even slower because the remaining challenges are harder. As in speech recognition, bridging the "semantic gap" between machine and human knowledge appears problematic. The context in all the varieties discussed above brought by humans to any classification task is much greater than what can be codified automatically from even the largest collections of training samples available to our community. Endowing fully automated systems with broad knowledge remains an elusive goal. Fortunately, in many applications it is not necessary to fully automate the task of document analysis. This may be the case when the focus is on a relatively few high-value documents (perhaps just one). The computer can play the role of an assistant to help the user acquire information that would otherwise remain inaccessible. While such documents could be collected and returned to a central repository for scanning and batch processing in the traditional manner, it may be advantageous to exploit the information immediately and *in situ*.

There are pronounced differences between human and machine cognitive abilities. A divide-and-conquer strategy for visual recognition can partition difficult domains into components that are relatively easier for both human and machine (Table I). Humans excel in gestalt tasks, like object-background separation. They apply to recognition a rich set of contextual constraints and superior noise-filtering abilities. They can also easily read degraded text (e.g., CAPTCHA's [104]) on which the best optical character recognition systems produce only gibberish. On the other hand, the study of psychophysics reveals that humans have limited memory and poor absolute judgment [105].

Computers can perform many tasks faster and more accurately. They can store thousands of images and the associations between them, and never forget a name or a label. They can compute geometrical properties like higher-order moments whereas a human is challenged to determine even the centroid of a complex figure. Spatial frequency and other kernel transforms can be easily computed to differentiate similar textures. Computers can count thousands of connected components and sort them according to various criteria (size, aspect ratio, convexity). They can quickly measure lengths and areas, and flawlessly evaluate multivariate conditional probabilities, decision functions, logic rules, and grammars. Nevertheless, computer vision systems have difficulty in recognizing "obvious" differences and they do not generalize well from limited training sets

We are *not* advocating here exploratory data analysis in feature space [106, 107], but operator interaction with displayed document images or parts thereof. Although we cannot clearly separate human interaction during training and testing (because when a human helps the system during classification time, it can be viewed as training) we attempt to categorize interaction as: (1) Human-initiated or Machine-initiated; (2) Durable or Ephemeral. *Durable Interaction* immediately alters some system parameters and therefore affects how the system deals with new data. *Ephemeral Interaction* merely labels new patterns or modifies the results of classification.

3.1 Examples of Human-Initiated Interaction

The most common example of human-initiated interaction is *labeling* training patterns. Another example is *word completion* on touch-screen devices (word

HUMAN	**MACHINE**
Dichotomies	
	Multi category classification
Figure-ground separation	
Part-whole relationships	
Salience	
	Non-linear high dimensional classification boundaries
Extrapolation from limited training samples	
Broad context	
	Precise mesaurement of individual features
	Enumeration
	Store and recall *many* labeled reference patterns
	Accurate estimation of statistical parameters
	Application of Markovian properties
	Estimation of decision functions from training samples
	Evaluation of complex sets of rules
Gauging *relative* size and intensity	
Detection of *significant* differences between objects	
	Computation of geometric moments
	Orthogonal spatial transforms (e.g. wavelets)
	Connected components analysis
	Sorting and searching
	Rank-ordering items according to a criterion
	Additive *white* noise, salt and pepper noise
Colored noise; Texture	
Non-linear feature dependence	
	Determination of local etrema in high-dimensions
Global optima in low dimensions	

Table 1. Comparison of relative strengths of human and machine in diverse aspects of visual pattern recognition

completion is seldom used with regular keyboards because it tends to distract the operator). Such interaction has also proved its value in the *vectorization* of engineering drawings and maps. We briefly describe these three applications.

Labeling Training Patterns in OCR

Nowadays all manual labeling of documents, or parts of documents, is carried out with computer display of digitized material, and can therefore be considered interactive. Indeed, considerable ingenuity has been applied to provide interfaces that speed up the process and reduce mislabeling. Commercial OCR firms strive to improve successive releases of their recognition systems by accumulating millions of labeled characters. If everything is keyed, it is human-initiated interaction. If they first OCR the training documents and only correct the errors, then the interaction is machine-initiated. It is *ephemeral* because the *current* classifier does not benefit from the newly labeled patterns.

Most OCR systems also provide at least limited facilities for additional training in the field for new shapes and new classes. If necessary, the operator can separate document segments set in different typefaces or written by different individuals. Entering only part of a document may help a recognition system designed with this in mind to fine-tune the classification algorithms. The underlying assumption is that if the remainder of the document(s) is from the same source, then the adjusted parameters will yield more accurate recognition. Training is not limited to characters: for example, a table-location algorithm can be trained via multi-parameter optimization [108].

Mobile Text Entry

It is clear that one bottleneck in mobile interactive document analysis is text entry. Without scanning or a regular keyboard, the alternatives are (1) virtual keyboard on a touch sensitive screen, (2) finger-operated keypad on arm or thigh (perhaps incorporated in the operator's clothing), and (3) automatic speech recognition. We believe that the stylus is the most appropriate solution, because in addition to text entry it can also mediate the graphical communication essential in other phases of document image analysis.

The virtual keyboard was invented in the seventies to avoid having the operator shift constantly between pointing device and keyboard while digitizing maps and line-drawings. It consisted of a picture of a keyboard that could be shifted to the area of the drawing being vectorized. Current virtual keyboards usually appear in a fixed partition of the touch-sensitive screen of a handheld device. Edwards' survey of input interfaces in mobile devices covers most of the relevant issues [109]. Data input is usually a local operation, so it makes little difference whether the device is networked or not.

Important considerations for stylus data entry are speed, operator comfort, and ramp-up time. The first two factors are influenced by the amount of space allocated to the keyboard, to the recognized or keyed text, and to control functions. The third factor depends heavily on the keyboard layout. The QWERTY layout, developed to prevent binding of type bars in mechanical typewriters, is suboptimal even for typing, and even more so for one-handed stylus entry.

Ancona gives a good overview of alternative keyboards and word-completion algorithms [110]. An upper bound on the speed of individual character entry is imposed by Fitts' Law, which is a nonlinear relationship between pointing time and the distance and size of the target. The relevant distance is that between the screen areas ("keys") corresponding to consecutive letters. The letter transition frequency is given by a language model. It is possible to reduce the average distance by having multiple keys for common symbols, but this decreases the size of the keys. Several researchers have optimized keyboard designs according to various language models [111, 112, 113, 114]. The computed speeds hover about 40 words per minute, but actual text entry is much slower.

The speed increase obtainable by word completion depends on the language model. Ancona (*op. cit.* [110]) demonstrates a keyboard with separate keys for the ten most common words (with a cumulative word frequency of 28%). After each tap on the screen, the ten most likely words appear in the selection area of the screen. If the correct word is included, it can be selected with one additional tap. If not, another letter is tapped, which brings up ten new words. With a vocabulary of 13,000 words, the expected number of taps per word was 3.3. The performance of word-completion systems depends on how well the stored lexicon is matched to the user input. Multiple lexicons – for different languages and applications – can be either stored on board, or downloaded via a wireless connection.

Vectorization

Entering line art (maps and engineering drawings) manually is even more laborious and expensive than keying text. Manual vectorization was first conducted from hardcopy on a digitizing table. The operator traced the lines with cross-hairs under a magnifying glass with a MARK button. After the advent of large-size roller-feed scanners and bitmapped displays all service bureaus and in-house operations adopted on-screen vectorization. Vectorized lines could now be displayed with a different color, deviations between the manually entered line segments and the original bitmap became clearly visible, and the operator could zoom in on dense portions of the drawing.

If most of the labels on a drawing or map cannot be recognized by OCR because of poor document quality or unusual character shapes, it is still possible to rapidly mark their location and orientation, rotate them to horizontal, and move them to a single area of the screen [115]. This accelerates manual label entry (a single E-size drawing may contain over 3,000 alphanumeric symbols). Most such data-entry systems are part of GIS or CAD software designed for standard workstations, where all graphical operator interaction is mediated by the mouse. As demonstrated by Engelbart and colleagues at SRI long ago, direct-action devices, like a touch-sensitive stylus, would allow faster and more accurate interaction [116]. However, the aspects of interest here are the machine-initiated algorithms developed for semi-automated data entry.

To enter colored maps, different color layers are first separated according to RGB values. Vectorizing algorithms are manually initialized to a line segment or curve, and automatically follow that line at least to the next intersection point. Some systems also attempt to automatically recognize map and drawing symbols (e.g., for schools or resistors). If it fails, the operator overrides it. The character recognition software recognizes cleanly lettered labels (elevations, part numbers, resistor values), but leaves labels confused by overlaid line art or poor lettering to the operator.

These interactive systems (like CAVIAR, below) exhibit clear speed advantages over completely manual data entry, and are robust enough (unlike automated systems) for operational application. Although some of these systems are laboriously trainable, one key difference compared to CAVIAR is that no commercial system that we are aware of incorporates active algorithms (i.e., durable interaction) that take advantage of routine operator input.

3.2 Machine-Initiated Interaction

All trainable systems incorporate, by definition, durable interaction. Most such systems, however, are human-initiated: training is a preliminary, separate phase from the recognition, without regard to what can be correctly or incorrectly recognized without additional training. We believe that eventually all interaction in DIA should be *machine-initiated* and *durable*. In other words, the operator should not even have to look at data that the system had no trouble in classifying, and every interaction should be utilized by the system to improve subsequent classification. We therefore present some of our work outside of DIA on machine-initiated, durable interaction. Then we propose several phases in DIA where we see potential applications of similar types of interaction.

Machine-Initiated, Durable Interaction in CAVIAR Systems

CAVIAR (Computer Assisted Visual Interactive Recognition) is an interactive system for recognizing faces and flowers, both problems of a level of difficulty (i.e., current automated accuracy) comparable to document recognition [117, 118, 119, 120, 121]. Experiments on sizable databases of faces and flowers indicate that interactive recognition is more than twice as fast as the unaided human, and yields an error rate ten times lower than state-of-the-art automated classifiers. The benefit margin of interactive recognition increases with improved automated classification. Parsimonious human interaction throughout the interpretation process is much better than operator intervention only at the beginning and the end, e.g., framing the objects to be recognized or dealing with rejects. Furthermore, this interactive architecture has been shown to scale up: it can start with only a single sample of each class, and it improves as recognized samples are added automatically to the reference database (decision-directed adaptation).

The notions embodied in CAVIAR differ in fundamental ways from past efforts at mobile, interactive recognition. Whether such an approach can be equally effective in the domain of documents as it is for flowers and faces is unproven, and adapting CAVIAR to document analysis requires further research. There are, however, other projects that share similar goals and assumptions. The Army Research Laboratory's Forward Area Language Converter (FALCon) system provides mobile optical character recognition (OCR) and translation capabilities [122, 123], but, so far as we know, it has a traditional user interface. Research on camera-based document acquisition is growing [124, 125]. However, this work, like FALCon, treats the later processing stages as though they will be fully automated.

Camera-based systems for locating and recognizing text in traffic signs and providing translation services for visitors to foreign lands are somewhat similar [126, 127], but their interaction paradigm is less integrated into classification than CAVIAR's. Reading systems for the vision-impaired likewise focus on page-at-a-time processing, but offer an auditory user interface [128, 129]. A somewhat similar notion is recent work on developing tools to support forensic document analysis [130]. Forensic systems are, however, intended for off-line use by domain experts (as opposed to opportunistic document readers whose primary jobs lie elsewhere), and have no need for mobility.

Potential for Machine-Initiated Durable Interaction in DIA

We mention some DIA tasks where CAVIAR-like systems may prove advantageous. We focus on scenarios where automated algorithms work accurately only on exceptionally clean documents, but where a little interaction can quickly produce acceptable results on ordinary material.

Binarization. Most OCR algorithms are designed for binarized images, because all scripts avoid discrimination based on shades of gray or color. Therefore documents must be converted to binary images after digitization to 8-bit gray scale or RGB. Global binarization algorithms work only if the foreground and background reflectance are uniform throughout the document, which may not be the case if part of a folded documents suffers prolonged exposure to sunlight, or if there are dark areas around the edges of a photocopy. Local binarization algorithms set the threshold according to the distribution of reflectance in a window translated through the page. The threshold estimates of the relative density and configuration of the foreground (ink) and background invariably depend on explicit or implicit assumptions that hold only for a narrow class of documents. An operator can easily tell when binarization fails. Setting the appropriate window size for local algorithmic thresholding requires far less work than setting the threshold manually everywhere, and it is more robust than fully automated local thresholding.

Page segmentation. Column, paragraph and line segmentation are other instances where interaction may be effective. The first step is usually estimating global document skew. While accurate skew estimation and correction algorithms have been developed for printed matter, they do not work well on handwriting because the orientation of individual lines varies, the margins are not straight, there may be only a few words on a page, or there may be several columns of words or phrases at different angles. Humans can, however, judge skew remarkably well, and convey this information to the computer by a few well chosen stylus taps or by rotating a superimposed grid. After the computer-proposed skew correction and line finding is corrected, the occasional merged pair of lines – due to overlapping ascenders and descenders – can be likewise rapidly separated.

Word segmentation. This is relatively easy for printed text, except for extremely tightly-set, micro-justified print. In handwriting, however, large spaces may appear within words. Towards the end of a line, words are often squeezed together. In Arabic and other scripts, some inter-letter spaces are mandatory. Underlined groups of words can further complicate the task. Again, humans can usually spot missed word boundaries even in unfamiliar languages and scripts. If the writing lines are already properly segmented, then a simple interface can be designed to correct linked and broken words.

Character recognition. An operator can provide global assistance to the character recognition system. He or she may be able to recognize the language or script of a document, indicate the average slant, and (in Western scripts) the prevalent case. The operator may decide which of the available lexicons would provide the best language model, and the chosen lexicons can be automatically updated with entries from the processed documents that have been deemed correct. Humans can also tell where perfect accuracy is important, as in telephone numbers, email addresses, and proper nouns and, if recognition fails, enter them manually or select them from the top recognition candidates. Finally, if the typeface is entirely outside the machine's repertory, it can cluster the character images, so that the operator need to label only a representative member of each cluster [131].

Active Learning: Machine-Initiated Durable Interaction During Training

During the training of pattern classifiers it is often feasible to provide labels for the training patterns incrementally. The most 'informative' patterns can be chosen iteratively, and their labels queried. This learning paradigm, wherein the learner is allowed to choose the information to be acquired, is called *active learning* and has been shown to significantly reduce the labeling cost, while preserving the accuracy of the trained classifier [132, 133]. Although we are not aware of any formal application of active learning in DIA (as opposed to document categorization), the training samples are often augmented when classification errors arise. This practice is seldom documented.

4 Technology and Applications

In this section we briefly discuss recent technological advances that alter the landscape of OCR and DIA and open up new applications.

4.1 Cameras and Displays

Solid state sensors are more sensitive to light than film. Current digital consumer-grade cameras, PDA cameras, and even cell-phone cameras with tiny lenses have comparable spatial sampling rate, geometric fidelity, and higher photometric range than desktop scanners of just a few years ago. Top of the line camera-phones already provide 5 mega pixels in color, which is sufficient for most A4 pages. The effects of non-uniform illumination can be mitigating by taking calibration pictures. We can therefore expect that most document acquisition will soon take place with 2-D sensor arrays rather than linear sensor sweep [134, 135].

High quality portable document acquisition systems (first for law enforcement and military applications) will require personal OCR, DIA, and document interpretation support systems [136]. Current defense interests are mainly in foreign-language documents and non-Latin scripts. Since the person acquiring the document is likely to have some expertise or at least interest in its contents, and images are not acquired in large quantities, increased interaction seems appropriate, at least in preprocessing. Interaction will be enhanced by direct action which allows pointing faster and more accurately than with a mouse, but hampered by the miniature screen. A letter-size document is certainly not readable on any camera-back display. Zooming and scrolling on both directions is impractical. Perhaps the new textile based displays will provide a satisfactory interface. Another alternative for interaction is notebook-sized touch-sensitive displays like the Tablet-PC.

Another topic of rising interest is reading text in videos, including road signs from car-mounted cameras [137, 138, 139, 140]. Such text is often in color and exhibits more geometric and photometric distortion than text scanned from paper. Furthermore, there is less context of every kind.

4.2 Web-Wide Data Accessibility

Rapidly increasing storage and communication capacity has led to a qualitative change in the nature of document image collections available for experimentation. The information retrieval community is alread making good use of web-based document collections to evaluate diverse approaches in the contests of the *Text Retrieval Conferences* (TREC) *Genomics Track*, the *Knowledge Discovery and Data Mining Cup*, and the *Creative Assessment of Information Extraction in Biology.* Image test databases typically contain at least two orders of magnitude fewer documents than test collections for information retrieval, extraction, categorization, and screening (because about a million bytes must be processed per page image, versus a few thousand bytes for an encoded page).

Most DIA research has been based on *ad hoc* collections of documents assembled by the researchers themselves because they are rich in aspects relevant to their particular research task. Although the collection, annotation and documentation of such test databases is not a trivial task, we seldom see much reuse by different groups of researchers, except possibly in Chinese and Japanese character recognition. This is likely to change as more and more research data sets are posted on the web. Large enough benchmarks would allow each test to be run on new, but statistically representative, samples. This would help avoid tuning algorithms to the test set, which is an almost inevitable consequence of open test collections of limited size [141].

Most applications must contend with highly repetitive material (for example, some firms do nothing but convert telephone directories to computable readable form). Nevertheless, many researchers strive for diversity within the constraints imposed by their task. Although this approach tests the range of applicability of the algorithms, it would also be desirable to experiment with adaptation on large, relatively homogeneous sets of document images that can now be readily found on the web.

Collection tools for DIA research require some database of digital libraries with downloadable page images, and a search engine capable of searching the database (or the whole web). The first step is the location of one or more collections with images of the desired type. (It is not always easy to tell, just by looking at a display, which pages are in image format, and which pages are in coded format). Whether *partial* processing of documents, such as type categorization, script or language recognition, contrast enhancement, skew detection and removal, segmentation at various levels (e.g. paragraph, line, word), table spotting, etc., is valuable by itself may be open to question, but there are certainly a great many researchers and publications engaged in pursuing such relatively narrow goal because end-to-end document processing requires a large team with varied resources. It would therefore be valuable to develop tools for the extraction of document collections with specific characteristics including degree of homogeneity or heterogeneity from digital libraries, and appropriate specifications of standard formats for intermediate results.

A small-scale study was reported on the *Making of America* collection (part of Cornell University's Digital Library), which at the time comprised 267 monographs (books) and 22 journals (equaling 955 serial volumes) for a total of 907,750 pages, making it three orders of magnitude larger than the datasets traditionally used in document analysis research (e.g., the UW1 CD-ROM).

Two tasks were evaluated: optical character recognition and table detection. In the case of the former, the textual transcriptions provided by the digital library (primarily for retrieval purposes) were used as the ground-truth, while for the latter, a visual inspection of the pages purported to contain tables was conducted, enabling precision (but not recall) measurements [142].

4.3 Digital Libraries

The *Million Book Project* at Carnegie Mellon is already well over the half-way mark. The Google consortium plans to digitize over 10.5 million unique books. The non-profit *Open Content Alliance*, initiated by the Internet Archive and Yahoo, proposes to provide broad access to non-proprietary world culture on paper. The CMU project produces only page images, but Google is experimenting with commercial OCR systems in dozens of languages with a view to provide searchable text. The *European Library* offers access to both digital and non-digital resources of the 45 national libraries of Europe. Most current digitization projects produce only page images: browsing digital libraries accessible through university libraries suggests that only a small fraction of their content has been transcribed. Crane addresses the issues of scale and sampling of quasi-infinite collections [143].

These "cultural" collections, which convert old books to computer-readable form, represent only part of the growth of digital libraries. Equally important are specialized collections for research, assembled from journals, conference proceedings and reports that are already in computer readable form. Some well known examples are: *ArXiv* for Physics, *DML* for Mathematics, *CiteSeer* for Computer Science, and *Medline* for medicine, but there are growing collections in every field of study.

In addition to web-wide access to cultural and technical collections, there are many novel services. Among the most popular are music servers, genealogical searches, and software that allows organizing and sharing personal photographs. Newspapers, radio and television stations offer access to their archives, and specialized search engines have been developed for sound effects [144]. Some of these require audio interaction, while many game sites and some web-based educational laboratories need a haptic modality. Nevertheless, we do not believe that multimedia will diminish the importance of digital sources of conventional printed information, and of related technologies. Current developments at the intersection digital library development and DIA include research opportunities in digitizing, coding, annotating, disseminating and preserving library documents [145].

4.4 Interoperability

A simple ASCII or Unicode file may be sufficient output for experiments on isolated digits and characters. But how should we code the output of an equation recognition system? Most researchers use either a proprietary format or TeX [146, 147, 148]. Both lack a good transition to analytical and numerical equation processing tools like Mathematica and Maple. A similar quandary arises with table recognition. Again, we would like a format that allows a smooth transition to a database query language. We favor *Wang Notation*, which provides a layout-independent representation of the relations between hierarchical category headers and content cells [149, 150]. For archival circuit diagrams and engineering drawings, the natural choice seems to be one of the widespread CAD formats (like Spice, Synopsis, and AutoCad).

Most business documents now carry XML tags, which facilitate their interpretation by whatever community agrees to the underlying convention. XML tags allows automated processing of equivalent fields, regardless of what they are called on the document. For instance, one tag may specify to whom payment should be routed, regardless of whether the name field is called vendor, provider, supplier, or seller. Digital libraries have evolved elaborate conventions for tagging metadata, beginning with the Library of Congress MARC format, and migrating from SGML to XML and the Dublin and SDLIP Cores [151]. Perhaps it is finally time for our research community to agree at least on XML schemes for "interoperability" [152]. This will also eventually help to relieve us from the tedious task of reading technical articles, which will be delegated to indefatigable autonomous agents.

XML tags have no actual meaning or semantics. The notion of meaning appears to require some kind of shared understanding of a topic. Since many current attempts to formalize meaning are focused on *ontologies*, ontological engineering may play a part in the extraction of concepts from documents [153].

4.5 Document Storage

We keep defining new units to keep up with the size and speed of storage devices. We can store book-length files and high-resolution pictures on devices that hang on our key chain. Whereas document image compression used to be a popular field of research that led to impressive increases in the compression ratio of mainly-text images, there is no longer any need to compress documents at the "retail" level. Large archives are still compressed, but communication links are so fast that they are often decompressed *before* retrieval.

Merely digitizing or coding something does not guarantee permanent access. For instance, many records from WWII were kept on punched cards. Not only did the punch cards disintegrate, but the card readers have disappeared. Magnetic tape and disk and optical media have a relatively short life. Furthermore, the software required to read the coded data may be incompatible with computers of another generation. It is not uncommon for engineering drawings prepared on earlier Computer Aided Design systems to be rescanned and revectorized, simply because the CAD software can no longer run on any available computer. Diskettes, tape cassettes, and ZIP drives are already obsolete. Until recently, many organizations opted for archiving documents on microfilm or microfiche instead of digital media. However, at current storage costs, it is plausible to keep everything *on line*. When the server is replaced, everything is copied, so there is no need to worry about removable media forgotten in some cabinet. Is the solution to keep every document spinning for ever?

5 Conclusions

This survey can be regarded only as samples of research selected from a large population according to prior probabilities that correspond only to the authors' own research interests. It is far from exhaustive or representative, and different topics are covered at different depths. In spite of the plethora of citations, it suffers from small-sample effects. Some of our samples may be distorted or mislabeled. As a training set for predicting research, it is highly biased. Nevertheless, we take the opportunity to list our impressions, based on imperfect training and grossly suboptimal recognition, of where the field is heading, and of what research problems should be addressed to reach the next stage.

5.1 Trends

Increasing processing power and storage capability by over a factor of 1000 during the last decade allows much greater use of context of all types. This leads naturally to exploiting common-source language and style constraints, i.e., field classification in OCR and joint processing of multiple tables, forms, figures and other document components in DIA. Further decreases in error rates are unlikely without adaptive/dynamic field classification. Although none of the underlying ideas are new, they can now be tested without access to supercomputers, and perhaps even incorporated into commercial OCR engines.

The availability of useful OCR and DIA software and inexpensive scanners on desktop computer systems means that all necessary interaction, including labeling training material, adjusting parameters, proofreading and corrections, is likely to be carried out by folks who would rather be doing something else. We do not believe that interaction can be eliminated in the foreseeable future, i.e., that most tasks of interest can be fully automated. This increases the premium on transparent and effective interaction. The demands on users can be alleviated by systems capable of taking full advantage of machine-initiated, durable interaction.

OCR and DIA capability is migrating to consumer-grade cameras, pocket computers and even to camera-phones. Although good interfaces that replace page-displays and keyboards are still lacking, many users will find these devices convenient enough to accept a touch-sensitive screen-and-audio interface for casual document capture.

5.2 Open Problems

Although both style and language constraints have been extensively investigated, we found little or no research on combining style, language constraints and order-dependent shape context. How can these diverse constraints be combined optimally?

In interactive dynamic recognition systems, both operator interventions and classification results can permanently change some of the classification parameters. Therefore the overall accuracy of such systems depends both on the quality of the interaction and on the order of arrival of patterns to be classified. How should such systems be evaluated?

Interactive visual classification benefits from the availability of a visual model for mediating communications between the operator and the machine. How can such visual models be constructed for new visual recognition tasks?

In all OCR and DIA systems with which we are familiar, feature sets are constructed by trial and error. How can a complete OCR and DIA system for a new language, script, and page layout, be generated automatically, starting only with a collection of labeled pixel maps?

Acknowledgement. George Nagy acknowledges support from the National Science Foundation under Award #0414854.

References

1. Grother, P.: Handprinted Forms and Character Database, NIST Special Database 19, technical report, March. 1995.
2. McCarthy, J.: Notes on formalizing contexts. In Kehler, T., and Rosenschein, S., eds., *Proceedings of the Fifth National Conference on Artificial Intelligence*, pp. 555–560. Los Altos, CA, Morgan Kaufmann, 1986.
3. Veeramachaneni, S., Sarkar, P., Nagy, G.: Modeling Context as Statistical Dependence, in *Procs. Modeling and Using Context: 5th International and Interdisciplinary Conference* CONTEXT 2005, Paris, France, July 5-8, (2005). Lecture Notes in Computer Science, Volume 3554, pp. 515–528, Jul 2005.
4. Bouquet, P., Serafini L.: Comparing formal theories of context in AI. *Artificial Intelligence*, (2004). 155: pp. 1–67.
5. Modeling and Using Context: Second International and Interdisciplinary Conference, CONTEXT'99, Trento, Italy, September pp. 9–11, (1999), Proceedings Lecture Notes in Computer Science Vol. 1688 Bouquet, P.; Serafini, L.; Brezillon, P.; Benerecetti, M.; Castellani, F. (Eds.)
6. Modeling and Using Context: 4th International and Interdisciplinary Conference, CONTEXT 2003, Stanford, CA, USA, June 23-25, 2003, Proceedings Series: Lecture Notes in Computer Science, Vol. 2680 Blackburn, P.; Ghidini, C.; Turner, R.M.; Giunchiglia, F. (Eds.) (2003).
7. Modeling and Using Context: 5th International and Interdisciplinary Conference, CONTEXT 2005, Paris, France, July 5-8, 2005, Proceedings Series: Lecture Notes in Computer Science, Vol. 3554 Dey, A.; Kokinov, B.; Leake, D.; Turner, R. (Eds.) (2005).

8. Yannakoudakis, E., Angelidakis, G.: An insight into the entropy and redundancy of the English dictionary, *IEEE Transactions on Pattern Analysis and Machine Intelligence*, 10(6), 960–970, 1988.
9. Suen, C.Y.: N-gram statistics for natural language understanding and text processing, *IEEE Transactions on Pattern Analysis and Machine Intelligence*, 1(2), 164–172, 1979
10. Katz, S. M. : Estimation of probabilities from sparse data for the language model component of a speech recognizer, *IEEE Transactions on Acoustics, Speech and Signal Processing*, 35(3):400–401, March 1987.
11. Rice, S., Nagy, G., Nartker T.: Optical Character Recognition: An Illustrated Guide to the Frontier, Kluwer Academic Publishers, Boston/Dordrecht/London, 1999.
12. Hull, J.J., Srihari S.N.: Experiments in Text Recognition with Binary N-Gram and Viterbi Algorithms, *IEEE Trans. Pattern Analysis and Machine Intelligence*, 4(5), 520–530, Sept. 1982.
13. Hull, J.J.: A hidden Markov model for language syntax in text recognition. In *Proceedings of the Eleventh Conference on Pattern Recognition*, volume 2, 124–127, 1992.
14. Hull, J.J.: Incorporating language syntax in visual text recognition with a statistical model. *IEEE Trans. Pattern Analysis and Machine Intelligence*, 18(12):1251–1256, 1996.
15. Nagy, G.: Teaching a Computer to Read, *Proc. 11th Int'l Conf. Pattern Recognition,* vol. 2, pp. 225–229, 1992.
16. Raviv, J.: Decision Making in Markov Chains Applied to the Problem of Pattern Recognition, *IEEE Trans. Information Theory,* VOL. IT-13, no. 4, 536–551, 1967.
17. Toussaint, G. T.: The use of context in pattern recognition,*Pattern Recognition,* Vol. 10, 189–204, 1978.
18. Shinghal, R., Toussaint, G.T., Experiments in text recognition with the modified Viterbi algorithm, *IEEE Trans. Pattern Analysis and Machine Intelligence* 1(2), 184–193, 1979.
19. Shinghal, R., Toussaint, G.T.: The sensitivity of the modified Viterbi algorithm to the source statistics, *IEEE Trans. Pattern Analysis and Machine Intelligence* 2(2), 1181–1184, 1980.
20. Sinha, R.M.K., Prasada, B.: Visual Text Recognition through Contextual Processing, *Pattern Recognition*, 20(5), 463–479, 1988.
21. Sinha, R.M.K., Prasada, B., Houle, G. F., Sabourin, M.: Hybrid Contextural Text Recognition with String Matching, *IEEE Trans. Pattern Anal. Mach. Intell.* 15(9), 915–925 (1993).
22. Rabiner, L.R. : A Tutorial on Hidden Markov Models and Selected Applications in Speech Recognition, *Proceedings of the IEEE*, 77(2), 257–286, 1989.
23. Gilloux, M., Leroux, M. Bertille J.M.: Strategies for Handwritten Words Recognition Using Hidden Markov Models, *Proc. Second Int'l Conf. Document Analysis and Recognition*, 299–304, 1993.
24. Kuo, S.S., Agazzi, O.E.: Visual keyword recognition using hidden Markov models, *Proc. of IEEE Computer Society Conference on Computer Vision and Pattern Recognition*, 329–334, 1993.
25. Nathan, K.A., Bellegarda, J.R., Nahamoo, D., Bellegarda, E.J.: On-Line Handwriting Recognition Using Continuous Parameter Hidden Markov Models, *Proc. Int'l Conf. Acoustics, Speech, and Signal Processing,* vol. 5, 121–124, 1993.

26. MacKay, D.J.C., Peto, L.: A hierarchical Dirichlet language model. *Natural Language Engineering,* 1(3):1–19, 1994.
27. Bazzi, I., Schwartz, R, Makhoul, J.: An Omnifont Open-Vocabulary OCR System for English and Arabic, *IEEE Trans. Pattern Analysis and Machine Intelligence,* vol. 21(6) 495–504, June 1999.
28. Feng, S., Manmatha, R., McCallum, A.: Exploring the Use of Conditional Random Field Models and HMMs for Historical Handwritten Document Recognition, *Proc. 2nd IEEE International Conference on Document Image Analysis for Libraries,* DIAL 2006, Lyon, France, April 2006.
29. Sarkar, P., Nagy, G.: Classification of Style-Constrained Pattern-Fields, *Proc. 15th Int'l Conf. Pattern Recognition*, 859–862, 2000.
30. Sarkar, P., Nagy, G.: Style Consistency in Isogenous Patterns, *Proc. Sixth Int'l Conf. Document Analysis and Recognition*, pp. 1169–1174, 2001.
31. Sarkar, P., Nagy, G.: Style Consistent Classification of Isogenous Patterns, *IEEE Trans. Pattern Analysis and Machine Intelligence,* 27(1), Jan. 2005.
32. Veeramachaneni, S., Nagy, G.: Analytical Results on Style-constrained Bayesian Classification of Pattern Fields, *IEEE Trans. Pattern Analysis and Machine Intelligence*, vol. 29(7) 1280–1285, July 2007.
33. Spitz, A.L.: An OCR based on character shape codes and lexical information, *Proceedings of the Third International Conference on Document Analysis and Recognition* (Volume 2) Volume 2, Page: 723, 1995.
34. Spitz, A. L., Maghbouleh, A.: Text Categorization using Character Shape Codes, *SPIE Symp on Electronic Image Science and Technology,* San Jose, pp. 174–181, 2000.
35. Ho, T.K., J.J. Hull, S.N. Srihari: A Computational Model for Recognition of Multifont Word Images, *Machine Vision and Applications 5,* 157–168, 1992.
36. Ho, T.K., J.J. Hull, S N. Srihari: A Word Shape Analysis Approach to Lexicon Based Word Recognition, *Pattern Recognition Letters 13*, 821-826, 1992.
37. Hong, T., Hull, J.J.: Visual Inter-Word Relations and Their Use in OCR Postprocessing, Proc. Third Int'l Conf. Document Analysis and Recognition, vol. 1, pp. 442–445, 1995.
38. Cesarini, F., Gori, M., Marinai, S., Soda, G.: INFORMys: A flexible invoice-like for reader system, *IEEE Trans. on Pattern Recognition and Machine Intelligence*, 20(7), 730–745, July 1998
39. Marinai, S., Marino, E., Soda, G.: Font Adaptive Word Indexing of Modern Printed Documents, *IEEE Trans. Pattern Recognition and Machine Intelligence* 28(8), 1187–1199, August 2006.
40. Shi, H., Pavlidis, T.: Font Recognition and Contextual Processing for More Accurate Text Recognition, *Proc. Fourth Int'l Conf. Document Analysis and Recognition*, vol. 1, pp. 39–44, 1997.
41. Zramdini, A.W., Ingold, R.: Optical Font Recognition from Projection Profiles, *Electronic Publishing* 6(3): 249–260 (1993).
42. Zramdini, A.W., Ingold, R.: Optical Font Recognition Using Typographical Features, *IEEE Trans. Pattern Analysis and Machine Intelligence*, 20(8), 877–882, Aug. 1998.
43. Bapst, F., Ingold, R.: Using Typography in Document Image Analysis. In *Proc. Raster Imaging and Digital Typography* (RIDT'98), Saint-Malo (France), pp. 240–251, 1998.

44. Srihari, S.N., Bandi, K., Beal, M.: A Statistical Model for Writer Verification, *Proc. Int. Conf. on Document Analysis and Recognition* (ICDAR-05) Seoul, Korea, August 2005.
45. Kawatani, T.: Character Recognition Performance Improvement Using Personal Handwriting Characteristics, *Proc. Third Int'l Conf. Document Analysis and Recognition,* vol. 1, pp. 98–103, 1995.
46. Veeramachaneni, S., Nagy, G.: Style Context with Second-Order Statistics, *IEEE Trans. Pattern Analysis and Machine Intelligence,* 27(1), Jan. 2005.
47. Andra, S.: Non-parametric approaches to style-consistent classification, Rensselaer Polytechnic Institute PhD dissertation, December 2006.
48. Andra, S., Nagy, G.: Combining Dichotomizers for MAP Field Classification, *Proceedings of International Conference on Pattern Recognition-XVIII*, Hong Kong, September 2006.
49. Widrow, B, Hoff, M.E.: Adaptive switching circuits, *1960 IRE WESCON Conv. Record,* Part 4, 96-104, 1960.
50. Aizerman, M.A., Braverman, E.M., Rozonoer, L.I.: The Robbins-Monroe process and the method of potential functions, *Automation and Remote Control* 26, 1882–1885, November 1965.
51. Tsypkin, Y. Z.: Adaptation, training, and self-organization in automatic systems, *Automation and Remote Control,* vol. 27, pp. 1652, January 1966.
52. Fu, K.S.: Learning techniques in pattern recognition systems, in Pattern Recognition (L.N. Kanal, ed.) Thompson Book Company, Washington, 1968.
53. Jain, A., Duin, R., Mao, J.: Statistical Pattern Recognition A Review. *IEEE Transactions on Pattern Analysis and Machine Intelligence*, 22(1):4–37, 2000.
54. Zadeh, L.A.: On the definition of adaptivity, *Proceedings of the IRE* 51, #3. 469–470, 1963.
55. Lendaris, G.G.: On the Definition of Self-Organizing Systems, *Proceedings of the IEEE 52*, 3, March, 1964
56. Nagy, G.: Pattern Recognition IEEE 1966 Workshop, *IEEE Spectrum*, pp. 92–94, February 1967.
57. Castelli, V., Cover, T.: On the exponential value of labeled samples, *Pattern Recognition Letters* 16, 105–111, 1995.
58. Castelli, V., Cover, T.: The relative value of labeled and unlabeled samples in pattern recognition with an unknown mixing parameter, *IEEE-Trans. Information Theory* 42(6), 2101–2117, 1996.
59. Scudder, H.J.: Probability of error of some adaptive pattern-recognition machines, *IEEE. Trans. Information Theory* IT-11, 363–371, July 1965.
60. Spragins, J.: Learning without a teacher, *IEEE Trans. Information Theory,* vol. IT-12, 223–229, April 1966.
61. Stanat, D.F.: Unsupervised learning of mixtures of probability functions, in Pattern Recognition (L.N. Kanal, ed.) Thompson Book Company, Washington, 1968.
62. Aizerman, M.A., Braverman, E.M., Rozonoer, L.I.: The probability problem of pattern recognition learning and the method of potential functions, *Automation and Remote Control* 25, 1175–1192, September 1964.
63. Braverman, E.M.: Experiments on machine learning to recognize visual patterns, translated from *Automat. i Telemekh.,* vol. 23, pp. 349–364, March 1962, *Automation and Remote Control,* vol. 23, 315–327, 1962.

64. Braverman, E.M.: The method of potential functions in the problem of training machines to recognize patterns without a trainer, *Automation and Remote Control,* vol. 27, 1748-1771, October 1966.
65. Dorofeyuk, A.A.: Teaching algorithm for a pattern recognition machine without a teacher, based on the method of potential functions, *Automation and Remote Control,* vol. 27, 1728–1737, October 1966.
66. Nagy, G.: State of the Art in Pattern Recognition,*Proceedings of the IEEE* 56, #5, 336–362, May 1968.
67. Jain, A.K.: Cluster Analysis, Chapter 2 in *Handbook of Pattern Recognition and Image Processing* (K-S Fu and T-Y Young, eds), Academic Press, NY 1986.
68. Ball, G.H.: Data analysis in the social sciences: What about the details? *Procs. Fall Joint Computer Conference,* pp. 533–560, Spartan Books, 1965.
69. MacQueen, J.: Some methods for classification and analysis of multivariate observations, *Proc. 5th Berkeley Symp on Statistics and Probability,* pp. 281-297, Berkeley, CA University of California Press, 1967.
70. Ball, G.H., Hall, D.J.: A clustering technique for summarizing multivariate data, Behavioral Science, 12, pp. 153–155, March 1967.
71. Linde, Y., Buzo, A., Gray, R.M. : An algorithm for vector quantization design, *IEEE Trans. Comm.* 28, 84–95, 1980
72. Gersho, A., Gray, R. M. : Vector Quantization and Signal Compression, The International Series in Engineering and Computer Science, 1991.91
73. Dubes, R., Jain, A.K.: Validity studies in clustering methodologies, *Pattern Recognition 11*, 235–254, 1979.
74. Jain, A.K., Dubes, R.: Algorithms for Clustering Data, Prentice Hall 1988.
75. Theodoridis, S., Koutroumbas, T.: Pattern Recognition, Academic Press, 1999.
76. Topchy, A., Jain, A.K., Punch, W.: Clustering Ensembles: Models of Consensus and Weak Partitions, *IEEE Trans. Pattern Analysis and Machine Intelligence*, 27(12), 1866–1881, Dec 2005.
77. Duda, R.O., Hart, P.E., Stork, D.G.: Pattern Classification. New York: John Wiley and Sons, 2001.
78. Nagy, G., Shelton, G.L. : Self-Corrective Character Recognition System, *IEEE Transactions on Information Theory* IT-12, #2, pp. 215–222, April 1966.
79. Baird, H.S., Nagy, G.: A Self-Correcting 100-Font Classifier, Document Recognition, *Proc., IS&;T/SPIE Symp. on Electronic Imaging: Science Technology*, San Jose, CA, February 6-10, 1994, L. Vincent and T. Pavlidis, eds., vol. 2181, pp. 106–115, 1994.
80. Breuel, T., Mathis, C.: Classification Using a Hierarchical Bayesian Approach, *Proc. 16th Int'l Conf. Pattern Recognition*, 40103–40106, Aug. 2002.
81. Sarkar, P., Baird, H.S., Zhang, X.: Training on Severely Degraded Text- Line Images, *Proc. Seventh Int'l Conf. Document Analysis and Recognition*, pp. 38–43, Aug. 2003.
82. Veeramachaneni, S., Nagy, G.: Adaptive Classifiers for Multisource OCR, *Int'l J. Document Analysis and Recognition,* 6(3), 154–166, Aug. (2004).
83. Marosi, I., Tóth, L.: OCR Voting Methods for Recognizing Low Contrast Printed Documents: in *Proc. 2nd IEEE International Conference on Document Image Analysis for Libraries,* DIAL 2006, Lyon, France, April 2006.
84. Gold, B.: Machine recognition of hand-sent Morse code, *IRE Trans. Information Theory,* vol. IT-5, pp. 17–24, March 1959.
85. Lucky, R. W.: Automatic Equalization for Digital Communication. *Bell Systems Technical Journal,* 44:547–588, 1965.

86. Lucky, R. W.: Techniques for adaptive equalization of digital communication systems. *Bell Systems Technical Journal,* 45:255–286, February 1966.
87. Nagy, G., Tolaba, J.: Nonsupervised Crop Classification through Airborne Multispectral Observations, *IBM Journal of Research and Development 16,* #2, pp. 138–153, March 1972.
88. Shahshani, B.M., Landgrebe, D.A.: Asymptotic improvement of supervised learning by utilizing additional unlabeled samples: normal mixture density case, *Proc. SPIE Vol. 1766, p. 143–155, Neural and Stochastic Methods in Image and Signal Processing,* Su-Shing Chen; Ed., 1992.
89. Dempster, A.P., Laird, M.M., Rubin, D.B.: Maximum Likelihood from Incomplete Data via the EM Algorithm, *J Royal Statistical Soc.*, vol. 39, no. 1, pp. 1–38, 1977.
90. Redner, R.A., Walker, H.F.: Mixture densities, maximum likelihood, and the EM algorithm, *SIAM Review* 26, 2, pp. 195–235, 1984.
91. Raudys, S., Jain, A.K.: Small Sample Size Effects in Statistical Pattern Recognition: Recommendations for Practitioners, *IEEE Trans. on Patt. Anal. and Machine Intell.,* 13(3), 252–264, 1991.
92. Casey, R.G., Nagy, G.: Recognition of Printed Chinese Characters, *IEEE Transactions on Electronic Computers EC-15,* #1, pp. 91–101, February 1966.
93. Liu, C.-L., Jaeger, S., Nakagawa, M.: On line recognition of Chinese characters: the State of the Art, *IEEE Trans. Pattern Analysis and Machine Intelligence* 26, 2, pp. 198–213, (2004).
94. Casey, R.G., Nagy, G.: Autonomous Reading Machine, *IEEE Transactions on Computers* C-17, #5, pp. 492–503, May 1968.
95. Casey, R.G., Nagy, G.: Advances in Pattern Recognition, *Scientific American* 224, #4, pp. 56–71, 1971.
96. Casey, R.G.: Text OCR by Solving a Cryptogram, *Proc. Eighth Int'l Conf. Pattern Recognition,* pp. 349–351, 1986.
97. Nagy, G., Seth, S., Einspahr, K., Meyer, T.: Efficient Algorithms to Decode Substitution Ciphers with Applications to OCR,*Proceedings of International Conference on Pattern Recognition*, vol. 8, 352–355, Paris, October 1986.
98. Ho, T.K., Nagy, G.: OCR with no shape training,*Proceedings of International Conference on Pattern Recognition-XV*, vol. 4, pp. 27–30, Barcelona, September 2000.
99. Ascher, R.N., Nagy, G.: A Means for Achieving a High Degree of Compaction on Scan-Digitized Printed Text, *IEEE Transactions on Computers* C-23, #11, pp. 1174–1179, October 1974.
100. Bottou, L. Haffner, P., Howard, P., Simard, P., Bengio, Y., LeCun, Y.: High Quality Document Image Compression with DjVu, *Journal of Electronic Imaging*, vol. 7, no. 3, pp. 410–425, July 1998.
101. Witten, I., Moffat, A., Bell, T.: Managing Gigabytes, Academic Press 1999.
102. Marinai, S., Marino, E., Soda, G.: Font Adaptive Word Indexing of Modern Printed Documents, *IEEE Trans. Pattern Recognition and Machine Intelligence* 28(8), pp. 1187–1199, August 2006.
103. Nagy, G.: Twenty Years of Document Image Analysis in IEEE PAMI, *IEEE Trans. Pattern Analysis and Machine Recognition* 22(1), 38–62, January 2000.
104. Baird, H.S., Lopresti, D. P., editors: Human Interactive Proofs, *Procs. Second International Workshop,* volume 3517 of LNCS, 2005.
105. Miller, G.: The magical number seven plus or minus two; some limits on our capacity for processing information. *Psychological Review*, 63:81–97, 1956.

106. Sammon, J.W.: Interactive pattern analysis and classification, *IEEE Trans. Computers* C-16, 594-616, July 1970.
107. Ho, T.K.: Exploratory Analysis of Point Proximity in Subspaces, *Proceedings of the 16th International Conference on Pattern Recognition,* Quebec City, August 11–15, 2002.
108. Cesarini, F., Marinai, S., Sarti, L., Soda, G.: Trainable table location in document images, *Proceedings of International Conference on Pattern Recognition-XVI*, Vol. 3, Quebec City, 236–240, 2002.
109. Edwards, J.. New interfaces: Making computers more accessible. IEEE *Computer*, pages 12–14, December 1997.
110. Ancona, M., Locati, S., Mancini, M., Romagnoli, A., Quercini, G.: Comfortable textual data entry for PocketPC: the WTX system. In Advances in Graphonomics, *Proceedings of International Graphonomics Symposium 2005*, Salerno, Italy, June 2005.
111. Langendorf, D.J.: Textware solution's Fitaly keyboard v1.0 easing the burden of keyboard input. *WinCELair Review,* February 1998.
112. Masui, T.: An efficient text input method for pen-based computers. In *Proceedings of the ACM Conference on Computer-Human Interaction*, pages 328–335, 1998.
113. James, C.L., Reischel, K.M.: Text input for mobile devices: Comparing model prediction to actual performance. In *Proceedings of the ACM Conference on Computer-Human Interaction,* pages 365–371, 2001.
114. Mackenzie, S., Soukore, W.: Text entry for mobile computing: Models and methods, theory and practice, *Human-Computer Interaction,* 17:147–198, 2002.
115. Nagy, G., Li, L., Samal, A., Seth, S., Xu, Y.: Integrated text and line-art extraction from a topographic map.*International Journal on Document Analysis and Recognition*, 2(4):177–185, June 2000.
116. English, W.K., Engelbart, D.C., Berman, M.L.: Display-selection techniques for text manipulation. *IEEE Transactions on Human Factors in Electronics,* HFE-8(1):5–15, March 1967.
117. Zou, J. Nagy, G.: Interactive visual pattern recognition,*Proceedings of the 17th International Conference on Pattern Recognition*, XVI, IEEE Computer Society Press, Vol. III, pp. 478–481, Aug. 2002.
118. Zou, J. Nagy, G.: Evaluation of model-based interactive ower recognition. In *Proceedings of the 17th International Conference on Pattern Recognition,* volume 2, pages 311–314, (2004).
119. Cha, S.-H., Evans, A., Gattani, A., Nagy, G., Sikorski, J., Tappert, C., Thomas, P., Zou, J.: Computer Assisted Visual Interactive Recognition (CAVIAR) technology. In *Proceedings of the IEEE Electro/Information Technology Conference, May 2005.* PDF
120. Nagy, G.: Interactive, Mobile, Distributed Pattern Recognition, *Proc. of the 13th International Conference on Image Analysis and Processing* ICIAP, Cagliari, Italy, LNCS 3617, 37–49, 2005.
121. Zou, J. Nagy, G.: Human-computer interaction for complex pattern recognition problems, to appear in Data Complexity in Pattern Recognition, Springer Verlag, Editors: Mitra Basu, Tin Kam Ho, Publication Date: Dec. 2006.
122. Holland, M., Schlesiger, C.: High-mobility machine translation for a battlefield environment. In *Proceedings of NATO/RTO Systems Concepts and Integration Symposium*, volume 15, pages 1–3, Monterey, CA, May 1998.

123. Swan, K.: FALCon: Evaluation of OCR and machine translation paradigms, August 1999. http://www.arl.army.mil/seap/reports/kreport.pdf. (accessed on 8/8/06)
124. Fisher, F.: Digital camera for document acquisition. In Symposium on Document Image Understanding Technology, Columbia, MD, 2001. http://www.dtic.mil/matris/sbir/sbir022/a038.pdf. (accessed on 8/8/06)
125. Jacobs, C., Simard, P.. Low resolution camera based OCR. International Journal on Document Analysis and Recognition. To appear. (2004)
126. Fujisawa, H., Sako, H., Okada, Y., Lee, S.: Information capturing camera and developmental issues. In*Proceedings of the Fifth International Conference on Document Analysis and Recognition* (ICDAR'99), pages 205–208, Bangalore, India, September 1999.
127. Yang, J., Chen, X., Zhang, J., Zhang, Y., Waibel, A.: Automatic detection and translation of text from natural scenes. In *Proceedings of the IEEE International Conference on Acoustics, Speech and Signal Processing* (ICASSP '02), May 2002.
128. Hedgpeth, T., Rush, M., Black, J., Panchanathan, S.: The iCare project reader. In *Procs. Sixth International ACM SIGACCESS Conference on Computers and Accessibility*, October (2004).
129. Peters, J.P., Thillou, C., Ferreira, S.: Embedded reading device for blind people: a user-centred design. In *Procs. IEEE Emerging Technologies and Applications for Imagery Pattern Recognition* (AIPR 2004), pages 217–222, (2004).
130. Srihari, S., Shi, Z.: Forensic handwritten document retrieval system. In *Procs. First International Workshop on Document Image Analysis for Libraries* (DIAL'04), pages 188-194, (2004).
131. Xu, Y., Nagy, G.: Prototype Extraction and Adaptive OCR, *IEEE Trans. Pattern Analysis and Machine Intelligence* Vol. 21, 12, pp. 1280–1296, Dec. 1999.
132. MacKay, D.: Information-based objective functions for active data selection. *Neural Computation,* Vol. 4, No. 4, pp. 590–604, 1992.
133. Freund, Y., Seung, H.S., Shamir, E., Tishby, N.: Selective sampling using the query by committee algorithm. *Machine Learning* 28, 133–168, 1997
134. Liang, J., Doerman, D., .Li, H.: Camera-based analysis of text and documents: a survey, *International Journal on Document Analysis and Recognition,* 7(2-3), 84–104, July 2005.
135. Pollard, S., Pilu, M.: Building cameras for capturing documents, *International Journal on Document Analysis and Recognition,* 7,(2–3), 123–137, July 2005.
136. Lopresti, D., Nagy, G.: Mobile Interactive Support System for Time-Critical Document Exploitation, *Procs. Symposium on Document Image Understanding,* College Park, MD November 2005.
137. Gandhi, T., Kasturi, R., Antani, S.: Application of planar motion segmentation for scene text extraction. In *Proc. of the ICPR*, 2000, I: 445–449.
138. Myers, G., Bolles, R., Luong, Q.-T., Herson, J.: Recognition of text in 3-D scenes. In *Proc. of the 4th Symp. on Document Image Understanding Technology*, pp. 23–25, 2001.
139. Wu, W., Chen, X., Yang, J.: Incremental Detection of Text on Road Signs from Video with Application to a Driving Assistant System, *Proceedings of ACM Multimedia 2004* (MM2004), pp. 852–859 (2004).

140. Yamaguchi, T., Maruyama, M., Miyao1, H., Nakano, Y.: Digit recognition in a natural scene with skew and slant normalization, *International Journal on Document Analysis and Recognition*, 7(2-3), 168–177, July 2005.
141. Salzberg, S.L.: On comparing classifiers: Pitfalls to avoid and a recommended approach, *Data Mining and Knowledge Discovery 1*, 317–327, 1997.
142. Lopresti, D., Zhou, J.: Document analysis and the World Wide Web, In *Proceedings of the Second IAPR Workshop on Document Analysis Systems,* pages 651–659, Malvern, PA, Oct. 1996.
143. Crane, G.: What Do You Do with a Million Books? *D-Lib Magazine* 12(3), ISSN 1082–9873, March 2006.
144. Rice, S.V., Bailey, S.M.: A Web Search Engine for Sound Effects, in *Proceedings of the 119th Convention of the Audio Engineering Society,* Paper #6622, New York, (2005) (PDF).
145. Nagy G., Lopresti, D.: Interactive Document Processing and Digital Libraries, *Proc. 2nd IEEE International Conference on Document Image Analysis for Libraries*, Lyon, France, IEEE Press, 2006.
146. Chou, P.A.: Recognition of equations using a two-dimensional stochastic context-free grammar. In W. A. Pearlman, editor, *Visual Communications and Image Processing* IV, vol. 1199 of SPIE Proceedings Series, 852–863, 1989.
147. Twaakyondo, H.M., Okamoto, M.: Structure analysis and recognition of mathematical expressions, *Proc. third Inter. Conf. on Document Analysis and Recognition,* ICDAR'95, Montral, Canada, pp. 430–437, 1995.
148. Zanibbi, R., Blostein, D., Cordy, J.R.: Recognizing Mathematical Expressions Using Tree Transformation, *IEEE Transactions on Pattern Analysis and Machine Intelligence*, 24 (11), 1455–1467, 2002.
149. Wang, X.: Tabular abstraction, editing, and formatting, PhD dissertation, University of Waterloo, Canada, 1006.
150. Embley, D., Lopresti, D., Nagy, G.: Notes on Contemporary Table Recognition, Document Analysis Systems VII, 7th International Workshop, *Procs. DAS 2006*, Nelson, New Zealand, February 13–15, 2006, Horst Bunke, A. Lawrence Spitz (Eds.) LNCS 3872, pp. 164–175 Springer 2006.
151. Macgregor, G., McCulloch, E.: Collaborative tagging as a knowledge organisation and resource discovery tool, *Library Review,* Volume: 55 Issue: 5 Page: 291–300, 2006
152. Hitz, O., Robadey, L., Ingold, R.: Using XML in Document Recognition, In *Proc. Document Layout Interpretation and its Applications* (DLIA'99), Bangalore (India), 1999.
153. Tijerino, Y.A., Embley, D. W., Lonsdale, D. W., Nagy, G.: Towards Ontology generation from tables, *World Wide Web Journal* 8, 3, Springer, September 2005.

Cursive Character Segmentation Using Neural Network Techniques

Michael Blumenstein

School of Information and Communication Technology, Griffith University, Gold Coast campus, PMB 50 Gold Coast Mail Centre, Queensland 9726, Australia
M.Blumenstein@griffith.edu.au

Summary. The segmentation of cursive and mixed scripts persists to be a difficult problem in the area of handwriting recognition. This research details advances for segmenting characters in off-line cursive script. Specifically, a heuristic algorithm and a neural network-based technique, which uses a structural feature vector representation, are proposed and combined for identifying incorrect segmentation points. Following the location of appropriate anchorage points, a character extraction technique, using segmentation paths, is employed to complete the segmentation process. Results are presented for neural-based heuristic segmentation, segmentation point validation, character recognition, segmentation path detection and overall segmentation accuracy.

1 Introduction

The problem of automated handwriting recognition has endured for many decades. Active research still persists in order to pursue a satisfactory solution for recognizing off-line cursive handwriting. The motivating factors include commercial applications and scientific progress in an age-old artificial intelligence problem. One of the main impediments for progress has been the inherent variability in handwritten material [1].

Handwriting recognition itself is a mechanical process that transforms graphical human handwritten scripts into symbols that are stored on a computer system in the form of ASCII code or Unicode. One of the major problems in recognizing unconstrained cursive words is the process of segmentation [2], [3]. Segmentation refers to the method of separating the characters in a word, so that they may be used to assist in final word interpretation. Some systems use the method of over-segmentation to dissect the word at many intervals into primitives. The term "primitive" refers to an entire character or character components. Following initial over-segmentation, various techniques may be used to correctly assemble the primitives using contextual processing to recognise entire words. The removal of incorrect segmentation points from

M. Blumenstein: *Cursive Character Segmentation Using Neural Network Techniques*, Studies in Computational Intelligence (SCI) **90**, 259–275 (2008)
www.springerlink.com

over-segmented words is still a difficult problem. A solution to this problem would guarantee a higher success rate for handwritten word recognition. A number of segmentation techniques have been proposed in the literature, some of which are described below.

In [4], Bozinovic and Srihari attempt to locate possible segmentation points based on proximity to minima in the lower contour and the use of other rules that force segmentations in areas that are between two distant segmentation points. A technique proposed by Cheriet [5] for extracting "key letters" in cursive script analyses face-up and face-down valleys along with open loop regions. Cheriet employs background analysis to achieve segmentation.

Some of the more recent studies employing dissection or presegmentation include that of Han and Sethi [6] who proposed an algorithm for segmenting handwritten words based on a number of features such as crossing points, loops, concave and convex points. They reported that 50 real-world postal address images were segmented with an accuracy of 85.7%. Yamada and Nakano [7] reported a segmentation algorithm that segmented cursive words based on contour features. Reasonable recognition rates were obtained when the segmentation algorithm was used as part of a complete word recognition system.

Yanikoglu and Sandon [8] proposed a segmentation algorithm by evaluating a cost function to locate successive segmentation points along the baseline. The decision to segment at a particular point is made if the first minimum cost is located. The cost is calculated by summing the weights of four global characteristics or "style parameters" in the cursive script. The algorithm used a linear programming technique to obtain the weights of the features. The global characteristics included pen thickness, dominant slant, average character width and distance from the previous segmentation point. Finally, characters were extracted by finding the best angular line.

Eastwood *et al.* [9] proposed a neural-based technique for segmenting cursive script. In their research they trained a neural network with feature vectors representing possible segmentation points as well as "negative" features that represented the absence of a segmentation point. The feature vectors were manually obtained from training and test words in the CEDAR benchmark database. The accuracy of the network on a test set of possible segmentation points was 75.9%.

Dimauro *et al.* [10] proposed an advanced technique for segmenting cursive words as part of a recognition system to read the amounts on Italian bank cheques. The segmentation technique is based on a hypothesis-then-verification strategy. Initially, the entire word image is searched, and connected components are located within the image. Each "block" detected via this process is passed to a recogniser. If the block is rejected, a hypothesis is generated to split the block by using a "drop falling" algorithm. The algorithm employs a number of rules that analyse the background of the image to determine the first cutting point. They then employ a descending procedure that simulates a "drop-falling" process. The dropping procedure is guided by rules that take into account neighbouring pixels and a regional analysis of the upper contour to form an appropriate segmentation path. The hypothesis is then verified by classifying the strokes that have originated as a result of segmentation. A nearest neighbour technique is employed for this process. If the stroke is classified with high confidence, the segmentation hypothesis is accepted. Otherwise, a different hypothesis is considered.

Nicchiotti *et al.* [11] presented a simple but effective segmentation algorithm. The algorithm is divided into three main steps. The first step is to detect possible segmentation points by analysing the minima in the lower contour and holes. The second step is to determine the cut direction of the segmentation point. The chosen direction is the one that contains the least number of black pixels. Finally, over-segmented strokes are merged back to the main character by some heuristic rules.

Xiao and Leedham [12] presented a knowledge-based technique for cursive word segmentation. Based on connected component analysis, those components that contain more than one character are over-segmented based on a face-up or face-down region. Then over-segmented components are merged into a single character based on the knowledge of the character structure.

In this chapter, an existing neural-based segmentation technique [13] is enhanced to validate prospective segmentation points. The existing technique first uses a Feature-based Heuristic Segmenter (FHS) [14] to over-segment the handwriting. Following this, a neural confidence-based module is used to evaluate a prospective segmentation point by obtaining a fused value from three neural confidence values: segmentation point validation (SPV), left character validation (LCV) and centre character validation (CCV). The segmentation technique has two advantages. Firstly, it can reduce the number of missed segmentation points and hence increase the overall character/word recognition rate in subsequent processing. Secondly, since the number of segmentation points is optimised directly following over-segmentation, it can reduce the processing time of later stages.

The enhancements to the existing segmentation technique include an Enhanced Heuristic Segmenter (EHS) that employs ligature detection and a neural assistant for obtaining better prospective segmentation points. In addition, the neural confidence-based module is improved by using 1) a recently proposed feature extraction technique [15] for processing relevant features, 2) a single character classifier for the recognition of left characters and centre characters and 3) a segmentation path detection-based character extraction technique [16].

The remainder of the chapter is broken down into 4 sections. Section 2 describes the enhanced neural-based segmentation technique. Section 3 provides experimental results, followed by discussion in Section 4. Finally, conclusions are drawn in Section 5.

2 Enhanced Segmentation Technique

This section presents some enhancements to the neural-based segmentation technique. The new heuristic segmenter, EHS, employs two new attributes - ligature detection and a neural assistant. The first component was investigated since the former segmenter, FHS, could not effectively locate prospective segmentation points that were located under over-lapped strokes. The second feature, the neural assistant, uses a hybrid strategy that combines a character classifier and heuristic rules to over-segment the handwriting. Figure 1 shows an overview of the EHS algorithm.

i. Boundary and connected component analysis
ii. Estimation of global features (average character width, average stroke width, word height)
iii. Assign a prospective segmentation point at:
- Minima on the lower contour
- Ligatures

iv. Remove segmentation points at holes
v. Add more segmentation points by the neural assistant

Fig. 1. Overview of EHS algorithm

The improved neural confidence-based module uses a newly proposed feature extraction technique, the Modified Direction Feature (MDF) for SPV, LCV and CCV. LCV and CCV use a single classifier for character recognition and a Segmentation Path Detection (SPD) technique is used to extract characters for the recognition process. Figure 2 illustrates an overview of the entire neural-based segmentation technique. In the following sub-sections, further details of ligature detection, the neural assistant, MDF, neural confidence calculation/fusion and SPD are provided.

2.1 Ligature Detection

A ligature is a small stroke that is used to connect joined/cursive characters. One of the major features of a ligature is that it is usually located within the "middle-region" of handwritten words spanning an area down to the word baseline. Hence, a baseline detection technique can be used to identify this middle region. In this work, a modified vertical histogram is generated based on the middle region of the handwriting to locate possible ligatures.

Baseline Detection

Small strokes in a word image may extend above or below the main body of handwriting. Such letter components are called ascenders and descenders

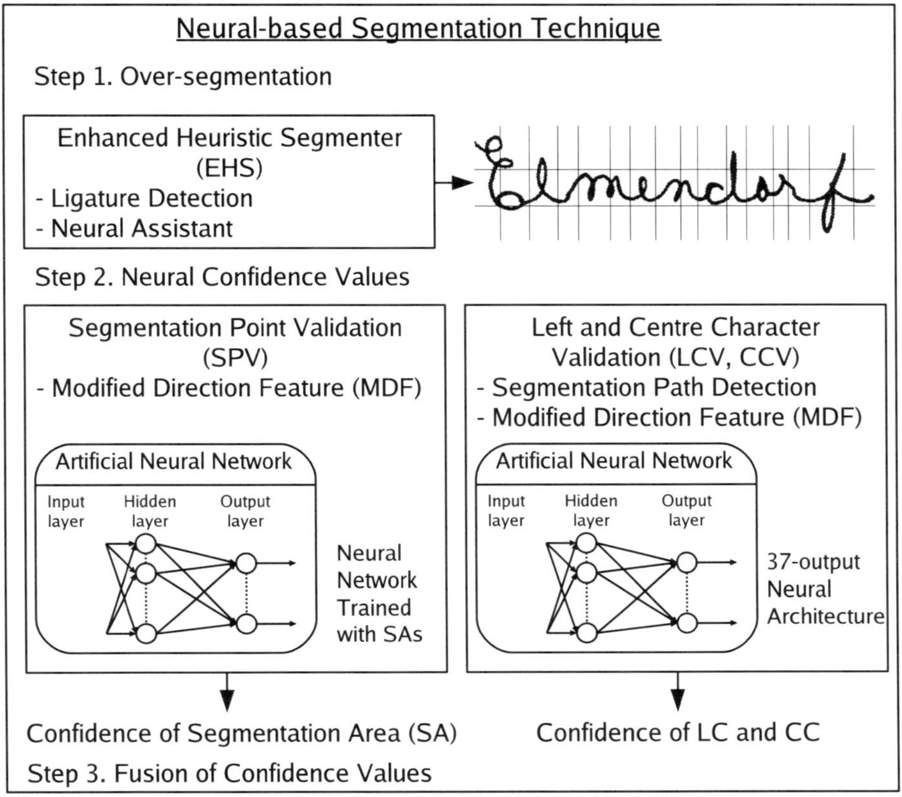

Fig. 2. Overview of the improved, neural-based segmentation technique

respectively. Examples of letters that contain such strokes are: ‘f’, ‘j’, ‘g’, ‘T’ etc. Hence the letters that contain ascenders or descenders may overlap parts of characters in the main body that do not contain such strokes. In order to over-segment the word image more accurately, it is necessary to remove ascenders and descenders before the actual segmentation process. In this research, the technique calculates the average vertical value of the maxima and minima of the upper and lower contours respectively. Outlier maxima and minima values are removed based on this average value. Finally, baselines are estimated by the average of the remaining maxima and minima.

Modified Vertical Histogram

The second step in the ligature detection algorithm is to analyse the middle region and to locate ligatures. One common approach is the use of vertical (density) histogram analysis. The analysis is based on the vertical distribution of foreground pixels. The histogram is drawn by a projection of the total number of foreground pixels in each column of the word image. Areas with low pixel density are then identified as possible segmentation points. Figure 3 illustrates an example vertical histogram; the vertical histogram is formed based on the middle region of the word "Top".

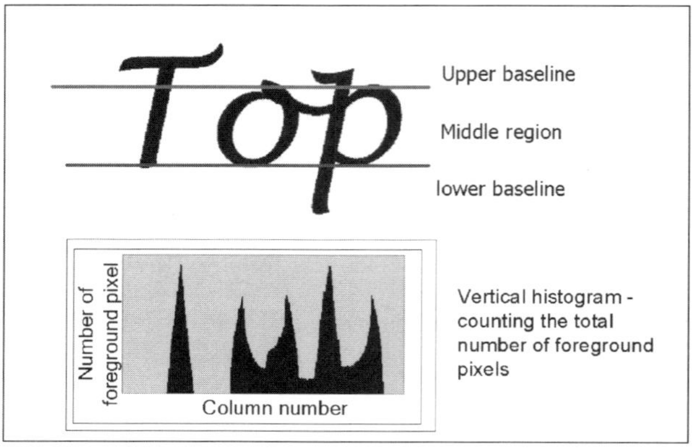

Fig. 3. Vertical histogram analysis

Figure 3 illustrates that there are an excessive number of "low" density regions. This is because the vertical histogram is not adequate to distinguish the difference between "holes" and "ligatures". In this research, a modified vertical histogram was developed to improve the accuracy of ligature location. Figure 4 shows the modified vertical histogram of the word shown in Figure 3.

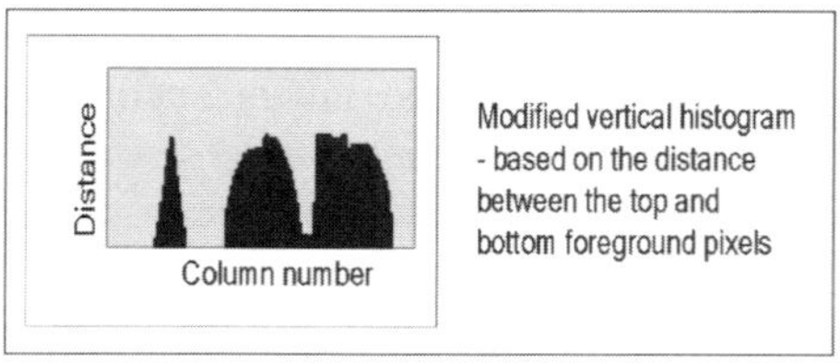

Fig. 4. Modified vertical histogram analysis

The concept of the modified vertical histogram is formed by calculating the distance between the top and bottom foreground pixels for each column in a word image. As may be seen from Figure 4, the ligature region is clear and hence easy for the segmenter to detect. One weakness of the modified vertical histogram is that it is not suitable for characters with overlapped strokes. But in this research, since the overlapped strokes are removed in most cases (i.e. the modified vertical histogram is formed from the middle region), the advantage of the modified vertical histogram can then be maximized.

Ligatures are located using the modified vertical histogram and a heuristic based on the average stroke width. Regions with distance values smaller than the average stroke width are defined as ligatures.

2.2 Neural Assistant

The neural assistant uses a character classifier and some extra heuristics to generate additional segmentation points following regular feature-based segmentation point assignment. Regions between two successive prospective segmentation points are extracted and processed by MDF in order to obtain a confidence value. Additional segmentation points are added based on the confidence value and the distance between the two prospective segmentation points. Experimental results in [16] showed that the classifier could be effectively used to distinguish character and non-character components, and hence could provide appropriate assistance in the current step.

2.3 Modified Direction Feature (MDF)

Recent work has shown that the Modified Direction Feature (MDF) enhances the character recognition process and outperforms some popular feature extraction techniques such as the Transition Feature (TF) [15]. This work demonstrated the superiority of MDF for describing patterns based on their contour or boundary. This prompted an investigation to determine the feasibility of employing MDF for SPV, LC and CC recognition to enhance the overall segmentation process. The details of MDF have been described in [15].

2.4 Neural Confidence Calculation and Fusion

Segmentation Point Validation (SPV)

Following heuristic segmentation it is necessary to discard "incorrect" segmentation points while preserving the "correct" points in a cursive word. This is achieved by calculating a number of confidence values for each prospective segmentation point (PSP) generated by the heuristic segmenter. For SPV, a neural network is trained with features extracted from segmentation areas (SAs) originally located by the heuristic algorithm. The neural network verifies whether each particular area is or is not characteristic of a segmentation point [14]. If an area is positively identified as a segmentation point, the network outputs a high confidence (>0.5). Otherwise the network will output a confidence close to 0.1. In this research, the MDF extraction technique was used to describe the segmentation area.

Left and Centre Character Classification

For this step, additional neural networks trained with handwritten characters (upper case and lower case) are required to confirm the first neural network's output. The network(s) is/are presented with areas immediately centred on/adjacent to each segmentation point. Area width is calculated based upon average character width. If for example, the area immediately to the left of the PSP proves to be a valid character, the network will output a high confidence (LC) for that character class. At the same time, if the area immediately centred on the segmentation point provides a high confidence for the reject neuron (CC), then it is likely that the PSP is a valid segmentation point. The "reject" output of the neural network is specifically trained to recognise non-character patterns (i.e. joined characters, half characters or unintelligible primitives). If this neuron gives a high confidence, this will usually indicate that the particular area being tested is a good candidate for a segmentation point. Otherwise, if any valid characters are given a high confidence (in the centre character area), it is unlikely that that particular area should be segmented. The procedure of SPV, LC and CC validation is illustrated in Figure 5. Fusion of character and segmentation point confidences is detailed in the next sub-section and in [13].

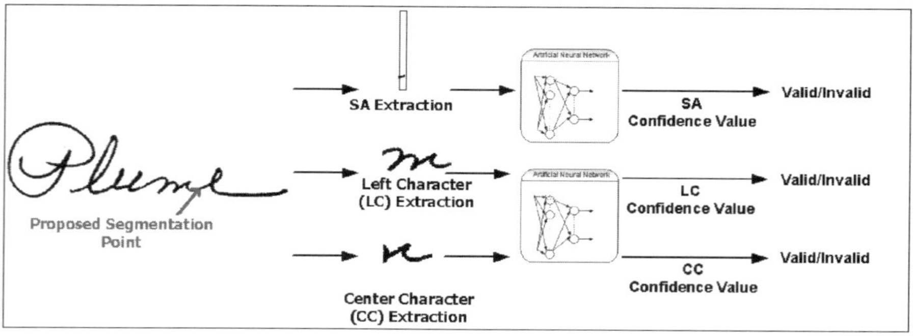

Fig. 5. Overview of SA, LC and CC extraction and validation

Confidence Fusion

A Correct Segmentation Point (CSP) is found:

if $f_{SPV\_Ver}(ft1) >= 0.5$ AND
$f_{LCC\_Ver}(ft2)$ is a high character confidence AND
$f_{CCC\_Ver}(ft3)$ is a high non-character confidence;

$$f_{CSP}(ft1, ft2, ft3) = f_{SPV\_Ver}(ft1) + f_{LCC\_Ver}(ft2) + f_{CCC\_Ver}(ft3)$$

where, $f_{SPV\_Ver}(features)$ - Confidence value from the Segmentation Point Validation neural network. $f_{LCC\_Ver}(features)$ - Left Character Confidence (LCC) value from the character neural network. $f_{CCC\_Ver}(features)$ - Centre Character Confidence (CCC) from the character neural network (reject neuron output).

An Incorrect Segmentation Point (ISP) is found:

if $f_{SPV\_Ver}(ft1) < 0.5$ AND
$f_{LCC\_Ver}(ft2)$ is a high non-character confidence AND
$f_{CCC\_Ver}(ft3)$ is a high character confidence;

$$f_{ISP}(ft1, ft2, ft3) = (1 - f_{SPV\_Ver}(ft1)) + f_{LCC\_Ver}(ft2) + f_{CCC\_Ver}(ft3)$$

where, $f_{SPV\_Ver}(features)$ - Confidence value from Segmentation Point Validation neural network. $f_{LCC\_Ver}(features)$ - Left Character Confidence value from character neural network (reject neuron output). $f_{CCC\_Ver}(features)$ - Centre Character Confidence value from character neural network (highest confidence from 36 character neuron outputs).

Finally, the outcome of fusion is decided by the following equation:
$f(confidence) = max(f(CSP), f(ISP))$

Enhancements to Classification Procedure

Building on previous work, two novelties are introduced to enhance the LC and CC classification rate. Firstly, instead of using the Transition Feature (TF) [17] for incorporation of character confidences into the segmentation technique, the neural network was trained on feature vectors produced by MDF. Secondly, in previous work, two separate neural networks were trained for both upper case and lower case characters. This introduced the problem of deciding upon when to use the lower case or upper case networks. Hence, in order to bypass this issue, lower case and upper case characters were combined into a single network containing 37 outputs. The configuration was similar to that undertaken in previous work [18], where upper and lower case characters that were similar in appearance were grouped in the same class i.e. 'c' and 'C' would share one output class. The only exception was that in this case a reject neuron was also added. The reject neuron was trained to fire when a non-character component was presented to the network (as described above).

2.5 Character Extraction by Segmentation Paths

Previously in the neural confidence-based segmentation technique, LC and CC were extracted using vertical dissections based on the x-coordinates of PSPs provided by the heuristic segmenter (mentioned earlier). It was found that this simplistic scheme was inadequate for the purpose of extracting overlapping and tightly coupled characters in cursive script. The reason being that in some cases, characters would be imprecisely split. This section details a novel character extraction procedure based on the segmentation points output by the heuristic segmentation algorithm.

Segmentation Path Detection (SPD)

The first step of extracting characters using SPD is to measure the ascenders and descenders of the word image. As mentioned earlier, ascenders and descenders are strokes that extend above or below the middle zone or main body of a handwriting sample. Next, the main body of the image is equally divided up into 4 sections, namely sections 1, 2, 3, and 4 (See Figure 6). Based on the x-coordinate of a segmentation point, SPD performs backward traversal. Once a foreground (black) pixel is encountered, the system checks whether the location of the black pixel is below section 1. The line at the bottom of section 1, in Figure 6, is called the "best-fit" line.

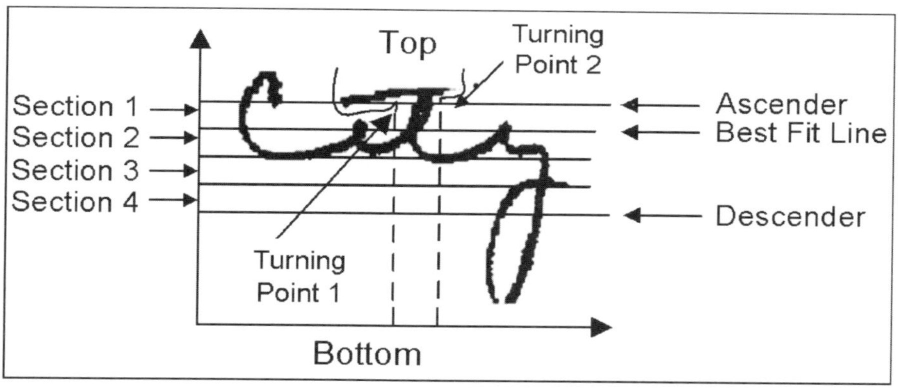

Fig. 6. Word sample sections and segmentation path generation

The "best-fit" line is used as a threshold position, which informs the algorithm whether or not an alternate extraction path should be detected. If the encountered black pixel is below the "best-fit" line, then this pixel, along with all connected foreground pixels are ignored. However, if this black pixel exists on or above the "best-fit" line, this is considered to be the starting point of an overlapping stroke. This pixel is called the "turning point". Commencing from this turning point, a path directed around the overlapping stroke is explored. The algorithm attempts to investigate the right hand side of the turning point. If it is possible to reach the top row of the image, then the extraction path is found. Otherwise, if the traversal to the right hand side is blocked, then the algorithm returns to the turning point, and traverses towards the left hand side. As shown in Figure 6, both left-hand and right-hand segmentation paths of the character 't' are detected. Once an extraction path is located, all pixel coordinates are stored for the purpose of character extraction.

3 Experimental Results

A number of experiments were conducted in this research. Experiments were first conducted to compare the performance of the EHS and FHS algorithms. Further experiments were conducted to compare the performance of MDF and a simple Density Feature (DF) extraction technique for SPV and subsequently the accuracy of character classification for LC and CC. In addition, the performance of character extraction was evaluated using SPD and finally the overall neural confidence-based segmentation technique for validating prospective segmentation points, was tested.

Segmentation performance is measured based on three types of segmentation errors: "over-segmentation", "missed" and "bad" metrics. "Over-segmentation" refers to a character that has been divided into more than three components. A "missed" error occurs when no segmentation point is found between two successive characters. The "bad" error refers to a segmentation point that could not be used to extract a character perfectly, but might still be used for the purpose of character separation.

3.1 Handwriting Database and Neural Network Configuration

The training and testing patterns for this work were obtained from handwritten words contained in the CEDAR benchmark database [19], specifically the "/train/cities/BD" and "/test/cities/BD" directories respectively.

The classifiers used in this research were feed-forward Multi-layered Perceptrons (MLPs) trained with the resilient back-propagation (BP) algorithm. For experimental purposes, the architectures were modified varying the number of inputs, outputs and hidden units.

3.2 EHS and SPV Segmentation Performance

Table 1 shows the segmentation performance of FHS and EHS. The results are based on the 1031 segmentation points that existed between joined, cursive characters contained in the CEDAR words used for testing.

Results for SPV are presented below in tabular form. Table 2 presents top results comparing MDF and DF using a total of 32028 segmentation patterns for training and 3162/4854 patterns for testing.

Table 1. Segmentation performance of EHS and FHS (1031 segmentation points)

	Segmentation Error Rates		
	Over-segmented [%]	Missed [%]	Bad [%]
FHS	4.07	4.07	6.99
EHS	2.72	2.42	4.56

Table 2. SPV rates with a BP-MLP

	Test Set Recognition Rate [%]			
	3162 Patterns		4854 Patterns	
	DF	MDF	DF	MDF
1-Output	81.21	**82.19**	80.61	**81.15**
2-Outputs	N/A	81.97	N/A	**81.15**

3.3 Character Classification Results

This sub-section lists character classification results using a single neural network trained with both upper and lower case characters in addition to a reject neuron (for non-character patterns). In total, 25830 characters were used for training and 3179 for testing. As the number of reject patterns in the above training set represented a large proportion of the data, it was decided that the number of reject patterns be halved in subsequent experiments to demonstrate the effect on the recognition rate. As a result of this procedure, the training set subsequently contained 20464 characters, with the test set remaining constant. Table 3 lists results using both configurations.

Table 3. Character recognition rates with a BP-MLP

	Test Set Recognition Rate [%]	
	All reject patterns	Half of reject patterns
Total Test Set	**67.54**	64.39
Reject Patterns only	**78.49**	70.1
Characters Only	50.29	**54.83**

3.4 Segmentation Path Results

Experimental results are displayed below for correct character extraction employing the SPD technique proposed above. Table 4 displays the percentage of words where characters were all successfully extracted whilst including errors introduced by the heuristic segmenter. Table 4 also shows the percentage of words where characters were all correctly extracted without the interference of incorrect segmentation points (ISPs). The latter is an ideal situation and supposes that all segmentation points are correct.

Table 4. Character extraction rates using SPD

	Character Extraction Rate [%]	
	Including ISPs	Excluding ISPs
317 Words	78.9	95.27

3.5 Performance of the Neural-based Segmentation Technique

The errors of the enhanced neural-based segmentation technique are calculated based on the number of correct segmentation points obtained in the word samples. The total number of segmentation points in the 317 test word samples is 1718. Only 1031 segmentation points that existed between joined/cursive characters were chosen for testing purposes. The reason for this is to test the segmenter on its ability to separate cursive character components. Table 5 shows the overall results of the enhanced neural-based segmentation technique and the existing neural-based segmentation technique using 317 testing words.

Table 5. Overall results of the neural-based segmentation technique (1031 segmentation points)

	Segmentation Error Rates		
	Over-segmented [%]	Missed [%]	Bad [%]
Existing Technique	7.08	2.33	10.86
Experiment 1 (FHS)	8.73	0.1	8.63
Experiment 2 (EHS)	7.37	0.1	6.79

4 Analysis and Discussion of Results

4.1 Analysis of EHS Over-segmentation

The introduction of ligature detection to locate prospective segmentation points hidden by large horizontal strokes or overlapping characters proved quite successful. As may be seen from Table 1, EHS performed fairly well on the test set with only 2.42% of "missed" errors being generated.

Two problems were found during the inspection process. The first problem arises when segmenting very noisy characters. Since the enhanced heuristic algorithm was heavily dependent on contour analysis, heavy noise that was inherent around the handwriting could cause serious errors. One of the solutions to this problem was additional pre-processing.

The second problem that was observed related to the neural assistant. The main problem was incorrect classification. However, overall the classification rate was acceptable based on the current character classifier's recognition accuracy (approx. 89%).

The missed segmentation points were due to the neural assistant misrecognising two joined characters as a single character. This type of error is very hard to deal with, since when two characters are tightly coupled, the ligature cannot be detected. One solution is to employ a better neural classifier or incorporate more heuristic rules. However, in some cases the missed segmentations may be recovered when the neural-based segmentation technique is employed, which uses the centre area associated with each segmentation point. Figure 7 provides some sample handwriting with segmentation points found by EHS.

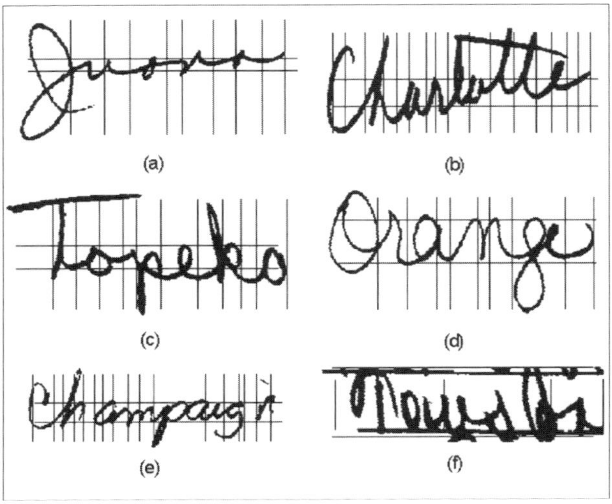

Fig. 7. Sample word images segmented by the enhanced feature-based heuristic segmenter. (a), (b), (c) successful words. (d), (e), (f) unsuccessful words.

Although neural classifiers may contribute problems in some instances, their use in the described segmenter was very beneficial, because it could introduce more segmentation points without using complex heuristics.

4.2 SPV Discussion

As may be seen from Table 2, in comparing the recognition rates when using DF and MDF, the MLP trained with MDF patterns produces a slightly higher recognition rate. The small increase in recognition rates demonstrates that the MDF is comparable with DF for small, uncomplicated patterns. When a two-output neural network was used (the first neuron indicated a "correct" segmentation and the second indicated an "incorrect" one), the recognition rates on both MDF data sets either remained constant or decreased nominally in comparison with the single-output MLP. A comparison was not directly possible in this case with the DF dataset.

4.3 Character Classification

The use of a 37-output neural architecture was considered an important step for the overall segmentation process. With the current configuration, although the recognition rate was not excessively high, it is possible to classify both lower and upper case characters with a single network.

As may be seen from Table 3, the results for recognizing reject patterns is nearly 80% when using all available patterns for training. This is a favourable outcome, as the LC and CC depend on this confidence for correct segmentation. Conversely, the character recognition rate is substantially lower, however it may be seen that when half of the reject patterns are removed for training, a higher character recognition rate is achieved. This indicates that the slight disproportion between characters and reject patterns may be leading to a bias during training.

4.4 SPD Discussion

As may be seen in Table 4, the results for correct character extraction are most favourable. The result of 78.9%, using the x-coordinates produced by the heuristic segmenter is encouraging. Upon improving the segmenter further, the success of the character extractor may approach the ideal rate of 95.27%.

4.5 Analysis of Neural-based Segmentation

As may be seen from Table 5, the segmentation technique was successful at discarding bad segmentation points as well as recovering "missed" segmentation points by adding them at large gaps between points in words based on the average character width. Both experiments (using FHS and EHS) recorded the same "missed" error of only 0.1%, which is a very promising result. Furthermore, the results also showed that the enhanced heuristic segmenter was able to produce better inputs to increase overall segmentation results.

The reason for the higher "bad" errors by the neural-based segmentation technique as compared to those obtained by the enhanced heuristic segmenter is because some "missed" errors are turned into "bad" ones. This is due to the technique recovering "missed" segmentation points based on the average character width. In some cases, it could not perfectly locate the character boundary (using SPD) and hence contributed to the "bad" error. Although the "over-segmentation" error went up slightly as compared to previous work, it is possible to recover this at a later stage.

Another reason for the increase of the segmentation performance is related to the use of the MDF and the segmentation path-based character extraction technique (SPD). Since 'clean' characters can be extracted and MDF provides better features for the single classifier, the performance of LCV and CCV are improved.

5 Conclusions and Future Work

This chapter describes an improved neural-based segmentation technique for cursive words. The technique included an enhanced heuristic segmenter to over-segment handwriting in addition to the use of an MDF extraction technique for SPV, LCV and CCV. The enhanced heuristic segmenter provided better inputs to the subsequent neural validation process. Encouraging results were obtained that can increase the overall performance of a segmentation-based handwriting recognition system.

In the future, EHS will be used to locate prospective segmentation points from the training set facilitating re-training and testing of the SPV classifier for further enhanced performance. The above-mentioned technique will also be tested on a larger dataset to validate the improvements proposed. Finally, a new character extraction technique that uses the direction feature on the character's boundary will be investigated.

References

1. Casey, R. G.: Moment Normalization of Handprinted Characters. IBM Journal of Research Development (1970) 14:548–557.
2. Casey, R. G., Lecolinet, E.: A Survey of Methods and Strategies in Character Segmentation. IEEE Trans. Pattern Analysis and Machine Intelligence (1996) 18:690–706.
3. Lu, Y., Shridhar, M.: Character Segmentation in Handwritten Words - An Overview. Pattern Recognition (1996) 29:77–96.
4. Bozinovic, R. M., Srihari, S. N.: Off-Line Cursive Script Word Recognition. IEEE Trans. Pattern Analysis and Machine Intelligence (1989) 11:68–83.
5. Cheriet, M., Reading Cursive Script by Parts. Proceedings of the Third International Workshop on Frontiers in Handwriting Recognition (IWFHR-3), Buffalo, New York, May 25-27,(1993) 403–408.
6. Han, K., Sethi, I. K.: Off-line Cursive Handwriting Segmentation. Proceedings of the 3rd International Conference on Documents Analysis and Recognition (ICDAR '95), (1995) 894–897.
7. Yamada, H., Nakano, Y.: Cursive Handwritten Word Recognition Using Multiple Segmentation Determined by Contour Analysis. IEICE Transactions on Information and Systems (1996) E79-D:464–470.
8. Yanikoglu, B., Sandon, P. A. Segmentation of Off-Line Cursive Handwriting using Linear Programming. Pattern Recognition (1998) 31:1825–1833.
9. Eastwood, B., Jennings, A., Harvey, A.: A Feature Based Neural Network Segmenter for Handwritten Words. Proceedings of the International Conference on Computational Intelligence and Multimedia Applications, Gold Coast, Australia, (1997) 286–290.

10. Dimauro, G., Impedovo, S., Pirlo, G., Salzo, A.: An Advanced Segmentation Technique for Cursive Word Recognition. In: Lee S W (ed) Advances in Handwriting Recognition, World Scientific Publishing, (1998) 255–264.
11. Nicchiotti, G., Scagliola, C., Rimassa, S. A: Simple and Effective Cursive Word Segmentation Method. Proceedings of the 7th International Workshop on the Frontiers of Handwriting Recognition (IWFHR-7), (2000) 499–504.
12. Xiao, X., Leedham, G.: Knowledge-based Cursive Script Segmentation. Pattern Recognition Letters (2000) 21:945–954.
13. Blumenstein, M., Verma, B. K.: Analysis of Segmentation Performance on the CEDAR Benchmark Database. Proceedings of the International Conference on Document Analysis and Recognition (ICDAR '01), Seattle, USA, (2001) 1142–1146.
14. Blumenstein, M., Verma, B.: A New Segmentation Algorithm for Handwritten Word Recognition. Proceedings of the International Joint Conference on Neural Networks (IJCNN '99), Washington D.C., (1999) 878–882.
15. Blumenstein, M., Liu, X. Y., Verma, B.: A Modified Direction Feature for Cursive Character Recognition. Proceedings of the International Joint Conference on Neural Networks (IJCNN '04), Budapest, Hungary, (2004) 2983–2987.
16. Cheng, C. K., Liu, X. Y., Blumenstein, M., Muthukkumarasamy, V.: Enhancing Neural Confidence-Based Segmentation for Cursive Handwriting Recognition. 5th International Conference on Simulated Evolution And Learning (SEAL '04), (2004) Busan, Korea, SWA-8, CD-ROM Proceedings.
17. Gader, P. D., Mohamed, M., Chiang, J.-H.: Handwritten Word Recognition with Character and Inter-Character Neural Networks. IEEE Trans. Systems, Man, and Cybernetics-Part B: Cybernetics (1997) 27:158–164.
18. Blumenstein, M., Verma, B., Basli, H.: A Novel Feature Extraction Technique for the Recognition of Segmented Handwritten Characters. Proceedings of the Seventh International Conference on Document Analysis and Recognition, (ICDAR '03), Edinburgh, Scotland, (2003) 137–141.
19. Hull, J. J.: A Database for Handwritten Text Recognition. IEEE Trans. Pattern Analysis and Machine Intelligence (1994) 16:550–554.

Multiple Hypotheses Document Analysis

Tatsuhiko Kagehiro and Hiromichi Fujisawa

Central Research Laboratory, Hitachi, Ltd., Tokyo 185-8601, JAPAN
{tatsuhiko.kagehiro.tx,hiromochi.fujisawa.sb}@hitachi.com

Summary. Document layout analysis is a tough task for a document analysis and recognition system, especially when there are many variations in the layouts. Often, layout analysis requires character recognition, while character recognition requires layout analysis beforehand. It's a Catch-22, or a "chicken-and-egg" problem. This chapter discusses this kind of dilemma and presents a two-part solution that first analyzes the layout and then, using a hypothesis-driven approach, segments the numerical character line. Basically, the approach is to first generate multiple hypotheses based on low-level image processing, then to conduct layout analysis and create many candidates based on each of the hypotheses. Finally, the correct candidate is selected by the results of content recognition. Probabilistic verification is used to select the candidates that are input to the recognition module, with parameters that are learned from samples in advance. The second part of the solution, which relies on a hypothesis-driven approach for the segmentation of the numerical character line will also be presented. As a test case, these solutions were applied to the Japanese postal address recognition system. They show how tough problems involved in analyzing the surface images of mail pieces can be solved. The hypotheses-driven approach manages every possible variation, including writing orientation, printed or hand-written, address-block location, character size, and so on.

1 Introduction

OCR (Optical Character Recognition) has many uses in the automation of work in offices and factories. About 30 years ago, OCR was developed to read numeral characters written in fixed areas, and these recognition results are used to generate helpful information to automate office and factory work. As time went on, the readable character position became more variable, and various character sets (Alphabets, Kanji, etc.) could also be read by OCR. OCR continues to expand and become more adaptable, and it is hoped that in future OCR will be able to recognize documents correctly and read more types of documents.

When the design of the document is known in advance, OCR can get the positions of the characters in the input image. An example of these documents

T. Kagehiro and H. Fujisawa: *Multiple Hypotheses Document Analysis*, Studies in Computational Intelligence (SCI) **90**, 277–303 (2008)
www.springerlink.com

are fixed forms. In this case, it is not necessary to analyze the layout of the document; OCR can segment the character patterns by using a priori knowledge, and recognize each character. But when OCR is used to read more varied types of documents, the positions and layout of characters can change with each type. The document types shown below are itemized by their degree of format complexity, from type 1 to type 4.

Type 1: Fixed form
Type 2: Business or name card, Cover page of paper, Contents page of book
Type 3: Address surface of mail, Cover page of magazine, Newspaper
Type 4: Handwritten memorandum, Signed board

For type 1, examples of fixed forms include those used by bank tellers and various office forms. OCR machines can read the contents of fixed forms correctly.

In type 2, the character position is strongly restricted, but not completely fixed. In this case, an OCR system can analyze the layout correctly. A little recognition error can occur and in a real workflow humans must correct these errors.

In type 3, there is a common rule in the document layout, but this rule has some variations. Type 3 documents make a good target for document analysis research, so many papers have been published [18][19][21][22][23][12]. The destination address on mail is an especially good candidate for automatic recognition, because current mail sorting machines read the destination address less than perfectly. Rejected mail requires humans to attach the correct address. The process described here represents a real solution for the use of OCR in the real world.

As for type 4, it is difficult for OCR to recognize handwritten memos and signed boards. Because humans write memos according to their own inspiration or memory, there is no format for the character allocation and sometimes no structure for the layout. There is a big demand for automatic recognition of signed boards in the real world, because they often contain very important information. But in order for OCR to read signed boards two different modules in the layout analysis must be developed completely, one to detect the position of the signed board from the scene image, the other to analyze the layout on the board. In order to recognize the kinds of documents in type 4, it will be necessary to implement a human vision approach to the computer [25].

2 Approaches to Layout Analysis

Layout analysis has a very important role in document recognition. If the layout cannot be analyzed correctly, it is impossible to understand the contents of the document. But on the other hand, the contents of the document provide useful information for analyzing the layout. This is a "Catch-22" or the so-called "chicken-and-egg" problem. Humans can analyze the layout and understand the content in parallel, but it is difficult for the computer to imitate the human process, because computer procedures are generally sequential.

There are two approaches to document analysis, top-down and bottom-up. The top-down approach tries to estimate the layout based on a model, and extract the features which are needed to validate the model [5][16]. The bottom-up approach first extracts various features from the input image, and then tries to match these features to the model. Accordingly, if it is possible to estimate the layout in advance, the top-down approach is more effective and reliable. The bottom-up approach is generally expensive, but this approach has advantage to not output the big error, if the model of the layout is not known in advance [10].

There is an optimal approach for each type of document. The top-down approach is most effective for documents with a predictable layout structure (for example, business cards, cover pages of papers and content pages of books). Common-sense knowledge of the layout is given to the OCR system in advance, and the positions of characters are matched to this layout knowledge. After that, the contents of the document are read using the information about the layout. In this case, it is necessary to dynamically match the character positions to the layout information, because the character positions can be somewhat ambiguous [14][30].

If the supposition of the layout is difficult, or if variations of the common-sense approach can be expected, then the bottom-up approach is more effective. But a complete bottom-up approach requires large calculation costs, because all characters in the document must be recognized before the layout analysis. So the real OCR is designed to use the positions and sizes of each character pattern to analyze the layout. Usually, a rectangle circumscribed to each pattern is used to express position and size. In this method, the rectangles of the character patterns are extracted first, and then close rectangles are merged to create the character lines or blocks. This method does not require huge calculation costs and can use the layout knowledge while merging the pattern rectangles.

3 ROI in Document

There are two types of office documents. For one type, it is necessary to understand the entire contents of the document; the other requires understanding of only the important part of the document. This important part is called ROI (Region Of Interest). In fact, OCR systems are often designed to read only the ROI in documents, because the content of the ROI has sufficient information to automate the office task. For example, the ROI of an application form might be the name and address of the applicant, and the ROI of a piece of mail is the destination address and name. In the case of a scientific paper, the title, authors, and affiliations constitute a ROI. In these cases, it is not necessary to understand the entire contents of the document. But in order to select the ROI, the whole structure of the layout must be analyzed.

The multiple hypotheses approach is an effective way to extract the ROI from a document. This approach creates a number of candidates by making some hypotheses in cases in which the layout analysis process can not decide on only one answer. If the post-process can get other useful information, the number of candidates is reduced. Finally, the most valid candidate is selected as the correct answer. The candidates are evaluated under each hypothesis which was applied during the candidate creation. For example, if the size of characters is unknown, a number of ROI candidates are created by some hypotheses regarding character size. After that, each candidate is evaluated on the condition of each character size which was used by the hypotheses. This approach can control the balance between calculation cost and recognition accuracy. If it is necessary to improve the recognition accuracy without increasing the calculation cost, unusable candidates must not be created and more effective information must be used to evaluate the candidates in each step. An OCR which adopts multiple hypotheses can be implemented quickly, because no one process in the system is responsible for outputting a completely correct answer, and recognition accuracy is increased while each process is revised step by step.

4 Target Documents for Discussion

In this chapter, the layout analysis of mail addresses is used as an example in order to discuss layout analysis and ROI extraction strategies. The layout of mail addresses is loosely restricted by common-sense knowledge. The ROI of a piece of mail is the destination address, which is needed to automate the mail-sorting task. But it is not easy to determine the ROI on a given piece of mail. It can be in a variety of positions, it can have one of two character types (printed and handwritten), there can be variations of the mail direction, and in the countries of East Asia, the character lines can be either horizontal or vertical. In addition there is the stamp area, the return address area, and sometimes advertisements on the mail, all of which create heavy obstacles to ROI extraction.

The multiple hypotheses approach is used to analyze the layout and read the destination information. After binarization of the input image, each process creates some candidates; post-processing then evaluates the candidates. In the final step, one candidate is selected as the correct answer by checking the destination point in a real address data set. This approach can control the balance between calculation cost and recognition accuracy, because the number of the candidates in each process can be controlled by setting various parameters.

The character segmentation method which uses the multiple hypotheses approach will be introduced later. Address characters on mail can be written by machine or hand, and in either horizontal or vertical lines in the Asian countries. There can be various interpretations of the character segmentation results, because Kanji characters consist of some simple patterns, many of which can be read as other characters. In order to solve this problem, the multiple hypotheses approach creates candidates of the segmented patterns, and selects the best candidate as a correct answer.

5 Ambiguity of Mail Address Recognition

The technique of mail destination address recognition is important for automation of the mail-sorting task. But it is difficult for a machine to understand the destination address if it does not know the layout of the address in advance [20][11]. The layout of the address has several variations and much ambiguity, so the design of the address recognition system is much more difficult and complex.

In order to recognize the destination address, it is necessary to extract the area of the destination address characters. This area is called the address-block. If many mail samples are observed, it is possible to observe that there are variations in the address-block position, direction, and printed or handwritten characters, along with the return address area and possible advertisements. Some parameters for extracting the address-block can not be fixed uniformly, because of the variations and ambiguities. Many papers have been published on the topic of analyzing the layout of mail pieces [2][19][4][5][6].

After getting the address-block, it is necessary to read the character lines in the address-block. Current character recognition and segmentation modules can hardly output the correct answer. Character recognition has been studied for several years, however many researchers are continuing to improve the technique [24][29][26][27], and various approaches to character segmentation have been reported [17][1][15][13]. For example, the first of these approaches applies the similarity of the characters for recognition, the second uses the peripheral information of the character pattern, and the last employs the linguistic information of the character strings [8][9].

The character segmentation of Japanese street numbers is an especially difficult problem. In Japan, the street numbers can be written in either horizontal or vertical lines. Furthermore, the Kanji numerals "One", "Two" and "Three" are expressed as simple horizontal lines, with the number of horizontal lines corresponding to the Kanji numerals. Thus, "One" is one horizontal line, "Two" is two horizontal lines, and "Three" is three horizontal lines. Kanji numerals written in a vertical line introduce much ambiguity into the character segmentation. In this case, it is not effective to apply linguistic information, because there are various street numbers in real addresses. So the Kanji numerals must be segmented using information about the figures and their positions.

In this chapter, two methods which can solve the above mentioned problems are described. One is the address-block extraction method, the other is the character segmentation method for Kanji numerals in a vertical line. These methods adopt the multiple hypotheses approach and use a Bayesian rule for the validation of each candidate. The Bayesian rule can combine several different confidence values, and calculate the total confidence value. If the likelihood ratio can be observed from many learning samples, the Bayesian rule can be implemented to solve real problems. In the following sections we describe a useful method which applies the Bayesian rule to real problems [31][28].

6 Address-Block Extraction from Mail Pieces

6.1 Background of Address-Block Extraction

In order to read a mail destination address correctly, it is first necessary to accurately extract the destination address lines. Because the lines of a destination address on Japanese mail can be written in various ways, extracting them is very difficult.

Moreover, the existence of the return address lines and of advertisements on pre-printed mail sometimes makes the extraction much more difficult. In this chapter, we call the area occupied by the destination address lines the "address-block".

There are several conventional methods for extracting the address-block [4][5]. For example, in the case of handwritten mail, there is a technique for estimating the address-block by analyzing the positional relationship between character lines and ruled lines. When the address-block has ruled lines, this method is very useful, but it essentially has no power to extract an address-block without ruled lines. In the case of large-sized mail, another technique extracts an area having a uniform grey value as an address-block [6]. This technique is effective for extracting an address-block on an attached address label. However, it is not enough to use techniques that can only be applied to specific types of address-blocks.

The proposed method, which will be described later, can be adapted to various address-block types. This method extracts several candidates for the address-block by using multiple hypotheses on types and sorts them according to a confidence value. To calculate the confidence value of a candidate, an evaluation based on a Bayesian rule is executed by using a type-hypothesis. Address-block candidates are read by an address character recognition module one by one until one candidate is accepted as an address; the result is then regarded as the destination address of the mail. This recognition procedure can effectively read the destination address of mail [7].

6.2 Variations of Address-block Types in Japanese Mail

There are various types of address-block in Japanese mail. We have classified address-blocks into several types by human observation. That is, the type of mail envelope is either portrait or landscape, the direction of the written address is either vertical or horizontal, and the address is either printed or handwritten. Note that a landscape mail envelope with vertical writing is very uncommon. The total number of types is therefore six, abbreviated as follows:

P-PV: (Printed, Portrait, Vertical writing)
P-PH: (Printed, Portrait, Horizontal writing)
P-LH: (Printed, Landscape, Horizontal writing)
H-PV: (Handwritten, Portrait, Vertical writing)
H-PH: (Handwritten, Portrait, Horizontal writing)
H-LH: (Handwritten, Landscape, Horizontal writing)

The six types of address-blocks are shown in Figure 1. We assume that the mail image is scanned in portrait orientation so that the stamp is always located in the upper left. The resolution of the image is 200 dpi.

The rectangles in Figure 1 show the address-blocks. It is difficult for a simple method to extract the address-block, because its position and area are different in each type (as shown in Figure 1). Since the type of address-block cannot be known in advance, it is not possible to adjust the extraction process for each type. Moreover, return addresses and pre-printed advertisements on the mail sometimes make address extraction even more difficult.

6.3 Mail Destination Address Reading System

To read a mail destination address, two operations must be performed: extraction of the address-block and reading of the address lines in the address-block. The address-block extraction should be a bottom-up approach, because it is performed without the knowledge of the address-block type. The address-block extraction consists of the following four steps:

1. Removal of noise by pre-processing
2. Extraction of objects
3. Creation of address-block candidates
4. Evaluation and sorting of the address-block candidates

Figure 2 shows the flow of these steps. In the pre-processing step, noise and underlines are removed from the input binary image by simple image processing. Next, connected components are derived from the image. All processing after this step is performed using connected components.

In the object extraction step, the objects (a stamp, a postal code, character lines, etc.) that may commonly exist in the mail image are extracted. The extraction uses common knowledge of all address-block types such as knowledge

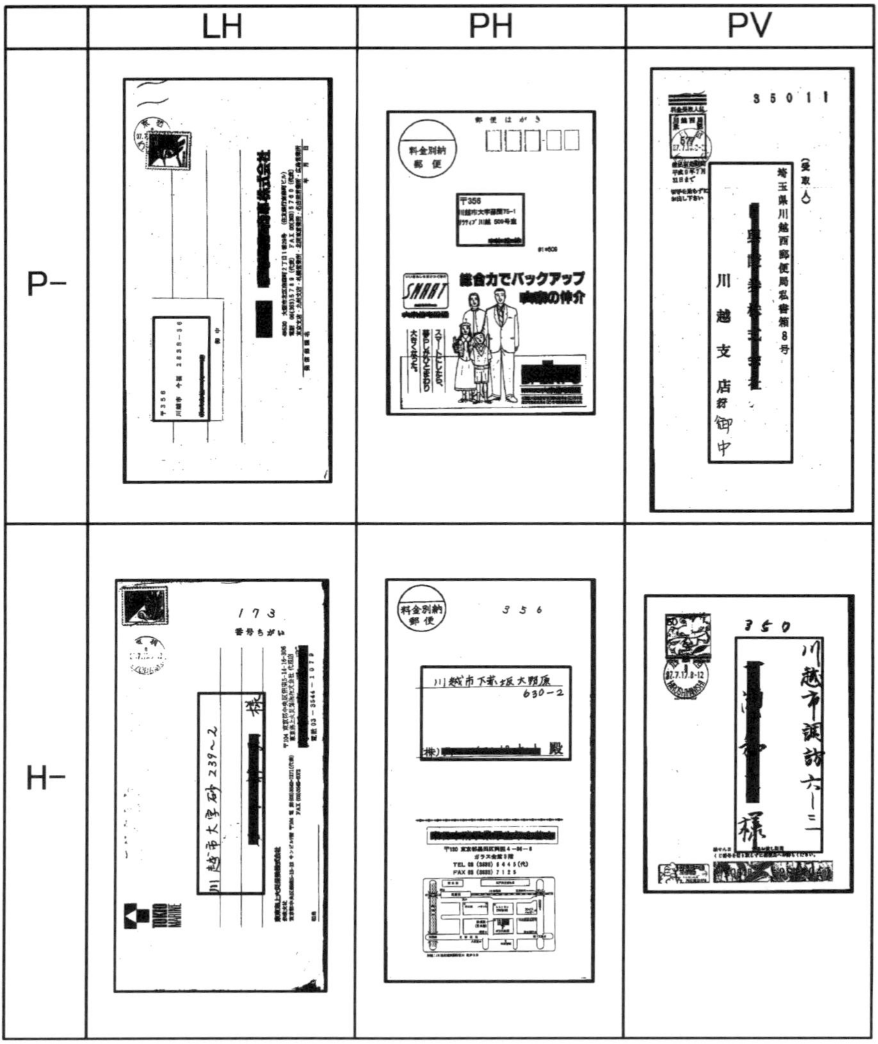

Fig. 1. Address-block Type Variations

of the position and the size of each object. In the address-block candidate creation step, several character lines extracted in the object extraction step are combined, and several candidates for the address-block are generated. This step creates several proper candidates for each mail type, because the position and the area of the correct address-block are different for each type. In the next step, one correct answer will be selected from the candidates even if there is no knowledge of the address-block type.

In the address-block candidate evaluation step, a confidence value expressing the likelihood that a candidate is the correct address-block is calculated according to a Bayesian rule. The confidence value is calculated by using features detected in the candidate. One candidate will have several confidence values corresponding to all address-block types. The number of confidence values for one candidate is therefore the same as the number of types. Next, the candidates are sorted in order of their maximum confidence values. In the address-lines reading step, the sorted candidates are analyzed one by one, using character recognition. When the recognition result is an address, it can be regarded as the destination address of the mail. Candidate evaluation using a Bayesian rule is described in detail in the next section.

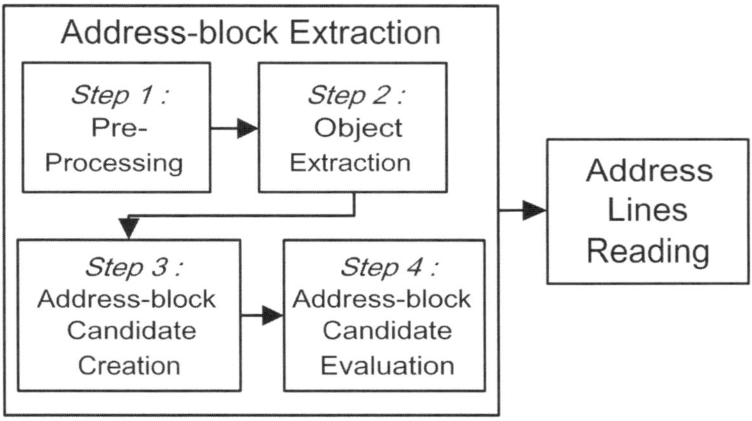

Fig. 2. Mail Destination Address Reading System

6.4 Segmentation of Character Lines by Multiple Hypotheses

Some character lines are extracted by merging some closely connected components of the patterns. These merge rules have different threshold values in response to the layout of the destination address. We call the segmentation setting for printed character the "small" character setting, and the segmentation setting for handwritten characters the "large" character setting. Figure 3 shows the character-line extraction results for each setting. In Figure 3, the printed character lines (Input Image A) are extracted correctly by the small character setting, but incorrectly by the large character setting. The handwritten character lines (Input Image B) are extracted correctly by the large character setting, but incorrectly by the small character setting. The character lines segmentation for all mail pieces can not be processed correctly by one-half of these settings.

The layout of the mail address lines is not known in advance. It is difficult to judge the layout of the destination address, because there also are advertisement and return address areas. Since the layout of the destination

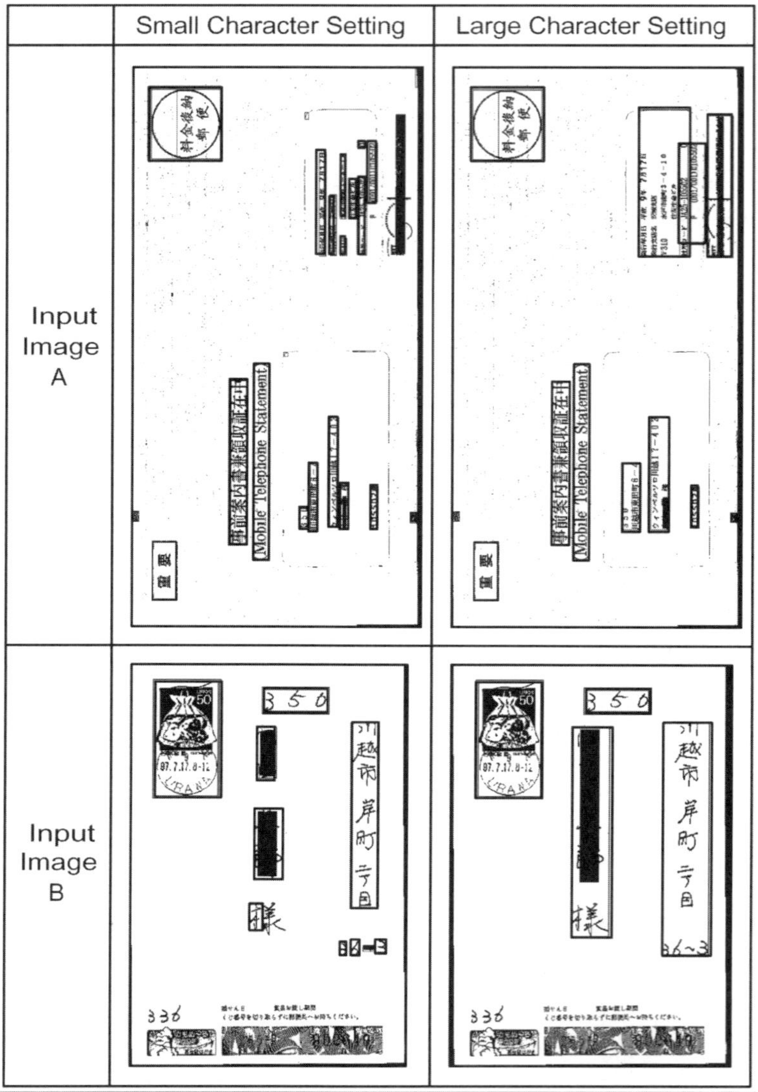

Fig. 3. Character Line Segmentation Results

address can not be decided, the segmentation settings can not be changed in response to the layout. Character lines segmentation is therefore executed twice by both settings, as shown in Figure 4. After the character lines are segmented by the multiple hypotheses, some candidates are outputted to the post-process. The correct character lines segmentation is found among these candidates wherever possible.

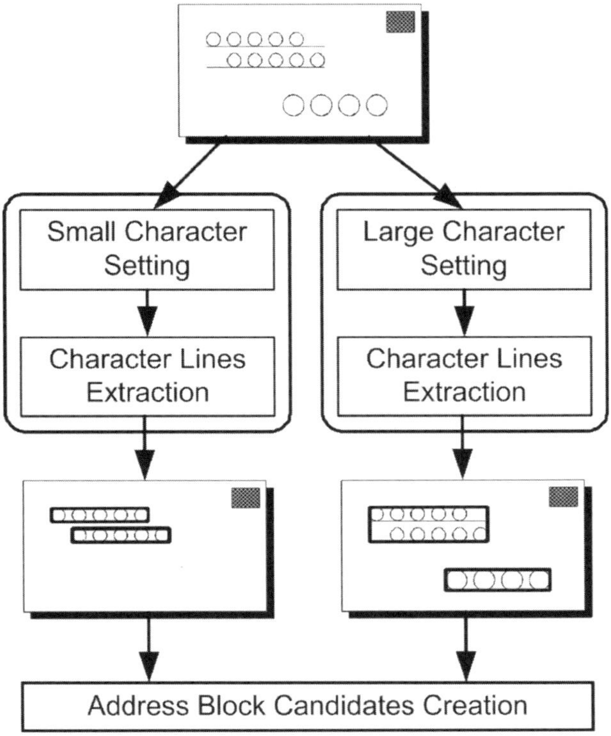

Fig. 4. Character Line Segmentation by Multiple Hypotheses

6.5 Address-Block Evaluation by Bayesian Rule

6.5.1 Calculation of Confidence Value

In the address-block candidate evaluation step, the confidence value is calculated from the following features that can be observed from the candidate.

1. Averages of height and width of character lines
2. Variances of height and width of character lines
3. Area of address-block candidate
4. Position of the candidate

Since the standard values of the features change according to the address-block type, a single set of parameters for the candidate cannot be determined. To cope with this problem, several address-block dictionaries are prepared according to each type, as shown in Figure 5. The optimal confidence values of all candidates in each type can thus be calculated. However, because the address-block type is unknown in advance, a dictionary corresponding to a certain type cannot be selected. Accordingly, an address-block candidate for each type is assumed, and a confidence value for each type is calculated independently as each dictionary is selected and referenced. Several confidence values per candidate are thus computed and are compared with each other to select the maximum value among them. This maximum value is regarded as a representative confidence value for the candidate, because the confidence value computed from the dictionary of the corresponding type should be the highest and all confidence values for an incorrect candidate should be low.

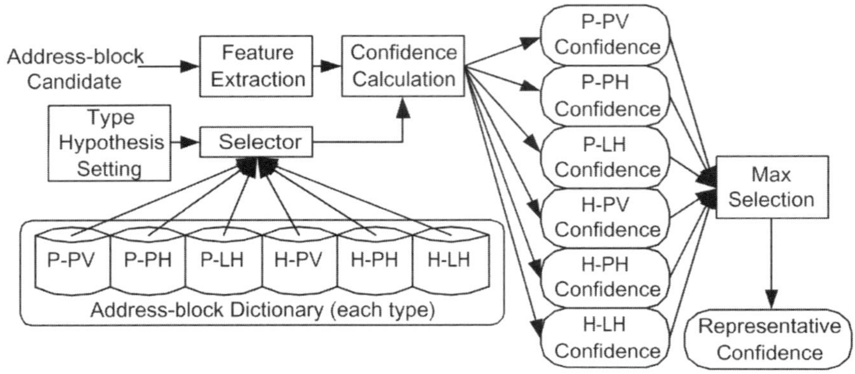

Fig. 5. Address-Block Evaluation

6.5.2 Address-Block Dictionary and Bayesian Rule

A Bayesian rule is applied to calculate the confidence value for the address-block candidate. The confidence value is calculated using the candidate features according to the Bayesian rule given in Eq. (1):

$$P(H_c|e_1, e_2, e_3, \cdots, e_n) = \frac{\frac{P(H_c)}{P(\bar{H}_c)} \prod_{k=1}^{n} L(e_k|H_c)}{1 + \frac{P(H_c)}{P(\bar{H}_c)} \prod_{k=1}^{n} L(e_k|H_c)} \tag{1}$$

$$Likelihood : L(e_k|H_c) = \frac{p(e_k|H_c)}{p(e_k|\bar{H}_c)} \tag{2}$$

where c is the address-block type, H_c is the hypothesis of the correct candidate for type c, e_1, e_2, e_3, ... e_n are the observed features of a candidate, $P(e_k|H_c)$ is a posterior probability of e_k supposing hypothesis H_c, and $L(e_k|H_c)$ is the likelihood ratio for e_k supposing hypothesis H_c as shown in equation 2.

We define $P(H_c|e_1, e_2, e_3, ...e_n)$ as a confidence for a candidate, where e_1, e_2, e_3, ... e_n are observed features.

To obtain $P(H_c|e_1, e_2, e_3, ...e_n)$, the likelihood ratio $L(e_k|H_c)$ is required in advance. The likelihood ratio is estimated from learning data, and is stored as shown in Figure 5. It is referenced while the confidence values are calculated. The likelihood ratio should also be calculated independently according to each type. The learning data for each type of address-blocks is therefore collected from real mail images in advance.

To estimate the likelihood ratio $L(e_k|H_c)$, $P(e_k|H_c)$ and $P(e_k|\bar{H}_c)$ must be known. $P(e_k|H_c)$ is a posterior probability of e_k supposing hypothesis H_c. We consider that a posterior probability can generally be estimated from the occurrence frequency of e_k in the case of H_c. Thus, to estimate $P(e_k|H_c)$ and $P(e_k|\bar{H}_c)$, in the address-block candidate creation step, many address-block candidates are created from a learning set of mail images. Next, all candidates are judged to be correct or incorrect by human observation. Features of candidates are extracted from the address-block candidates, and the frequencies of features in correct and incorrect candidates are calculated. These frequencies are used to calculate probabilities, $P(e_k|H_c)$ and $P(e_k|\bar{H}_c)$.

Note that $P(H_c)$ and $P(\bar{H}_c)$ are prior probabilities of H_c and $\bar{H}_c$, and they can be obtained by calculating the ratios of the number of correct and incorrect candidates to the total number of candidates.

6.6 Experimental Results

6.6.1 Evaluation of Address-block Extraction

To evaluate the proposed address-block extraction method, we measured its accuracy rate. Two datasets were prepared: a set of 500 pieces of printed mail and a set of 500 pieces of handwritten mail. We considered 1341 learning samples that included both types of mail. This dataset is different from the evaluation dataset mentioned above. Two methods were compared in terms of their accuracy in the extraction of the address-block:

(A) Conventional heuristic method
(B) Proposed Bayesian rule method

In method (A), the parameter set was adjusted heuristically based on the learning data. In the evaluation of both methods, the type of test mail was not known in advance. We compared the cumulative extraction rate for the address-block in mails of the P-LH and H-PV types.

Figure 8 shows the cumulative extraction rates for the two methods. Cumulative extraction rate is calculated from the frequency of candidate rankings, including correct ones. Hereinafter, the cumulative extraction rate from the top to the 5th candidate is called the "top-five correct rate".

In the case of the P-LH type, the top-one correct rate for method (B) was improved by 16 points up to 79% in comparison with method (A). The top-three, top-four, and top-five correct rates for methods (A) and (B) became equivalent, that is, 95%.

In the case of type H-PV, the top-one correct rate for method (B) was improved 11 points up to 90% in comparison with method (A). The top-five correct rate for (B) is still larger, that is, 91%, than for method (A).

A general experiment with all types of mail confirms that the proposed method can extract the top-five candidates, including the correct address-block, in 94% of printed mail cases and in 89% of handwritten mail cases.

Figure 6 and 7 show the input images of the P-LH and H-PV formats, and the address-block extraction results obtained by this proposed method. The rectangles in these images are the extraction results of the character lines, postal code, and stamp. In Figure 6, the first candidate is the correct answer. The others consist of some small character lines, not the destination address. These incorrect candidates are located at a valid position in the H-PV format, but the confidence value is not large because none of the other information, apart from position, is valid for H-PV format. In Figure 7, the first candidate is the correct answer, and the second candidate is the part of the return address area. It is possible to put the correct candidate in a superior position to all other candidates, if the confidence value is calculated from features of each candidate.

6.7 Evaluation of Address Reading

We measured the read rate of a destination address using the proposed method. The number of test mail samples was 13778 for printed mail and 7265 for handwritten mail. In the destination address reading system mentioned in Section 6.3, the read rates of two systems using methods (A) and (B) respectively were compared.

In the case of the printed mail, the read rates for methods (A) and (B) were equivalent, 84.5%. In the case of the handwritten mail, the read rate for method (B) was improved by 0.61pt to 67.5% in comparison with method (A).

As shown by this evaluation, the read rate improvement using method (B) for handwritten mail is larger than that for printed mail. This is because the parameter set for method (A) was heuristically adjusted for printed mail but not for the handwritten mail. If we had used a parameter set adjusted for handwritten mail, the improvement for handwritten mail would have been larger than that for printed mail. Therefore if it were possible to judge in advance whether the mail is printed or handwritten, an optimal parameter set for method (A) could be selected according to the judgment. However, it is difficult to predict the mail type in advance. Method (B), on the other hand, can cope with both printed and handwritten cases without prior knowledge of types.

6.8 Conclusion of Address-Block Extraction

A method for extracting the address-block from a mail image in various formats was developed and tested. In terms of the address-block extraction

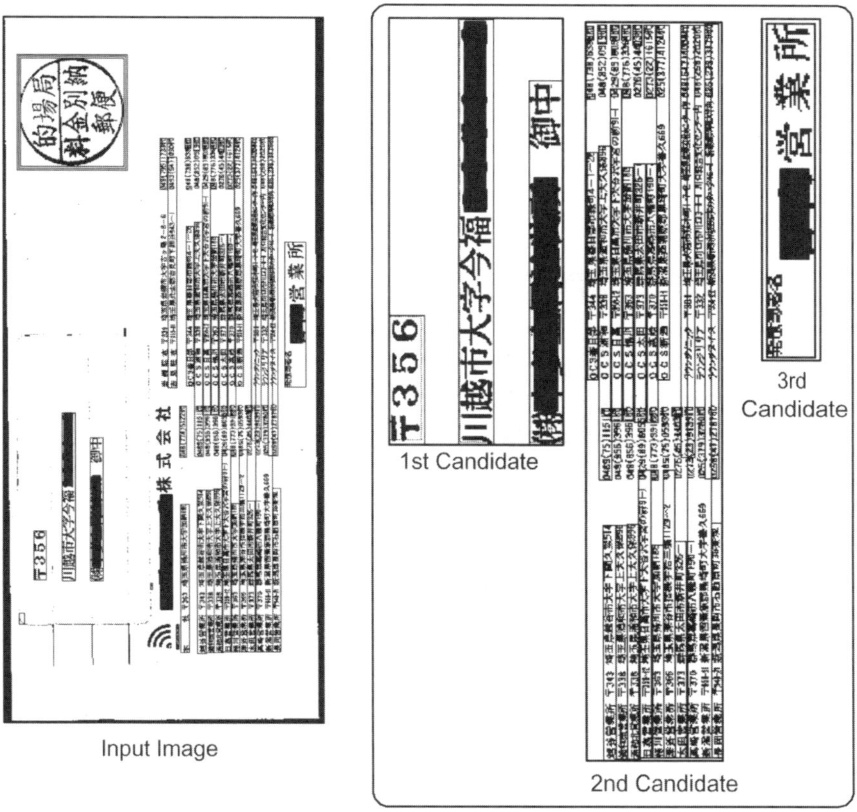

Fig. 6. Address-block Extraction Results (Printed)

accuracy of this method, the top-five correct rate was 94% for printed mail and 89% for handwritten mail. In the case of an address reading system using this method, the read rate for printed mail was 84.5% and that for handwritten mail was 67.5%. It is thus concluded the developed method can be applied to future areas such as signboard extraction and document analysis, if sufficient learning data is collected in advance.

7 Segmentation of Handwritten Kanji Numerals Integrating Peripheral Information

7.1 Background of Character Segmentation

We have developed a new method for segmenting handwritten Kanji numerals written vertically. The segmentation of Kanji numerals written vertically is difficult, because part of one Kanji numeral pattern can be read as another

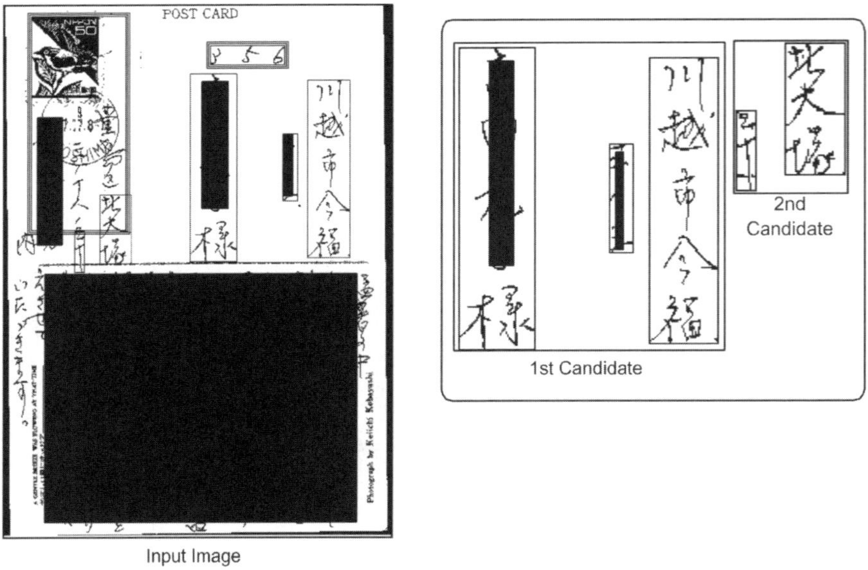

Fig. 7. Address-block Extraction Results (Handwritten)

Kanji numeral character. Figures 9 (a)-(d) show Kanji numerals and hyphens written vertically. In Figure 9 (a), there are three Kanji numerals and two hyphens, showing the Kanji numeral "three", hyphen, "two", hyphen, and "one" from top to bottom. The Kanji numeral "one" is expressed by one horizontal stroke, "two" is expressed by two horizontal strokes; and "three" is expressed by three horizontal strokes. Thus, if two horizontal strokes are written in a vertical direction, the pattern is ambiguous; they can be read as a pair of one's or as a single two.

Three kinds of information can generally be used to segment characters [8].

(1) peripheral information like size, shape, and relative positions of patterns
(2) similarity of a candidate character pattern(s) to a valid character
(3) word matching score

In the case of Kanji numeral segmentation, information types (1) and (2) can be used effectively, but type (3) information can not be used because unlike a string of characters, the numeral sequence has no contextual information. Type (1) information is effective for character segmentation but is not sufficient by itself. Type (2) information is particularly useful for valid segmentation, but it is risky to use it alone because segmentation in some cases is highly ambiguous, as described above.

Our method uses both information types (1) and (2) to resolve the ambiguities. The first problem we faced was the fact that the peripheral features

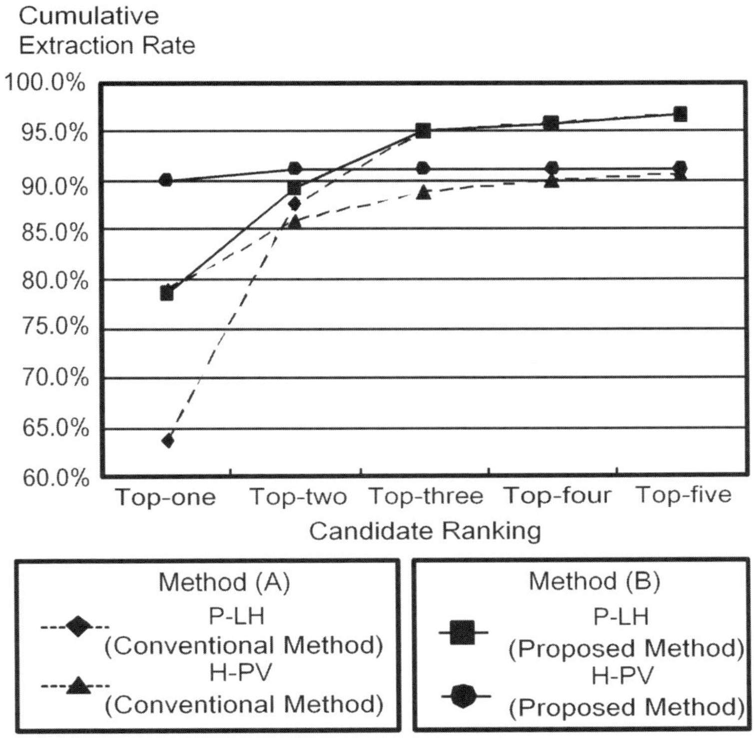

Fig. 8. Address-Block Extraction Accuracy

included in type (1) information, for example width, height, aspect ratio, and gap, do not have the same degree of usefulness for ambiguity resolution. Further, the degree of usefulness depends on the character category. To overcome these problems, we applied the Bayesian rule to obtain confidence values for each potential reading. These confidence values enable us to increase the segmentation accuracy.

7.2 Conventional Character Segmentation Method

An input image is analyzed by a pre-segmentation module that creates multiple candidates of segmented character patterns. In Figure 10, there are pre-segmented patterns of Kanji numerals. The hypotheses of the pre-segmentation are represented in terms of a graph (or a network), and one of the paths from the initial node to the terminal node is the result of segmentation. To select the optimum path among all possible paths, the links between nodes, which represent pre-segmentation candidates, are evaluated. The similarity of each pre-segmented pattern is used in the conventional segmentation method. The valid patterns (links) have a high similarity value, and the invalid

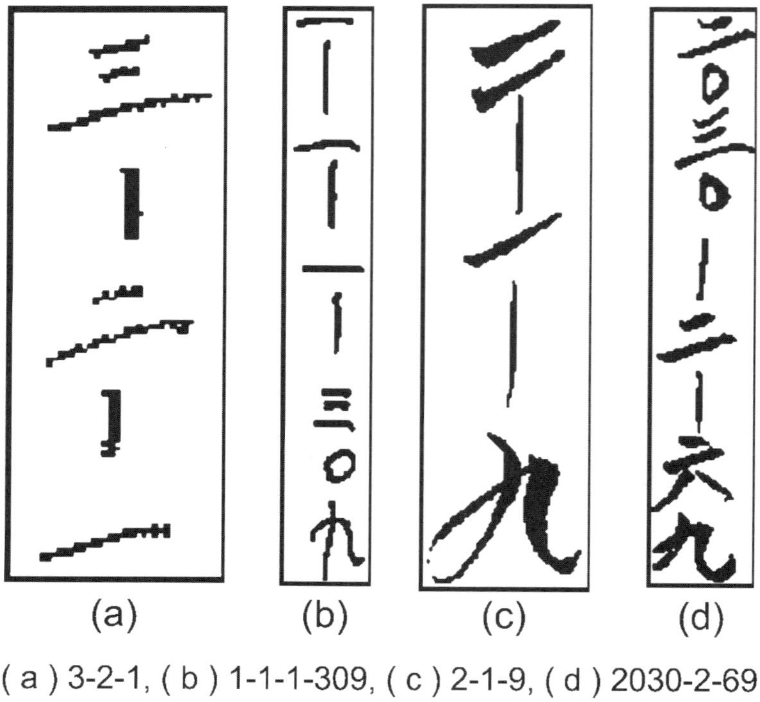

Fig. 9. Kanji Numerals Written Vertically

patterns (links) have a low value. In Figure 10, one of the pre-segmented patterns is inputted into the character classification module, then the character category and similarity value of 0.9 are obtained. In this case, the similarity is high because the inputted pattern has a valid shape. The evaluation of the links between nodes eventually identifies the optimum path. This search for the optimum path can be done through dynamic programming.

However, many pre-segmented patterns of Kanji numerals could be a part of several Kanji numeral patterns. As shown in Figure 10, the pre-segmented patterns consist of one horizontal line, two horizontal lines, and three horizontal lines. All patterns can be classified as the Kanji numerals, or with a high degree of similarity. Therefore, all evaluation values are very high, so it is difficult to determine the correct path from these pre-segmentation hypotheses in the case of Kanji numerals.

7.3 Information Concerning Character Category and Peripheral Features

7.3.1 Character Category Information

When the peripheral features of Kanji numerals "three", hyphen, "two", hyphen, and "one" (Figure 9(a)) are determined, the aspect ratio of the Kanji

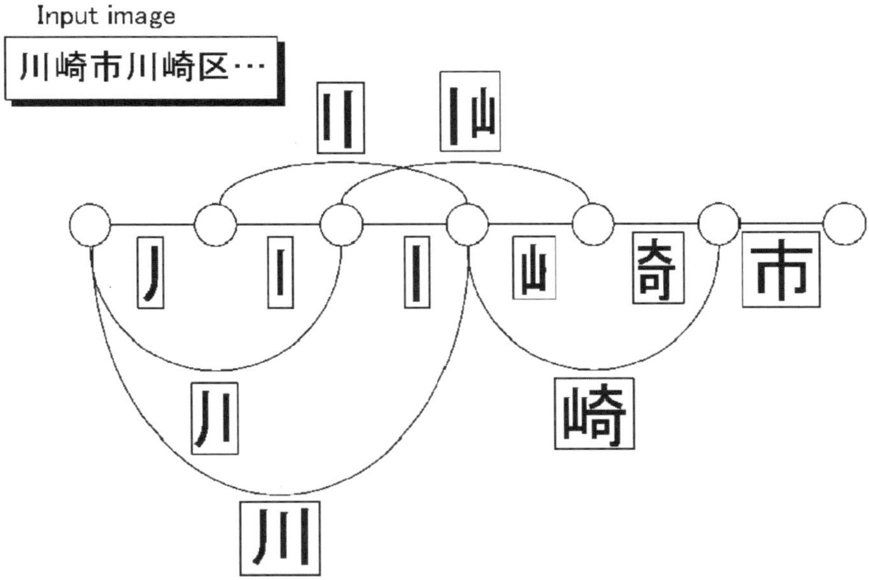

Fig. 10. Character Segmentation by Multiple Hypotheses Method

numeral "three" is about 1.0. In contrast the width of Kanji numeral "one" is much greater than the height. Also, the Kanji numeral "three" consists of three connected components, while the Kanji numeral "one" consists of one. In this way, the valid values of the peripheral features differ depending on the character categories.

The usefulness of a feature also differs depending on the character categories. For example, the usefulness of the spaces in front of and behind the characters is different for Kanji numerals "three" and "one" because the Kanji numeral "one" has the same shape as one part of the Kanji numeral "three", but a single horizontal line can be distinguished as either "one" or as one part of Kanji numeral "three" by using information concerning the spaces. Therefore, for the Kanji numeral "one", the information concerning the spaces in front of and behind characters is very important, and the usefulness of this feature is high.

On the other hand, the similarity given by the character classifier is not important and is not useful for Kanji numeral "one" because the similarity values for "one" and the one part of "three" is large in both cases. However, this information is very useful for other character categories.

Thus, we have to determine the usefulness of specific features for each character category, and the degrees of usefulness can then be used to calculate the link evaluation values.

7.3.2 Usefulness of Features

The degree of feature usefulness can be presented as a likelihood ratio that is given by equation 3,

$$L(e_k|H_c) = \frac{p(e_k|H_c)}{p(e_k|\bar{H}_c)} \tag{3}$$

where

c : character category assumed for a candidate
H_c : the positive hypothesis of the link being a pattern of character category c assumed for a candidate
e_k : one of the measured features, or the evidence of the hypothesis
$P(e_k|H_c)$: the probability of e_k supposing event H_c

The likelihood ratio is then evaluated for the assumed category. We computed the conditional probability $P(e_k|H_c)$ in terms of the histogram of e_k when event H_c occurred. To make this histogram, features e_k are extracted from all pre-segmented patterns, then classified into the corresponding character categories, and further divided into two classes, namely the correct and incorrect segmentation classes. So $P(e_k|H_c)$ can then be obtained from the histogram. By using the above formula, likelihood ratio $L(e_k|H_c)$ can be obtained from this $P(e_k|H_c)$.

Figure 11 shows the likelihood ratio graph of a feature extracted from the pre-segmented character patterns. This graph shows the likelihood distribution in the case of the character category of Kanji numeral "one"; the horizontal axis is the feature value of space in front of and behind the character, and the vertical axis is the likelihood ratio. Increasing the space value increases the likelihood ratio, demonstrating that the feature of the space in front of and behind characters is useful.

Thus, after extracting many features from pre-segmented patterns, we selected the most useful features for character segmentation. These features are shown below.

(A) Pattern height (normalized by line width)
(B) Pattern width (normalized by line width)
(C) Pattern aspect ratio
(D) Spaces in front of and behind characters
(E) Number of connected components in pattern
(F) Similarity of character classifier

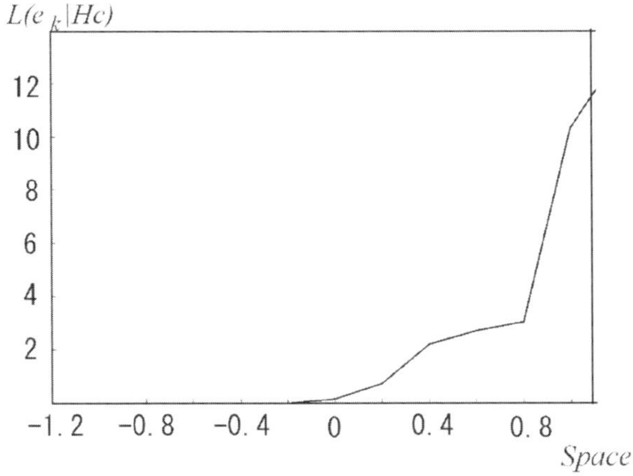

Fig. 11. Likelihood Ratio

7.4 Integrate Likelihood Ratios

The method we used to calculate the confidence value of a link by using the likelihood ratios of features is described below. We define this confidence value as the a posteriori probability $P(H_c|e_1, e_2, e_3, ..., e_n)$, where $e_1, e_2, e_3, ..., e_n$ is a list of measured features, or the evidence of the hypothesis. In this chapter, $e_1, e_2, e_3, ..., e_n$ are the (A)-(F) features listed above. The confidence value can be transformed into a computable formula by applying the Bayesian rule (Bayes theorem), as shown in equation 4.

$$P(H_c|e_1, e_2, e_3, \cdots, e_n) = \frac{\frac{P(H_c)}{P(\bar{H}_c)} \prod_{k=1}^{n} L(e_k|H_c)}{1 + \frac{P(H_c)}{P(\bar{H}_c)} \prod_{k=1}^{n} L(e_k|H_c)} \tag{4}$$

where

c : character category assumed for a candidate
H_c : positive hypothesis of category c given the candidate
$e_1, e_2, e_3, ...e_n$: measured features of the candidate
$L(e_k|H_c)$: likelihood ratio for e_k for correct category c

The usefulness of each feature is represented by the likelihood ratio. Thus, the calculated confidence value can be used for the link evaluation.

7.5 Implementation

7.5.1 Learning Stage

In the learning stage, many sample images are pre-segmented where candidate character patterns, including over-segmented partial patterns, are extracted. Peripheral features and similarity values are extracted from all of the pre-segmented patterns, and the histogram of these features is computed as described in 7.3.2 The sampled versions of the conditional density functions are then calculated, and the likelihood ratios $L(e_k|H_c)$ are pre-computed in a non-parametric way. The calculated $L(e_k|H_c)$ of each feature is stored and used in the recognition stage.

7.5.2 Recognition Stage

When an image is to be recognized, the pre-segmentation module performs segmentation as described above, and the features for each candidate are gathered. Among the features, there is a category index c given by a character classifier, and the confidence is computed for the assumed category. The computation process is depicted in Figure 12. Selection of the best path is then made using these confidence values.

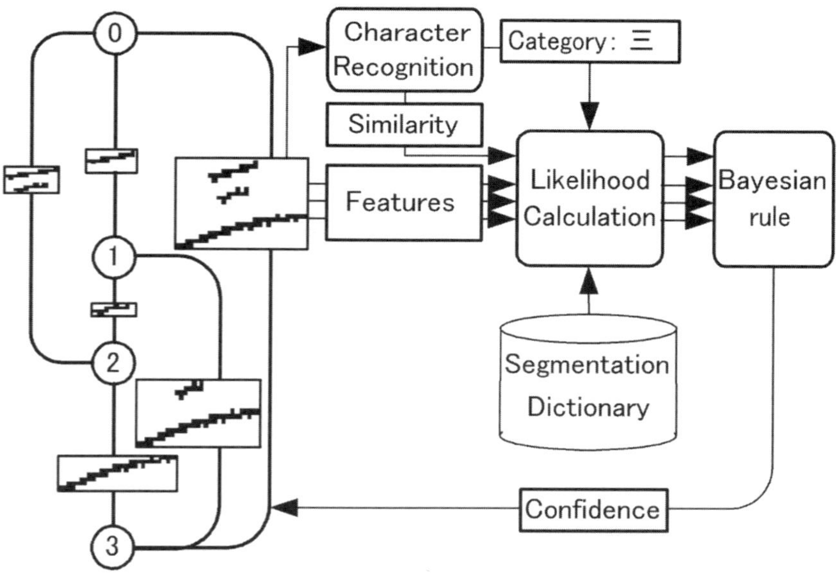

Fig. 12. New Character Segmentation Method for Street Numbers

7.6 Experimental Results

In order to measure the character segmentation accuracy, 208 sample images of character strings which were written by the general public were used for the experiment. There was no noise pattern in this input data. In the learning step 1288 samples of correct character patterns and 5082 samples of incorrect character patterns were used. This learning data set is different from the above 208 sample images. The likelihood ratio was created from the learning data set. In these samples, characters were written vertically with a ball-point pen, a writing brush, or a fiber-tipped pen.

In this experiment, we compared three methods as follows.

M1: Use similarity for confidence
M2: Use similarity and peripheral information uniformly for confidence
M3: Use similarity and peripheral information in response to character category for confidence

Table 1 shows the character segmentation and the recognition accuracy. In this experiment, a correct result for character segmentation means that all characters in the character line were segmented correctly. A correct recognition result means that all characters in the character line were recognized correctly.

Table 1. Accuracy of character segmentation and recognition

	Character segmentation	Recognition
M1	33%	30%
M2	73%	62%
M3	87%	71%

Table 1 shows that the character segmentation accuracy of M3 increased by 54 points compared to M1, and 14 points compared to M2. This means that peripheral information is effective for the character segmentation of Kanji numerals. Changing the importance level of the peripheral information in response to the character category is also effective.

With M3, character recognition accuracy increased by 41 points compared to M1, and 9 points compared to M2. The increment value of character recognition is less than character segmentation, because it is difficult to use linguistic information with the street numbers, despite the correct character segmentation results.

Figure 13 shows one example of segmentation results. In the Figure, M1 segmented each simple pattern as the characters in ImageA, ImageB, and ImageC. This means that each simple pattern has a high similarity value, because each can be read as Kanji numerals. M2 segmented those patterns which are separated by wide gaps from other patterns as shown in Figure 13 ImageB. M3 was able to segment the character patterns correctly, because

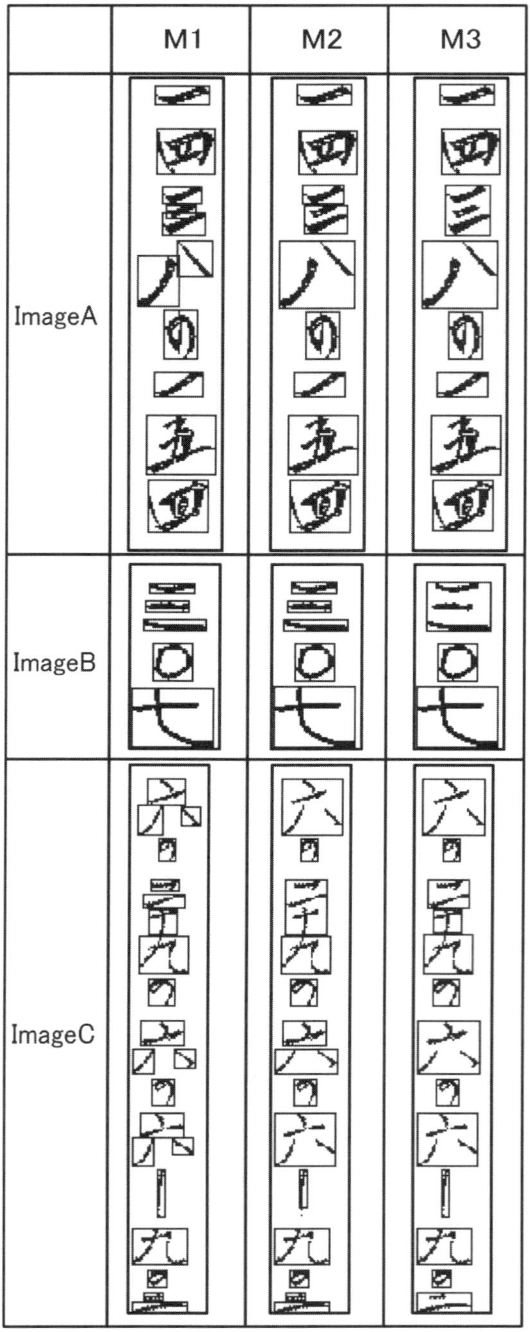

Fig. 13. Character Segmentation Results

it used information about the size, aspect, and gaps for the confidence value in response to the character category. Especially, ImageC in Figure 13 shows that M3 can solve the difficult character segmentation problem when there are many simple patterns and various sizes.

8 Conclusion

There are many approaches to document analysis, and these approaches can be divided in two types, top-down and bottom-up. Each of these two types has its good and bad points, and the most effective approach is different according to the kind of document in question. The top-down approach is best suited to well-known layouts, and the bottom-up approach is better when the layout is unknown.

OCR systems which read only the ROI (Region Of Interest) in a document can be effectively used for office automation. In this case, there are many methods which adopt the multiple hypotheses approach. If the multiple hypotheses approach is used for an OCR system, it is easy to design the whole system and adjust the balance between recognition accuracy and calculation costs.

In this discussion, a recognition system for mail destination addresses was introduced as a case in which the multiple hypotheses approach was used successfully. We described an effective method which can solve the ambiguity and variation in destination address recognition. Specifically, an address-block extraction method and character segmentation of the street numbers were introduced. These methods create a number of candidates using multiple hypotheses, and evaluate each candidate by a Bayesian rule. These evaluated candidates are then sorted by confidence value, and one correct answer is finally selected.

Future OCR systems will probably include expanded target fields and increased recognition accuracy. This will enable many more tasks in the office to be automated, freeing humans from simple, repetitive labor. Many researchers are currently working on character recognition in real scenes. For example, an embedded OCR was developed for mobile phones. We believe that the most difficult problems facing OCR will gradually be solved, and that eventually OCR systems will be able to recognize whole characters which are written anywhere in the world.

References

1. Kobayashi, Y., Yamada, K., Tsukumo, J.: A Segmentation Method for Hand-Written Japanese Character Lines Based on Transitional Information. Proc. ICPR (1992) 487–491
2. Kamimura, T., et al.: Determing Address Format Using Layout Rules for Reading Japanese Hand-written Mail. (in Japanese) IEICE PRU95-107 (1995) 25–30

3. Yeh, P.S., et al.: Address Location on Envelopes. Pattern Recognition (1987) Vol. 20, No. 2, 213–227
4. Wolf, M., et al.: Form-Based Localization of the Destination Address Block on Complex Envelopes. Proc. ICDAR (1997) 908–913
5. Wolf, M., et al.: Fast Address Block Location on Handwritten and Machine Printed Mail-piece Images. Proc. ICDAR (1997) 753–757
6. Yu, B., et al.: Address Block Location on Complex Mail Pieces. Proc. ICDAR (1997) 897–901
7. Sako, H., et al.: On The Rejection Ability Required in Multiple Hypothesis Techniques. Proc. IWFHR (2000) 123–132
8. Casey, R.G., et al.: A Survey of Methods and Strategies in Character Segmentation. IEEE Trans. on PAMI (1996) Vol. 18, No. 7, 690–706
9. Fujisawa, H., et al.: Segmentation Method for Character Recognition. Proc. IEEE (1992) Vol. 80 No. 7, 1079–1092
10. Gyohten, K., Sumiya, T., Babaguchi, N., Kakusho, K., Kitahashi, T.: A Multi-Agent Based Method for Extracting Characters and Character Strings. IEICE TRNS. INF. & SYST. (1996) VOL. E79-D, NO. 5
11. Bruckner, T., Suda, P., Block, H.U., Maderlechner, G.: In-house Mail Distribution by Automatic Address and Content Interpretation. SDAIR96 67–75
12. Srihari, S.N., Srikantan, G., Hong, T., Grom, B.: A General-Prupose Japanse Optical Character Recognition System. Proc. Conference on Document Recognition (1996) SPIE Symposium
13. Cohen, E., Hull, J.J., Srihari, S.: Control Structure for Interpreting Handwritten Addresses. IEEE TRANS. PAMI. (1994) VOL. 16, NO. 10, 1049–1055
14. Shinjo, H., Nakashima, K., Koga, M., Marukawa, K., Shima, Y., Hadano, E.: A Method for Connecting Disappeared Junction Patterns on Frame Lines in Form Documents. Proc. ICDAR (1997) 667–670
15. Dunn, C.E., Wang, P.S.P.: Character Segmentation Techniques for Handwritten Text - A Survey. Pattern Recognition Methodology and Systems, Proceedings., 11th IAPR International Conference (1992) 577–580
16. Tnag, Y.Y., Ma, H., Liu, J., Li, B.F., Xi, D.: Multiresolution Analysis in Extraction of Reference Lines from Documents with Gray Level Background. IEEE TRANS. PAMI. (1997) VOL. 19, NO. 8
17. Koga, M., Kagehiro, T., Sako, H., Fujisawa, H.: Segmentation of Japanese Handwritten Characters Using Perioheral Feature Analysis. Proc. ICPR (1998) 1137–1141
18. Sawaki, M.: Text-Line Extraction and Character Recognition of Document Headlines With Graphical Designs Using Complementary Similarity Measure. IEEE TRANS. PAMI. (1998) VOL. 20, NO. 10
19. Yen, P., Antoy, S., Litcher, A., Rosenfeld, A.: Address Location on Envelopes. Pattern Recognitoin (1987) Vol. 20, No. 2, 213–227
20. Srihari, S.N.: Integration of Hand-Written Address Interpretation Technology into the United States Postal Service Remote Computer Reader System. Proc. ICDAR (1997) 892–896
21. Palumbo, P.W., Srihari, S.N., Soh, J., Sridhar, R., Demjanenko, V.: Postal Address Block Location in Real Time. Computer (1992) 34–42
22. Whichello, A.P., Yan, H.: Locating Address Blocks and Postcodes in Mail-Piece Images. Proc. ICPR (1996) 716–720
23. Nakajima, N., Tsuchiya, T., Kamimura, T., Yamada, K.: Analysis of Address Layout on Japanese Handwritten Mail. Proc. ICPR (1996) 726–731

24. Liu, C., Koga, M., Fujisawa, H.: Gabor Feature Extraction for Character Recognition: Comparison with Gradient Feature. Proc. ICDAR (2005) Vol. 1, 121–125
25. Koga, M., Mine, R., Kameyama, T., Takahashi, T., Yamazaki, M., Yamaguchi, T.: Camera-based Kanji OCR for Mobile-phones: Practical Issues. Proc. ICDAR (2005) Vol. 2 635–639
26. Liu, C., Mine, R., Koga, M.: Building Compact Classifier for Large Character Set Recognition Using Discriminative Feature Extraction. Proc. ICDAR (2005) Vol. 2 846–850
27. Liu, C., Sako, H.: Class-specific Feature Polynomial Classifier for Pattern Classification and its Application to Handwritten Numeral Recognition. Pattern Recognition (2006) Vol. 39 669–681
28. Kagehiro, T., Koga, M., Sako, H., Fujisawa, H.,: Address-Block Extraction by Bayesian Rule. Proc. ICPR (2004) Vol. 2 582–585
29. Ikeda, H., Furukawa, N., Koga, M., Sako, H., Fujisawa, H.: A Context-free Grammar-based Language Model for String Recognition. Int. J. Comput. Proc. Oriental Lang. (2002) Vol. 15, No. 2 149–163
30. Shinjo, H., Hadano, E., Marukawa, K., Shima, Y., Sako, H.: A Recursive Analysis for Form Cell Recognition. Proc. ICDAR (2001) 694–698
31. Kagehiro, T., Koga, M., Sako, H., Fujisawa, H.: Segmentation of Handwritten Kanji Numerals Integrating Peripheral Information by Bayesian Rule. Proc. MVA (1998) 439–442

Learning Matching Score Dependencies for Classifier Combination

Sergey Tulyakov and Venu Govindaraju

Center for Unified Biometrics and Sensors, University at Buffalo, Amherst, NY 14228, USA tulyakov@cedar.buffalo.edu,venu@cubs.buffalo.edu

Summary. Two types of matching score dependencies might be observed during the training of multiple classifier recognition system - the dependence between scores produced by different classifiers and the dependence between scores assigned to different classes by the same classifier. Whereas the possibility of first dependence is evident, and existing classifier combination algorithms usually account for this dependence, the second type of dependence is mostly disregarded. In this chapter we discuss the properties of such dependence and present few combination algorithms effectively dealing with it.

1 Introduction

The integration of recognition algorithms into a single document processing system might involve different available modules suitable for a single task. For example, we might possess few character or word recognition algorithms which all can be used in the system. One possible approach is to test these algorithms and to choose the one with the best performance. But practice shows that better approach is to try to use all available algorithms and to combine their outputs in order to achieve a better performance than any single algorithm. The combination problem consists in learning the behavior of given algorithms and deriving best possible combination function.

We assume that both the combined algorithms and the result of combination are classifiers. Thus a finite number of classes are distinguished in the problem, and the task is to find a class, which corresponds most to the input. As examples, classes might be a character set, a word lexicon, a person list, etc. Usually classifiers output the numeric matching scores corresponding to each class, and we will assume that these scores are available for combination. The combination algorithm is a function producing a final combined score for each class, and the final classifier selects class with the best combined score.

The purpose of this chapter is to investigate the different scenarios of combining classifiers, to show the difficulties in finding the optimal combination

S. Tulyakov and V. Govindaraju: *Learning Matching Score Dependencies for Classifier Combination*, Studies in Computational Intelligence (SCI) **90**, 305–332 (2008)
www.springerlink.com

algorithms, and to present few possible approaches to combination problems. Generally, the classifier combination problem can be viewed as a construction of postprocessing classifier operating on the matching scores of combined classifiers. For many classifier combination problems, though, the number of classes or the number of classifiers and, consequently, the number of matching scores is too big, and applying generic pattern classification algorithms is difficult. Thus some scores are usually discarded from combination algorithm, or simplifying assumptions on score distributions are made and used in the combination algorithm. Though the dependency between classifiers is usually learned by the combination algorithms, the dependency between scores assigned to different classes by the same classifier is discarded. In this work we will show that accounting for score dependencies is essential for proper combination of classifiers. The theory will be complemented by the experiments we perform on handwritten word recognizers and biometric person matchers.

2 Problem Description

Though the general theory presented in this chapter can be applied to any classifier combination task, we will mostly focus on two particular applications: handwritten word recognition and biometric person authentication. As a result, we are making few assumptions about combined classifiers. First we assume that each classifier assigns a matching score for each class, and we use these scores for combination. It would be convenient to call these classifiers 'matchers' or 'recognizers', in contrast to the general notion of classifiers making a decision and thus having selected class as their only output. Second, we assume that we only combine a small number of given matchers; in fact, for both applications we consider combinations of two matchers. Thus we separate ourselves from the so called 'classifier ensembles' having potentially large number of dynamically generated classifiers. Finally, we assume that the number of classes is large and may be variable. Indeed, the number of possible handwritten words defined by the corresponding lexicon or the number of enrolled persons in biometric database can be both large and variable. To be more specific, we describe both applications next.

2.1 Handwritten Word Recognizers

We consider the application of handwritten word recognizers in the automatic processing of United Kingdom mail. The destination information of the mail piece will usually contain the name of the postal town or county. After automatic segmentation of the mail piece image the goal of handwritten word recognizer is to match hypothesized town or county word image against a lexicon of possible names. Provided lexicon contains 1681 entries.

We use two handwritten word recognizers for this application: Character Model Recognizer (CMR)[1] and Word Model Recognizer (WMR)[2]. Both recognizers employ similar approaches to word recognition: they oversegment the word images, match the combinations of segments to characters and derive a final matching score for each lexicon word as a function of character matching scores. Still, the experiments (see Table 1) reveal that these matchers produce somewhat complementary results and their combination might be beneficial.

Our data consists of three sets of word images of approximately same quality (the data was provided as these three subsets and we did not regroup them). The images were manually truthed and only those images containing any of the 1681 lexicon words were retained. The word recognizers were run on these images and their match scores for all 1681 lexicon words were saved. Note, that both recognizers reject some lexicon entries if, for example, the lexicon word is too short or too lengthy for presented image. We assume that in real systems such rejects will be dealt with separately (it is possible that the lexicon word corresponding to image truth will be rejected), but for our combination experiments we only keep scores of those lexicon words which are not rejected by any of the two recognizers. Thus for each image I_k we have a variable number N_k of score pairs (s_i^{cmr}, s_i^{wmr}), $i = 1, \dots, N_k$ corresponding to non-rejected lexicon words. One of these pairs corresponds to the true word of the image and we will call these scores 'genuine', and other 'impostor' score pairs correspond to non-truth words.

After discarding images with non-lexicon words, and images where truth word was rejected by any recognizer, we are left with three sets of 2654, 1723 and 1770 images and related sets of score pairs. We will refer to the attempt of recognizing word image as identification trial. Thus each identification trial has a set score pairs (s_i^{cmr}, s_i^{wmr}), $i = 1, \dots, N_k$ with one genuine score pair and $N_k - 1$ impostor pairs. The scores of each recognizer were also linearly normalized so that each score is in the interval $[0, 1]$ and bigger score means better match.

In order to get the general picture of the performance of considered recognizers we can count the numbers of identification trials where genuine score is better than all impostor scores of that trial. We summarized these counts in Table 1. The number of trials where first matcher (CMR) produced the genuine score bigger than all impostor scores is 3366, and second matcher (WMR) did the same 4744 times. Apparently, WMR has better performance, but still there are some identification trials ($5105 - 4744 = 361$), where CMR is correct and WMR is not. Since there is such distinction between recognizers, we strongly hope that their combination might achieve higher recognition rates.

Since our data was already separated into three subsets, we used this structure for producing training and testing sets. Each experiment was repeated three times, each time one subset is used as a training set, and two other sets are used as test sets. Final results are derived as averages of these three training/testing phases.

Matchers	Total # of trials	1st matcher is correct	2nd matcher is correct	Both are correct	Either one is correct
CMR&WMR	6147	3366	4744	3005	5105
li&C	5982	4870	4856	3937	5789
li&G	5982	4870	4635	3774	5731

Table 1. Numbers of identification trials with any matcher having best score for the correct class

2.2 Biometric Person Matchers

We used biometric matching score set BSSR1 distributed by NIST[3]. This set contains matching scores for a fingerprint matcher and two face matchers 'C' and 'G'. Fingerprint matching scores are given for left index 'li' finger matches and right index 'ri' finger matches. In this work we used both face matching scores and fingerprint 'li' scores and we do two types of combinations: 'li'&'C' and 'li'&'G'.

Though the BSSR1 score set has a subset of scores obtained from same physical individuals, this subset is rather small - 517 identification trials with 517 enrolled persons. In our previous experiments[4] we used this subset, but the number of failed identification attempts for most experiments was less than 10 and it is difficult to compare algorithms with so few negatives. In this work we use bigger subsets of fingerprint and face matching scores of BSSR1 by creating virtual persons; the fingerprint scores of a virtual person come from one physical person and the face scores come from another physical person. The scores are not reused, and thus we are limited to the maximum number of identification trials - 6000 and the maximum number of classes, or enrolled persons, - 3000. Some enrollees and some identification trials also needed to be discarded since all corresponding matching scores were invalid probably due to enrollment errors. In the end we split data in two equal parts - 2991 identification trials with 2997 enrolled persons with each part used as training and testing sets in two phases.

Table 1 shows the numbers of identification trials with genuine scores bigger than all impostor scores of that trial. The matchers now are more equal in strength and there is only a small number of trials where neither matcher correctly identified the genuine person.

3 Verification and Identification Tasks

Above described applications might include different operating scenarios. In one scenario the system generates a hypothesis of a true class of the input beforehand, and the task of the matchers is to verify if the input indeed of the hypothesized class. For example, a bank check recognition system might hypothesize about the value of the check based on the legal field, and numeric string recognition module must confirm that courtesy value coincides with the legal amount[5]. In biometric person verification systems a person presents a unique person identifier to the system, and biometric recognition module verifies if person's biometric scan matches the enrolled biometric template of claimed person's identity.

In another operating scenario a class of the input should be selected from a set of possible classes. Each lexicon word can be associated with a class for word recognition applications. In our considered application a set of UK postal town and county names serves as a lexicon for word recognizers. For biometric person recognition a set of classes can coincide with the set of enrolled persons. The task of recognizer in this scenario is to select the class, which is the true class of input signal. We will assume that we deal with so called 'closed set identification', where the true class of input is included in the set of possible classes; in contrast 'open set identification' might not include true class in this set, and input needs to be rejected in this case.

We will call the system operating in the verification mode as verification system, and system operating in identification mode as identification system. Correspondingly, the problem solved by matchers or their combinations in the first case will be called verification task, and in the second case - identification task. Note that there could also be other operating scenarios involving considered matchers; as an example we have given open set identification.

3.1 Performance Measures

Different modes of operation demand different performance measures. For verification systems the performance is traditionally measured by means of Receiver Operating Characteristic (ROC) curves or by Detection Error Trade-off (DET) curve. These curves are well suited for describing the performance of two-class pattern classification problems. In such problems there are two types of errors: the samples of first class are classified to belong to second class, and samples of second class are classified to be in first class. The decision to classify a sample to be in one of two classes is usually based on some threshold. Both performance curves show the relationship between two error rates with regards to a threshold (see [6] for precise definition of above performance measures).

In our case we will use ROC curves for comparing algorithm performance. If a matcher is used for verification task there are two classes: genuine if input belongs to the same hypothesized class, and impostor otherwise. The decision is traditionally based on the matching score of a recognizer assigned for hypothesis class.

For measuring performance of identification systems we will use ranking approach. In particular, we are interested in maximizing the rate of correctly identifying the input, first-rank-correct rate. If we look at identification task as a pattern classification problem, this performance measure will directly correspond to the traditional minimization of the classification error. Note that there are also other approaches to measure performance in identification systems[6], e.g. Rank Probability Mass, Cumulative Match Curve, Recall-Precision Curve. Though they might be useful for some applications, in our case we will be more interested in correct identification rate.

4 Verification Systems

The problem of combining matchers in verification systems can be easily solved with pattern classification approach. As we already noted, there are two classes: genuine verification attempts and impostor verification attempts. The hypothesis class of the input is provided before matching. Each matcher j outputs a score s^j corresponding to a match confidence between input sample and hypothesis class. Assuming that we combine M classifiers, our task is to perform two-class classification (genuine and impostor) in M-dimensional score space $\{s^1, \dots, s^M\}$. If the number of combined classifiers M is small, we will have no trouble in training pattern classification algorithm.

We employ the Bayesian risk minimization method as our classification approach[7]. This method states that the optimal decision boundaries between two classes can be found by comparing the likelihood ratio

$$f_{lr}(s^1, \dots, s^M) = \frac{p_{gen}(s^1, \dots, s^M)}{p_{imp}(s^1, \dots, s^M)} \tag{1}$$

to some threshold θ where p_{gen} and p_{imp} are M-dimensional densities of score tuples $\{s^1, \dots, s^M\}$ corresponding to two classes - genuine and impostor verification attempts. In order to use this method we have to estimate the densities p_{gen} and p_{imp} from the training data. For our applications the number of matchers M is 2 and the number of training samples is large (bigger than 1000), so we can successfully estimate these densities.

In our data each identification trial has one genuine and $N_k - 1$ impostor score pairs, so the total number of genuine score pairs is $T = K$ (K is the number of identification trials in the training set) and the total number of impostor score pairs is $T = \sum_{k=1}^{K} (N_k - 1)$. We approximate both densities as the sums of 2-dimensional gaussian Parzen kernels

$$\hat{p}(s^1, s^2) = \frac{1}{T} \sum_{t=1}^{T} \frac{1}{2\pi\sigma^2} e^{-\frac{(s^1 - s_t^1)^2 + (s^2 - s_t^2)^2}{2\sigma^2}}$$

where $\{s_t^1, s_t^2\}_{t=1,\dots,T}$ are the set of training score pairs. The window parameter σ is estimated by the maximum likelihood method on the training set[8] using leave-one-out technique. Note that σ is different for genuine and impostor density approximations.

For a given threshold θ we calculate the number of misidentified samples from the test data set of each class. The genuine samples (s^1, s^2) are misidentified as impostor samples if $\hat{f}_{lr}(s^1, s^2) = \frac{\hat{p}_{gen}(s^1,s^2)}{\hat{p}_{imp}(s^1,s^2)} < \theta$ (false rejects), and impostor samples misidentified as genuine if $\hat{f}_{lr}(s^1, s^2) \geq \theta$ (false accepts). Thus for each θ we calculate false reject and false accept rates, $FRR(\theta)$ and $FAR(\theta)$, and construct ROC curve, which is a graph of $FRR(\theta)$ versus $FAR(\theta)$. The resulting ROC curves for original matchers and for their combinations with likelihood ratio method are shown in Figures 1, 2 and 3.

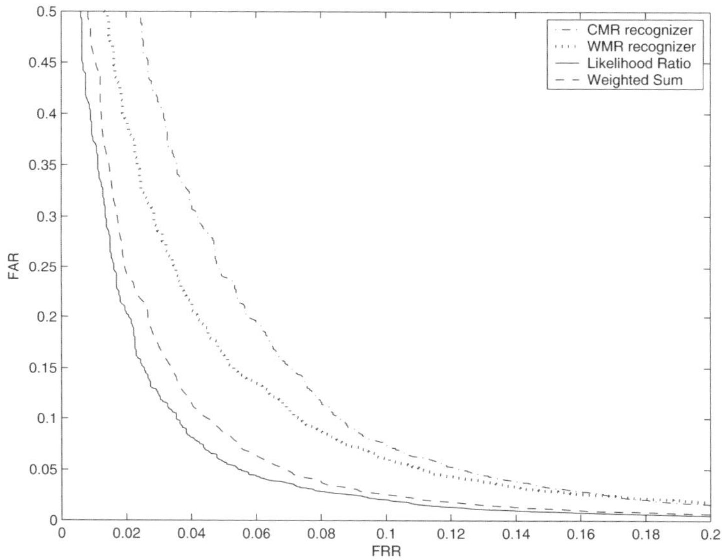

Fig. 1. ROC curves for two handwritten word recognizers (WMR and CMR) and their combinations by likelihood ratio and weighted sum methods

As we expected, the combination has better performance than any of the individual matchers. Biometric matchers are based on different modalities and thus better complement each other than word recognizers. This is indicated by the performance graphs: the improvement is bigger in the case of biometric matchers.

The likelihood ratio combination method is theoretically optimal for verification systems and its performance only limited by our ability to correctly estimate score densities. The density estimation is known to be a difficult task; working with many-dimensional data, having heavy tailed distributions or discreteness in the data can lead to very poor density estimates. In our experiments we had sufficient number of training samples in 2-dimensional space and the task was relatively easy, but still we had to make adjustments for the discreteness of fingerprint scores represented by the integer numbers in the range $0 - 350$.

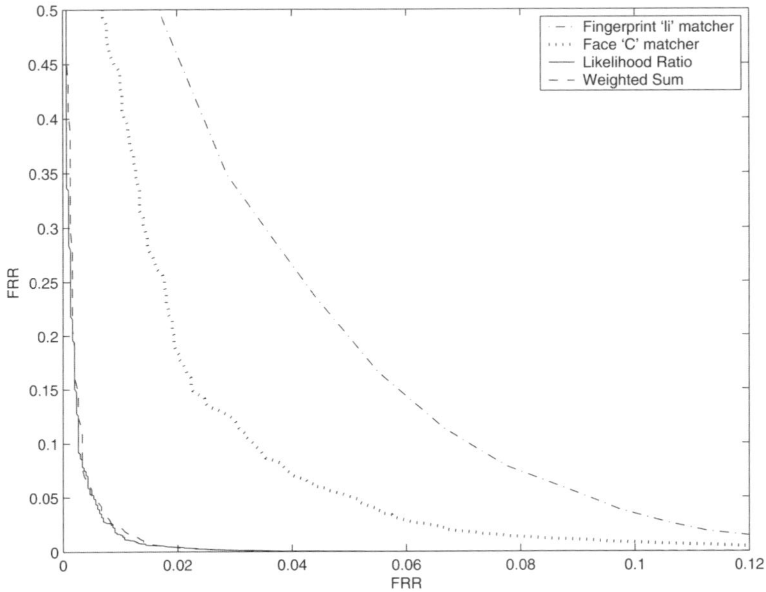

Fig. 2. ROC curves for two biometric matchers (fingerprint 'li' and face 'C') and their combinations by likelihood ratio and weighted sum methods

Since our problem is the separation of genuine and impostor classes, we could apply many existing pattern classification techniques. For example, support vector machines have shown good performance in many tasks, and can be definitely used to improve the likelihood ratio method. In [9] we performed some comparisons of likelihood ratio method with SVMs on an artificial task and found that on average (over many random training sets) SVMs do have slightly better performance, but for a particular training set it might not be true. The difference in performance is quite small and decreases with the increasing number of training samples. Also note that many pattern classification algorithms provide only a single decision boundary (separating hyperplane in the kernel mapped space for SVMs), and this effectively results in the single point of FAR-FRR plane instead of ROC curve. The advantage of likelihood ratio combination method is that we get the whole range of solutions by varying threshold parameter θ and which are represented by ROC curve.

5 Identification Systems

In identification systems a hypothesis of the input sample is not available and we have to choose the input's class among all possible classes. Denote N as the number of classes. The total number of matching scores available for combination now is MN: N matching scores for each class from each of M

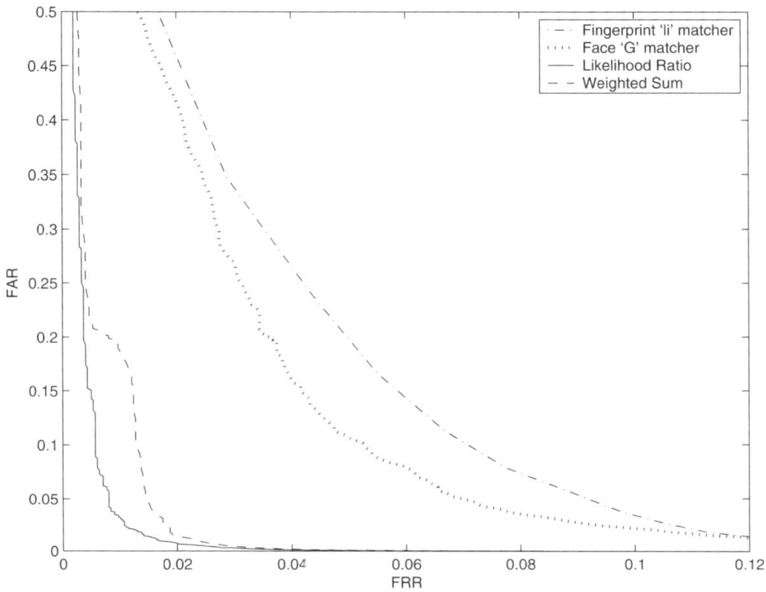

Fig. 3. ROC curves for two biometric matchers (fingerprint 'li' and face 'G') and their combinations by likelihood ratio and weighted sum methods

combined classifiers. If numbers M and N are not big, then we can use generic pattern classifiers in MN-dimensional score space to find the input's class among N classes. For some problems, e.g. digit or character recognition, this is an acceptable approach; the number of classes is small and usually there is a sufficient number of training samples to properly train pattern classification algorithms operating in MN score space.

But for our applications in handwritten word recognition and biometric person identification the number of classes is too big and the number of training samples is too small (there might be even no training samples at all for a particular lexicon word), so the pattern classification in the MN-dimensional score space seems to be out of the question. The traditional approach in this situation is to use some combination rules. The combination rule implies the use of some combination function f operating only on M scores corresponding to one class, $f(s^1, \dots, s^M)$, and it states that the decision class C is the one which maximizes the value of a combination function:

$$C = \arg \max_{i=1,\dots,N} f(s_i^1, \dots, s_i^M) \tag{2}$$

Note that in our notation the upper index of the score corresponds to the classifier, which produced this score, and lower index corresponds to the class for which it was produced. The names of combination rules are usually directly derived from the names of used combination functions: the sum function

$f(s^1, \dots, s^M) = s^1 + \dots + s^M$ corresponds to the sum rule, the product function $f(s^1, \dots, s^M) = s^1 \dots s^M$ corresponds to the product rule and so on.

Many combination rules have been proposed so far, but there is no agreement on the best one. It seems that different applications require different combination rules for best performance. Anyone wishing to combine matchers in real life has to test few of them and choose the one with best performance. Combination rules are also frequently used for verification problems to find the final score, which is compared with threshold and the decision is based on this comparison. But there is no real need to do it - the plethora of pattern classification algorithms is available for solving combinations in verification problems.

Our main interest in this chapter is to investigate the problem of finding the optimal combination function for identification systems. This problem appears to be much more difficult in comparison to combinations in verification systems.

5.1 Likelihood Ratio Combination Rule

As we already know, likelihood ratio function is the optimal combination function for verification systems. We want to investigate whether it will be optimal for identification systems. Suppose we performed a match of the input sample by all M matchers against all N classes and obtained MN matching scores $\{s_i^j\}_{i=1,\dots,N;j=1,\dots,M}$. Assuming equal prior class probabilities, the Bayes decision theory states that in order to minimize the misclassification rate the sample should be classified as one with highest value of likelihood function $p(\{s_i^j\}_{i=1,\dots,N;j=1,\dots,M}|\omega_i)$. Thus, for any two classes ω_1 and ω_2 we have to classify input as ω_1 rather than ω_2 if

$$p(\{s_i^j\}_{i=1,\dots,N;j=1,\dots,M}|\omega_1) > p(\{s_i^j\}_{i=1,\dots,N;j=1,\dots,M}|\omega_2) \tag{3}$$

Let us make an assumption that the scores assigned to each class are sampled independently from scores assigned to other classes; scores assigned to genuine class are sampled from M-dimensional genuine score density, and scores assigned to impostor classes are sampled from M-dimensional impostor score density:

$$\begin{aligned} & p(\{s_i^j\}_{i=1,\dots,N;j=1,\dots,M}|\omega_i) \\ & = p(\{s_1^1, \dots, s_1^M\}, \dots, \{s_{\omega_i}^1, \dots, s_{\omega_i}^M\}, \dots, \{s_N^1, \dots, s_N^M\}|\omega_i) \\ & = p_{imp}(s_1^1, \dots, s_1^M) \dots p_{gen}(s_{\omega_i}^1, \dots, s_{\omega_i}^M) \dots p_{imp}(s_N^1, \dots, s_N^M) \end{aligned} \tag{4}$$

After substituting 4 into 3 and canceling out common factors we obtain the following inequality for accepting class ω_1 rather than ω_2:

$$p_{gen}(s_{\omega_1}^1, \dots, s_{\omega_1}^M) p_{imp}(s_{\omega_2}^1, \dots, s_{\omega_2}^M) > p_{imp}(s_{\omega_1}^1, \dots, s_{\omega_1}^M) p_{gen}(s_{\omega_2}^1, \dots, s_{\omega_2}^M)$$

or

$$\frac{p_{gen}(s^1_{\omega_1}, \ldots, s^M_{\omega_1})}{p_{imp}(s^1_{\omega_1}, \ldots, s^M_{\omega_1})} > \frac{p_{gen}(s^1_{\omega_2}, \ldots, s^M_{\omega_2})}{p_{imp}(s^1_{\omega_2}, \ldots, s^M_{\omega_2})} \tag{5}$$

The terms in each part of the above inequality are exactly the values of the likelihood ratio function f_{lr} taken at the sets of scores assigned to classes ω_1 and ω_2. Thus, the class maximizing the MN-dimensional likelihood function of inequality 3 is the same as a class maximizing the M-dimensional likelihood ratio function of inequality 5. The likelihood ratio combination rule is the optimal combination rule under used assumptions.

Matchers	1st matcher is correct	2nd matcher is correct	Either one is correct	Likelihood Ratio Rule	Weighted Sum Rule
CMR&WMR	3366	4744	5105	4293	5015
li&C	4870	4856	5789	5817	5816
li&G	4870	4635	5731	5737	5711

Table 2. Correct identification rate for likelihood ratio and weighted sum combination rules

Table 2 shows the performance of this rule on our data sets. Whereas the combinations of biometric matchers have significantly higher correct identification rates than single matchers, the combination of word recognizers has lower correct identification rate than a single WMR matcher. This fact is rather surprising: the calculation of the combined scores by the likelihood ratio is exactly the same as we did for combinations in verification systems which gave us significant improvements in all cases (Figures 1, 2 and 3).

Few questions arise after reviewing the results of these experiments:

- If likelihood ratio combination rule was not able to improve correct identification rate of word recognizers, is there any other rule which will succeed?
- What are the reasons for the failure of seemingly optimal combination rule?
- What is the true optimal combination rule, and can we devise an algorithm of learning it from the training data?

In the rest of this chapter we will investigate these questions.

5.2 Weighted Sum Combination Rule

One of the frequently used rules in classifier combination problems is the weighted sum rule with combination function $f(s^1, \ldots, s^M) = w_1 s^1 + \cdots + w_M s^M$. The weights w_j can be chosen heuristically with the idea that better performing matchers should have bigger weight, or they can be trained to optimize some criteria. In our case we train the weights so that the number of successful identification trials on the training set is maximized. Since we have two matchers in all configurations we use brute-force method: we calculate the correct identification rate of combination function $f(s^1, s^2) = ws^1 + (1-w)s^2$ for different values of $w \in [0, 1]$, and find w corresponding to highest rate.

The numbers of successful identification trials on the test sets is presented in Table 2. In all cases we see an improvement over the performances of single matchers. The combination of word recognizers is now successful and is in line with the performance of other combinations of matchers.

We also investigated the performance of this method in the verification task. Figures 1, 2 and 3 contain ROC curves of the weighted sum rule used in verification task with the same weights as in identification experiments. In all cases we get slightly worse performance from the weighted sum rule than from the likelihood ratio rule. This confirms our assertion that the likelihood ratio is the optimal combination method for verification systems.

5.3 Explaining Identification System Behavior

The main assumption that we made while deriving likelihood ratio combination rule in section 5.1 is that the score samples in each identification trial are independent. That is, genuine score is sampled from genuine score distribution and is independent from impostor scores which are independent and identically distributed according to impostor score distribution. We can verify if this assumption is true for our matchers.

Matchers	$first_{imp}$	$second_{imp}$	$third_{imp}$	$mean_{imp}$
CMR	0.4359	0.4755	0.4771	0.1145
WMR	0.7885	0.7825	0.7663	0.5685
li	0.3164	0.3400	0.3389	0.2961
C	0.1419	0.1513	0.1562	0.1440
G	0.1339	0.1800	0.1827	0.1593

Table 3. Correlations between s_{gen} and different statistics of the impostor score sets produced during identification trials for considered matchers

Table 3 shows correlations between genuine score and some functions of the impostor scores obtained in the same identification trial. $first_{imp}$ column has correlations between genuine and the best impostor score, $second_{imp}$ and $third_{imp}$ consider second-best and third-best impostor scores, and $mean_{imp}$ has correlations between the mean of all impostor scores obtained in an identification trial and a genuine score. Non-zero correlations indicate that the scores are dependent. The correlations are especially high for word recognizers, and this might be the reason why the likelihood ratio combination rule performed poorly there.

The dependence of matching scores obtained during a single identification trial is usually not taken into account. One of the reasons might be that as a rule all matching scores are derived independently from each other: the same matching process is applied repeatedly to all enrolled biometric templates or all lexicon words, and the matching score for one class is not influenced by the presence of other classes or the matching scores assigned to other classes. So it might seem that the matching scores are independent, but it is rarely true. The main reason for this is that all matching scores produced during identification trial are derived using the same input signal. For example, a fingerprint matcher, whose matching score is derived from the number of matched minutia in enrolled and input fingerprint, will produce low scores for all enrolled fingerprints if the input fingerprint has only few minutiae.

The next three examples will illustrate the effect of score dependences on the performance of identification systems. In particular, second example confirms that if identification system uses likelihood ratio combination, then its performance can be worse than the performance of a single matcher.

5.3.1 Example 1

Suppose we have an identification system with one matcher and, for simplicity, $N = 2$ classes. During each identification attempt a matcher produces two scores corresponding to two classes, and, since by our assumption the input is one of these two classes (closed set identification), one of these scores will be genuine match score, and another will be impostor match score. Suppose we collected a data on the distributions of genuine and impostor scores and reconstructed score densities (let them be gaussian) as shown in Figure 4.

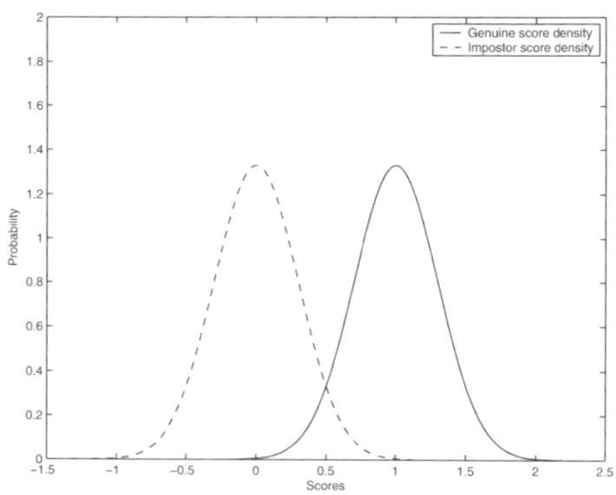

Fig. 4. Hypothetical densities of matching(genuine) and non-matching(impostors) scores

Consider two possible scenarios on how these densities might have originated from the sample of the identification attempts:

1. Both scores s_{gen} and s_{imp} are sampled independently from genuine and impostor distributions.

2. In every observed identification attempt : $s_{imp} = s_{gen} - 1$. Thus in this scenario the identification system always correctly places genuine sample on top. There is a strong dependency between scores given to two classes, and score distributions of Figure 4 do not reflect this fact.

If a system works in verification mode and we have only one match score to make a decision on accepting or rejecting input, we can only compare this score to some threshold. By doing so both scenarios would have same performance: the rate of false accepts (impostor samples having match score higher than threshold) and the rate of false rejects (genuine samples having match score lower than threshold) will be determined by integrating impostor and genuine densities of Figure 4 no matter what scenario we have. If system works in identification mode, the recognizer of the second scenario will be a clear winner: it is always correct while the recognizer of first scenario can make mistakes and place impostor samples on top.

This example shows that the performance of the matcher in the verification system might not predict its performance in the identification system. Given two matchers, one might be better for verification systems, and another for identification systems.

5.3.2 Example 2

Consider a combination of two matchers in two class identification system: one matcher is from the first scenario, and the other is from the second scenario. Assume that these matchers are independent. Let the upper score index refer to the matcher producing this score; s_i^j is the score for class i assigned by the classifier j. From our construction we know that the second matcher always outputs genuine score on the top. So the optimal combination rule for identification system will simply discard scores of first matcher and retain scores of the second matcher:

$$f(s^1, s^2) = s^2 \tag{6}$$

The input will always be correctly classified as $\arg\max_i s_i^2$.

Let us now use the likelihood ratio combination rule for this system. Since we assumed that matchers are independent, the densities of genuine $p_{gen}(s^1, s^2)$ and impostor $p_{imp}(s^1, s^2)$ scores are obtained by multiplying corresponding one-dimensional score densities of two matchers. In our example, impostor scores are distributed as a Gaussian centered at $(0, 0)$, and genuine scores are distributed as a Gaussian centered at $(1, 1)$. Figure 5(a) contains the contours of function $|p_{gen} - p_{imp}|$ which allows us to see the relative position of these gaussians. The gaussians have same covariance matrix, and thus the optimal decision contours are hyperplanes[7] - lines $s^1 + s^2 = c$. Correspondingly, the likelihood ratio combination function is equivalent to the combination function $f = s^1 + s^2$ (note, that true likelihood ratio function will be different, but if two functions have same contours, then their combination rules will be the same). Such combination improves the performance of the verification system relative to any single matcher; Figure 5(b) shows corresponding ROC curves for any single matchers and their combination.

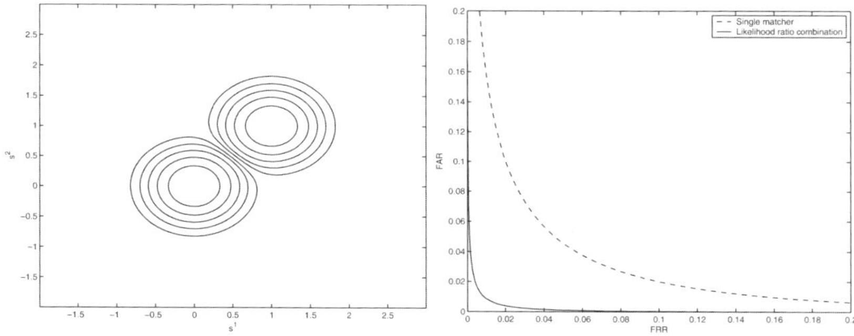

(a) (b)

Fig. 5. (a) Two-dimensional distributions of genuine and impostor scores for examples 2 and 3 (b) ROC curves for single matchers and their likelihood ratio combination

Suppose that (s_1^1, s_1^2) and (s_2^1, s_2^2) are two score pairs obtained during one identification trial. The likelihood ratio combination rule classifies the input as a class maximizing likelihood ratio function:

$$\arg\max_{i=1,2} \frac{p_{gen}(s_i^1, s_i^2)}{p_{imp}(s_i^1, s_i^2)} = \arg\max_{i=1,2} \; s_i^1 + s_i^2 \tag{7}$$

Let the test sample be $(s_1^1, s_1^2) = (-0.1, 1.0)$, $(s_2^1, s_2^2) = (1.1, 0)$. We know from our construction that class 1 is the genuine class, since the second matcher assigned score 1.0 to it and 0 to the second class. But the class 2 with scores $(1.1, 0)$, has combined score $s_2^1 + s_2^2 = 1.1 + 0 = 1.1$, which is bigger than combined score for class 1, $s_1^1 + s_1^2 = -0.1 + 1.0 + 0 = 0.9$. Hence class 2 has bigger ratio of genuine to impostor densities than class 1, and the likelihood ratio combination method would incorrectly classify class 2 as the genuine class.

Thus the optimal for verification system likelihood ratio combination rule (7) has worse performance than a single second matcher. On the other hand, the optimal for identification system rule (6) does not improve the performance of the verification system. Recall, that in section 5.1 we showed that if scores assigned by matchers to different classes are independent, then likelihood ratio combination rule is optimal for identification systems, as well as for verification systems. Current example shows that if there is a dependency between scores, this is no longer a case, and the optimal combination for identification systems can be different from the optimal combination for verification systems.

It seems that this example is analogous to our experiments with the combination of word recognizers. Our better performing word recognizer, WMR, has strong dependence between scores assigned to different classes (Table 3), and the resulting combination by likelihood ratio rule has worse performance than WMR's.

5.3.3 Example 3

The problem of finding optimal combination function for verification systems was a relatively easy task: we needed to approximate the densities of genuine and impostor scores and take their ratio. It turns out that the problem of finding optimal combination function for identification systems is considerably more difficult - we are not able to express it in such simple form. In fact, it is even difficult to construct an artificial example where we would know what this function is. Here we consider one such example.

Let X_{gen}, X_{imp} and Y be independent two-dimensional random variables, and suppose that genuine scores in our identification system are sampled as a sum of X_{gen} and Y: $\mathbf{s_{gen}} = \mathbf{x_{gen}} + \mathbf{y}$, and impostor scores are sampled as a sum of X_{imp} and Y: $\mathbf{s_{imp}} = \mathbf{x_{imp}} + \mathbf{y}$, $\mathbf{x_{gen}} \sim X_{gen}$, $\mathbf{x_{imp}} \sim X_{imp}$ and $\mathbf{y} \sim Y$, bold symbols here denote two-dimensional vector in the space (s^1, s^2). The variable Y provides the dependence between scores in identification trials; we assume that its value $\mathbf{y}$ is the same for all scores in one identification trial.

Let X_{gen} and X_{imp} have gaussian densities $p_{X_{gen}}(s^1, s^2)$ and $p_{X_{imp}}(s^1, s^2)$ as in the previous example and in the Figure 5(a). For any value of $\mathbf{y}$ conditional densities of genuine and impostor scores $p_{X_{gen}+Y|Y=\mathbf{y}}(s^1, s^2)$ and $p_{X_{imp}+Y|Y=\mathbf{y}}(s^1, s^2)$ are also gaussian and independent. As we discussed in the previous example, the likelihood ratio combination rule results in the combination function $f(s^1, s^2) = s^1 + s^2$, and this rule will be optimal for every identification trial and its associated value $\mathbf{y}$. The rule itself does not depend on the value of $\mathbf{y}$, so we can use it for every identification trial, and this is our optimal combination rule for identification system.

On the other hand, this rule might not be optimal for the verification system defined by the above score distributions. For example, if Y is uniformly distributed on the interval $0 \times [-1, 1]$, then the distributions of genuine and impostor scores $X_{gen}+Y$ and $X_{imp}+Y$ will be as shown in the Figure 6(a) and the optimal combination rule separating them will be as shown in the Figure 6(b). By changing the distribution of Y and thus the character of dependence between genuine and impostor scores we will also be changing optimal combination rule for verification system. At the same time, the optimal combination rule for identification system will stay the same - $f(s^1, s^2) = s^1 + s^2$.

If we knew only the overall score distributions as in the Figure 6(a) we would not have enough information to find the optimal combination function for identification system. If score vectors having distributions of Figure 6(a) are in its own turn are independent, then likelihood ratio combination of Figure 6(b) will be optimal for identification system. Or, if scores are generated

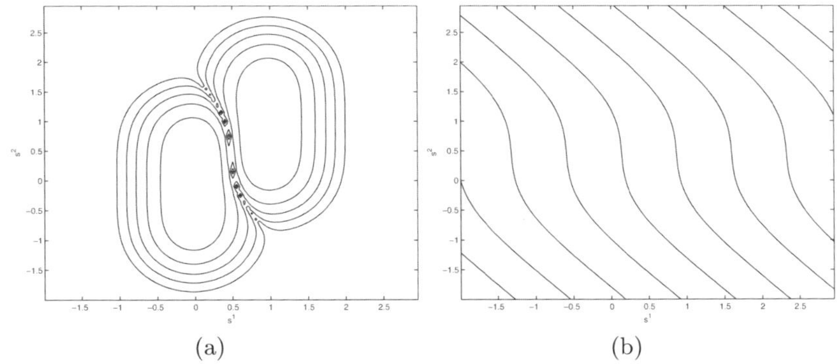

Fig. 6. (a) Two-dimensional distributions of genuine and impostor scores for example 3 (b) Contours of the likelihood ratio combination function

by the initial construction, linear combination function is the optimal one. Thus, there could be different optimal combination functions for identification systems with scores distributed as in the Figure 6(a), and the difference is determined by the nature of the score dependencies in identification trials.

6 Estimating Optimal Combination Function for Identification Systems

As we saw in the example 3 of the previous section, it is rather difficult to say from the training samples what is the optimal combination function for the identification system. The densities of genuine and impostor matching scores are of little help, and might be useful only if the scores in identification trials are independent. For dependent scores we have to consider the scores in each identification trial as a single training sample, and train the combination function on these samples.

This was precisely the technique we used to train the weighted sum rule for identification systems in section 5.2. For each training identification trial we checked whether the genuine score pair produced bigger combined scores than all impostor score pairs. By counting the numbers of successful trials we were able to choose the proper weights.

Though the weighted sum rule provides a reasonable performance in our applications, its decision surfaces are linear and might not completely separate generally non-linear score distributions. We might want our combination function to be more complex, trained with available training set and possibly approaching ideal optimal function when the size of the training set is increased. In this section we present two ideas on learning such combination functions. Since we do not know the exact analytical form of optimal combination function, the presented combination methods are rather heuristic.

6.1 Learning Best Impostor Distribution

The likelihood ratio combination function of section 5.1 separates the set of genuine score pairs from the set of all impostor score pairs. But we might think that for identification systems it is more important to separate genuine score pairs from the best impostor score pairs obtained in each identification trial. There is a problem, though, that we do not know which score pair is the best impostor in each identification trial. The best impostor score pair can be defined as one having biggest combined score, but the combination function is unknown.

To deal with this problem we implemented an iterative algorithm, where the combination function is first randomly initialized and then updated depending on found best impostor score pairs. The combination rule is based on the likelihood ratio function with the impostor density trained only on the set of found best impostor score pairs. The exact algorithm is presented below:

1. Make initialization of $f(s^1, s^2) = \frac{\hat{p}_{gen}(s^1,s^2)}{\hat{p}_{imp}(s^1,s^2)}$ by selecting random impostor score pairs from each training identification trial for training $\hat{p}_{imp}(s^1, s^2)$.
2. For each training identification trial find the impostor score pair with biggest value of combined score according to currently trained $f(s^1, s^2)$.
3. Update $f(s^1, s^2)$ by replacing impostor score pair of this training identification trail with found best impostor score pair.
4. Repeat steps 2-3 for all training identification trials.
5. Repeat steps 2-4 for predetermined number of training epochs.

The algorithm converges fast - after 2-3 training epochs, and found best impostor score pairs change little in the subsequent iterations. The trained combination function subsequently gets tested using a separate testing set. Table 4 (Best Impostor Likelihood Ratio method) provides the results of the experiments.

Matchers	Likelihood Ratio Rule	Weighted Sum Rule	Best Impostor Likelihood Ratio	Logistic Sum Rule	Weighted Sum + Ident Model
CMR&WMR	4293	5015	4922	5005.5	5025.5
li&C	5817	5816	5803	5823	5826
li&G	5737	5711	5742	5753	5760

Table 4. Correct identification rate for all considered combination methods

The method seems to perform well, but weighted sum combination rule is still better for word recognizers and biometric li&C matchers. This method is not able to fully account for the dependence of scores in identification trials, and the learning of the optimal combination function will not be probably achieved with it.

6.2 Sum of Logistic Functions

Generally, the matching score reflects the confidence of the match, and we can assume that if the score is bigger, then the confidence of the match is higher. When the scores are combined, the higher score should result in higher combination score. Thus, the combination function $f(s^1, s^2)$ should be monotonically nondecreasing in both of its arguments. One type of monotonic functions, which are frequently used in many areas, are logistic functions:

$$l(s^1, s^2) = \frac{1}{1 + e^{-(\alpha_1 s^1 + \alpha_2 s^2 + \alpha_3)}}$$

If $\alpha_1 \geq 0$ and $\alpha_2 \geq 0$, then $l(s^1, s^2)$ is monotonically nondecreasing in both of its arguments. Our goal is to approximate the optimal combination function as a sum of such logistic functions. The sum of monotonically nondecreasing functions will also be monotonically nondecreasing.

Suppose we have one identification trial and $\mathbf{s_1} = (s_1^1, s_1^2)$ and $\mathbf{s_2} = (s_2^1, s_2^2)$ are two score pairs of this trial. Let $\mathbf{s_1}$ be a genuine score pair, and $\mathbf{s_2}$ be an impostor score pair. Suppose also that we have some initial sum of logistic functions as our combination function. If both matchers gave a higher score to the genuine class and $s_1^1 > s_2^1$ and $s_1^2 > s_2^2$, then by our construction the combination score for genuine class will be higher than the combination score for impostor class. There is no need to do any modifications to our current combination function. If both matchers gave a lower score to the genuine class and $s_1^1 < s_2^1$ and $s_1^2 < s_2^2$, then we can not do anything - any monotonically nondecreasing function will give a lower combination score to the genuine class.

If one matcher gave a higher score to the genuine class and another matcher gave a higher score to the impostor class, we can adjust our combination function by adding corresponding logistic function to the current sum. For example, if $s_1^1 > s_2^1$ and $s_1^2 < s_2^2$ logistic function $l(s^1, s^2) = \frac{1}{1+e^{-(\alpha_1 s^1 + \alpha_3)}}$ will be increasing with respect to the first argument and constant with respect to the second argument. The input sample will be assigned genuine class since first matcher correctly identified it. We choose parameters α_1 and α_3 relative to the training sample:

$$l(s^1, s^2) = \frac{1}{1 + e^{-\frac{1}{h}\frac{1}{a-b}(s^1 - \frac{a+b}{2})}} \tag{8}$$

where $a = s_1^1$ and $b = s_2^1$, and h is the smoothing parameter. If a and b are close to each other, we get a steeper logistic function, which will allow us better separate genuine and impostor score pair. Similar logistic function is added to the current sum if second matcher is correct, and first is not: we replace s^1 by s^2 in equation (8), and $a = s_1^2$,$b = s_2^2$.

The overall training algorithm is similar to the training we did for best impostor likelihood ratio in the previous section:

1. Make initialization $f(s^1, s^2) = s^1 + s^2$, $n = 1$.
2. For each training identification trial and for each impostor score pair in this trial check if its combined score is higher than combined score of the genuine pair.
3. Update $f(s^1, s^2)$ by adding described above logistic function: $f(s^1, s^2) = \frac{1}{n+1}(nf(s^1, s^2) + l(s^1, s^2))$, $n = n + 1$.
4. Repeat steps 2-3 for all training identification trials.
5. Repeat steps 2-4 for predetermined number of training epochs.

The smoothing parameter h is chosen so that the performance of the algorithm is maximized on the training set. The convergence of this algorithm is even faster than the convergence of the best impostor likelihood ratio algorithm. Table 4 (Logistic Sum method) presents correct identification rate for this method.

The method outperforms weighted sum method for both biometric combinations, but not for the combination of word recognizers. This suggests that our heuristic was quite good, but still can be improved somehow. We can also see that the advantage of this method for second biometric combination outweighs its disadvantage for the combination of word recognizers, and thus we can consider it as the best combination rule so far.

7 Utilizing Identification Model

The previous two section investigated the usage of the so called combination rules in identification systems. We defined the combination rules by equation (2) and mentioned that such combination rules are a specific type of a classifiers operating in MN-dimensional score space and separating N classes, M is the number of classifiers. By considering the combinations of this restricted type we are able to significantly reduce the difficulty of training combination function, but at the same we might not get the best possible performance from our system.

We discussed this topic in length in [4] (see also the chapter on the review of combination methods). It turns out that besides two already mentioned types of combinations (combination rules of equation (2), *low complexity* combinations, and all possible N-class pattern classification methods in MN-dimensional score space, *high complexity* combinations) we can distinguish two additional types of classifier combinations in between. *Medium I complexity* combinations make the combination function class-specific:

$$C = \arg \max_{i=1,\dots,N} f_i(s_i^1, \dots, s_i^M) \tag{9}$$

while *medium II complexity* combinations remain class-generic and derive the combination score for each class not only from M scores assigned to this class but from potentially all available MN scores:

$$C = \arg\max_{i=1,\dots,N} f(s_i^1, \dots, s_i^M; \{s_k^j\}_{j=1,\dots,M;k=1,\dots,N;k\neq i}) \quad (10)$$

Generally, it is possible to use both medium I and medium II complexity type combinations for our applications, but we will concentrate on medium II complexity type. Since the combination functions of this type consider scores for all classes in order to derive a combined score for a particular type, we have a fair chance to properly learn the dependency between scores assigned to different classes, and train the combination function with this dependency in mind.

7.1 Identification Models

The goal of constructing an identification model is to somehow model the distributions of scores in identification trials. Better model will provide more information to the combination algorithm and result in better performance. We can use different heuristics in order to decide on which identification model might work best in a given application. For example, we might want the identification model to provide a good estimate for posterior class probability for a score from a current set of identification scores.

Consider our third example from the section 5.3. Recall, that genuine and impostor distributions are represented as sums of two random variables: $X_{gen} + Y$ and $X_{imp} + Y$. If each identification trial has many impostor samples, we can estimate the current value of Y as sum of all scores in this trial: $\hat{\mathbf{y}} = \sum_{i=1,\dots,N} \mathbf{s_i}$ (note, that the mean of X_{imp} is 0). The identification model in this case could state that instead of scores $\mathbf{s_i}$, we have to take their transformations: $\mathbf{s_i'} = \mathbf{s_i} - \hat{\mathbf{y}}$. If the combination rule is trained to use $\mathbf{s_i'}$ instead of $\mathbf{s_i}$, we will achieve near-optimal combination.

The identification model produced for this example is non-trainable, and it is only justified by the assumption that genuine and impostor scores are the sums of two random variables. If the assumption is not true, then the identification model might not perform well. In our research we are interested in designing general identification models which can be learned from the training data and which perform well for any applications.

There might be two approaches on using identification models as represented in Figures 7 and 8. In the first approach the identification model is applied to each score before the actual combination. Thus the score is normalized using identification model and the other identification trial scores. In the second approach identification model provides some statistics about current identification trial, and these statistics are used together with the scores in a single combination step. For our example, we can normalize scores $\mathbf{s_i'} = \mathbf{s_i} - \hat{\mathbf{y}}$ and use normalized score $\mathbf{s_i'}$ in subsequent combination. This will be a two step combination approach. Alternatively, we can use both $\mathbf{s_i}$ and $\hat{\mathbf{y}}$ as an input to the 1-step combination algorithm.

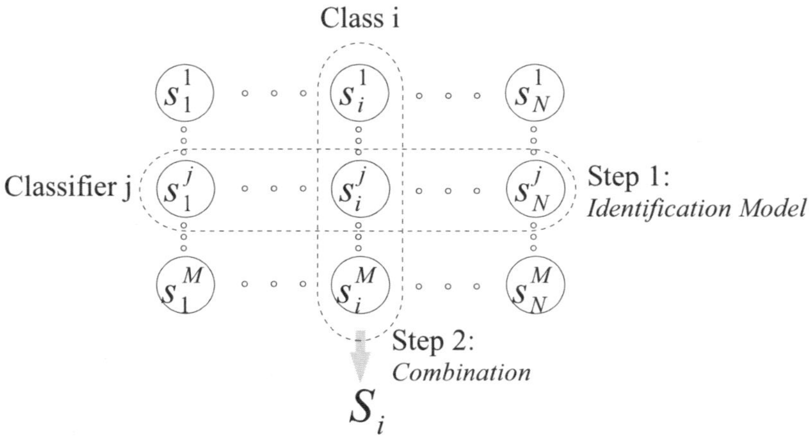

Fig. 7. 2-step combination method utilizing identification model

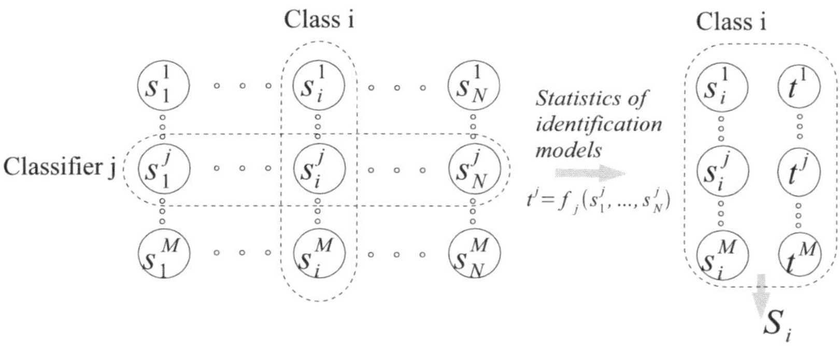

Fig. 8. 1-step combination method utilizing identification model

7.2 Related Research

We can list two general approaches in classifier combination research, which implicitly use the concept of identification model. These are the combination approaches based on rank information and combinations utilizing score normalization with current identification trial scores.

Rank based approaches replace the matching scores output by classifiers by their rank among all scores obtained in the current identification trial. Such transformation is performed for each classifier separately, and the ranks are combined afterward. T.K. Ho has described classifier combinations on the ranks of the scores instead of scores themselves by arguing that ranks provide more reliable information about class being genuine [10]. If there is a dependence between identification trial scores as for second matcher in our first example of section 5.3 (where the top score always belongs to the genuine class), then the rank of the class will be a perfect indicator if the class is genuine or not. Combining low score for genuine class with other scores as in the second example could confuse a combination algorithm, but the rank of the genuine class is still good, and using this rank should result in true classification. Brunelli and Falavigna [11] considered a hybrid approach where traditional combination of matching scores is fused with rank information in order to achieve identification decision. Saranli and Demirekler [12] provide additional references for rank based combination and a theoretical approach to such combinations.

Another approach for combinations, which might use the identification model, is a score normalization followed by some combination rule. Usually score normalization [13] means transformation of scores based on the classifier's score model learned during training, and each score is transformed individually using such a model. Such normalizations do not use the information about scores in identification trial, and the combinations using them can still be represented as a combination rule of equation (2). But some score normalization techniques indeed use a dynamic set of identification trial scores. For example, Kittler et al. [14] normalize each score by the sum of all other scores before combination. The combinations employing such normalizations are medium II complexity type combinations and can be considered as implicitly using an identification model.

Score normalization techniques have been well developed in the speaker identification problem. Cohort normalizing method [15, 16] considers a subset of enrolled persons close to the current test person in order to normalize the score for that person by a log-likelihood ratio of genuine (current person) and impostor (cohort) score density models. [17] separated cohort normalization methods into cohorts found during training (constrained) and cohorts dynamically formed during testing (unconstrained cohorts). Normalization by constrained cohorts followed by low complexity combination amounts to medium I combination types, since whole combination method becomes class-specific, but only one matching score of each classifier is utilized. On the other hand, normalization by unconstrained cohorts followed by low complexity combination amounts to medium II or high complexity combinations, since now potentially all scores of classifiers are used, and combination function can be class-specific or non-specific.

The related normalization techniques are Z(zero)- and T(test)- normalizations [17, 18]. Z- normalization is similar to constrained cohort normalization, since it uses impostor matching scores to produce a class specific normalization. Thus Z-normalization used together with low complexity combination rule results in medium I combination. T-normalization uses a set scores produced during single identification trial, and used together with low complexity combination rule results in medium II combination (note that this normalization is not class-specific).

Medium II combinations seem to be the most appropriate type of combinations for identification systems with large number of classes. Indeed, it is usually hard to train class-specific combination types of medium I and high complexity since the number of training samples for each class can be too small. As an example justifying medium II combinations in biometrics, [19] argued for applying T-normalizations in face verification competition. Ranks, T-normalization and many other investigated score normalization approaches are usually non-trainable. The concept of identification model implies that there is some training involved.

7.3 Identification Model for Weighted Sum

We will use the following idea for our identification model in this section. The confidence of a matching score is determined by the score itself and by the other scores in the same identification trial. If for a given score of a classifier there is another score in the same trial which is higher, then we have less confidence that the score belongs to the genuine class. Conversely, if all other scores are lower than a given score, we have more confidence that the score belongs to the genuine class.

The identification model in this case will consists in considering the following function of the identification trial scores: $sbs(s_i^j)$ - the best score besides score s_i^j in set of the current identification trial scores $\{s_i^j\}_{i=1,\dots,N}$ of classifier j:

$$sbs(s_i^j) = \max_{k=1,\dots,N;k\neq i} s_k^j \tag{11}$$

We use the 1-step identification model combination with weighted sum combination function. It means that instead of using only matching scores s_i^j, $j = 1,\dots,M$ for producing combined score S_i of class i, we will be using both s_i^j and $sbs(s_i^j)$. For two classifiers in our applications we will have the following combination function:

$$S_i = w_1 s_i^1 + w_2 sbs(s_i^1) + w_3 s_i^2 + w_4 sbs(s_i^2) \tag{12}$$

The number of considered input parameters for this method is two times bigger than the number of input parameters to the original weighted sum rule. We can still use the brute force approach to train the corresponding weights. Note, that though the number of weights is increased, the increase is rather small in comparison to the total number of classes (thousands). Thus we achieved the good trade-off between taking into consideration all scores produced by classifiers and the simplicity of training combination function.

The results of the experiments are presented in the Table 4 (Weighted Sum + Ident Model). The method outperforms all other methods for identification tasks. Note, that as in all our experiments, we used separate data sets for training weights and testing the trained method; thus the performance improvement is due not to more possibilities for training, but due to more complex combination function.

7.4 Identification Model for Verification Systems

Although most verification systems use only matching scores for one given class to make combinations and decisions on whether the class is genuine or impostor, there is an idea that the performance can be improved if the matching scores for other classes are taken into consideration. In fact, most of the cohort score normalization methods, which we referenced above, employ a superfluous set of matching scores for a cohort of a given class in order to make verification decision. These scores might be available naturally in identification system, but the verification system has to do additional matches to create these scores.

If the scores for other classes are available in addition to the score for a given class, they can provide significant amount of information to the combination algorithm. Indeed, as we discussed before, the matching scores are usually dependent and the dependence is caused by the quality of the input sample. Scores for other classes can implicitly provide us the information about the input sample quality. Consequently, we can view the application of identification model as score normalization with respect to the input sample.

The information supplied by the identification model can be considered as a predictor about the given score we consider in the verification task. We imply that this score is genuine, and the goal of the identification model is to check if this score is reasonable in comparison with scores we get for other, impostor, scores. Thus we can check the correlations of the genuine score with different functions of the impostor scores in order to find the statistics, which best predict the genuine score. Table 3 contains the correlation measurements for our matchers, and these measurements can be used to determine which statistics of impostor scores the identification model should include. In our experiments we considered first and second best impostor statistics. They seem to be good predictors according to Table 3, and, as an additional advantage, first few best scores are usually available due to the utilizing indexing methods in identification systems.

The application of the identification model in verification system is clear now. Instead of taking a single match score for a given class from a particular matcher, take few additional match scores for other class, and calculate some statistics from them. Then use these statistics together with a match score for a designated class in the combination. Since the likelihood ratio method is optimal for verification tasks, we use it here. If we employ the statistic of second best score from the previous section, our combination method will be written as

$$f_{lr}(s_i^1, \dots, s_i^M; \{s_k^j\}_{j=1,\dots,M;k=1,\dots,N;k\neq i}) = \frac{p_{gen}(s_i^1, sbs(s_i^1), \dots, s_i^M, sbs(s_i^M))}{p_{imp}(s_i^1, sbs(s_i^1), \dots, s_i^M, sbs(s_i^M))} \tag{13}$$

Note that we are dealing with the verification task, so we only produce the combined score for thresholding, and do not select among classes with arg max. Also, during our experiments we used a little different statistics than the statistics $sbs = \max_{k=1,\ldots,N;k\neq i} s_k^j$ from the previous section - we selected the second ranked score from $\{s_k^j, k = 1, \ldots, N; k \neq i\}$.

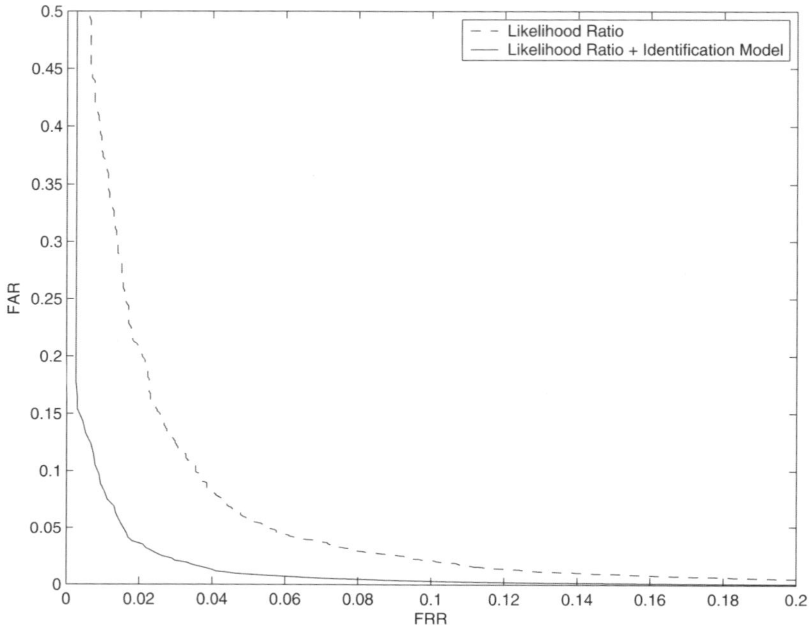

Fig. 9. The effect of utilizing identification model in the likelihood ratio combination function for handwritten word recognizers

Figure 9 contains the resulting ROC curve from utilizing identification model by equation 13 in the combination of word recognizers. Note, that this method performs significantly better than the original likelihood ratio method. We have also reported similar improvements for the biometric matchers before[20].

If we look at the verification task as the two class pattern classification problem in the M-dimensional score-feature space, then using identification model corresponds to expanding the feature space by the statistics of identification trials. The achieved improvements confirm the usefulness of these additional features.

8 Summary

In this work we considered combinations of handwritten word recognizers and biometric matchers. There can be different operating scenarios for the applications involving these matchers, and we considered two of them - verification and closed set identification. Different operating scenarios require different performance measures: ROC curves for verification problems, and correct identification rate for identification problems.

It turns out that for different scenarios we need to construct different combination algorithms in order to achieve optimal performance. This need is caused by the frequent dependence among scores produced by each matcher during a single identification trial. The optimal combination algorithm for verification systems corresponds to the likelihood ratio combination function. It can be implemented by the direct reconstruction of this function with genuine and impostor score density approximations. Alternatively, many generic pattern classification algorithms can be used to separate genuine and impostor scores in the M-dimensional score space, M is the number of combined matchers.

The optimal combination algorithm for the identification systems is more difficult to realize. We do not know how to express analytically the optimal combination function, and can only speculate on the heuristics leading to its construction. We described two possible approaches for approximating the optimal combination function in identification systems and compared them with traditionally used weighted sum combination method. The results are promising, but it is clear, that further development is needed in this area.

The concept of identification model provides a different point of view on the combinations in identification systems. The score dependence in identification trials can be explicitly learned in these models. The combination algorithm utilizing identification model uses more information about identification trial scores than traditional combination methods relying on a single match score for designated class. As a result it is possible to achieve significant improvements using these models.

References

1. Favata, J.: Character model word recognition. In: Fifth International Workshop on Frontiers in Handwriting Recognition, Essex, England (1996) 437–440
2. Kim, G., Govindaraju, V.: A lexicon driven approach to handwritten word recognition for real-time applications. Pattern Analysis and Machine Intelligence, IEEE Transactions on **19**(4) (1997) 366–379
3. Nist biometric scores set. http://www.nist.gov/biometricscores/ (2007)
4. Tulyakov, S., Govindaraju, V.: Classifier combination types for biometric applications. In: 2006 IEEE Computer Society Conference on Computer Vision and Pattern Recognition (CVPR 2006), Workshop on Biometrics, New York, USA (2006)
5. G. Kim, V. Govindaraju: Bank check recognition using cross validation between legal and courtesy amounts. Int'l J. Pattern Recognition and Artificial Intelligence **11**(4) (1997) 657–674
6. Bolle, R.M., Connell, J.H., Pankanti, S., Ratha, N.K., Senior, A.W.: Guide To Biometrics. Springer, New York (2004)

7. Theodoridis, S., K., K.: Pattern Recognition. Academic Press (1999)
8. Silverman, B.W.: Density estimation for statistics and data analysis. Chapman and Hall, London (1986)
9. Tulyakov, S., V., G.: Using independence assumption to improve multimodal biometric fusion. In: 6th International Workshop on Multiple Classifiers Systems (MCS2005), Monterey, USA, Springer (2005)
10. Ho, T.K., Hull, J.J., Srihari, S.N.: Decision combination in multiple classifier systems. Pattern Analysis and Machine Intelligence, IEEE Transactions on **16**(1) (1994) 66–75
11. Brunelli, R., Falavigna, D.: Person identification using multiple cues. Pattern Analysis and Machine Intelligence, IEEE Transactions on **17**(10) (1995) 955–966
12. Saranli, A., Demirekler, M.: A statistical unified framework for rank-based multiple classifier decision combination. Pattern Recognition **34**(4) (2001) 865–884
13. Jain, A., Nandakumar, K., Ross, A.: Score normalization in multimodal biometric systems. Pattern Recognition **38**(12) (2005) 2270–2285
14. Kittler, J., Hatef, M., Duin, R., Matas, J.: On combining classifiers. Pattern Analysis and Machine Intelligence, IEEE Transactions on (1998) 226–239
15. Rosenberg, A., Parthasarathy, S.: Speaker background models for connected digit password speaker verification. In: Acoustics, Speech, and Signal Processing, 1996. ICASSP-96. Conference Proceedings., 1996 IEEE International Conference on. Volume 1. (1996) 81–84 vol. 1
16. Colombi, J., Reider, J., Campbell, J.: Allowing good impostors to test. In: Signals, Systems & Computers, 1997. Conference Record of the Thirty-First Asilomar Conference on. Volume 1. (1997) 296–300 vol. 1
17. Auckenthaler, R., Carey, M., Lloyd-Thomas, H.: Score normalization for text-independent speaker verification systems. Digital Signal Processing **10**(1-3) (2000) 42–54
18. Mariethoz, J., Bengio, S.: A unified framework for score normalization techniques applied to text independent speaker verification. IEEE Signal Processing Letters **12** (2005)
19. Grother, P.: Face recognition vendor test 2002 supplemental report, nistir 7083. Technical report, NIST (2004)
20. Tulyakov, S., Govindaraju, V.: Identification model for classifier combinations. In: Biometrics Consortium Conference, Baltimore, MD (2006)

Perturbation Models for Generating Synthetic Training Data in Handwriting Recognition

Tamás Varga and Horst Bunke

University of Bern
Institute of Computer Science and Applied Mathematics (IAM)
Neubrückstrasse 10, CH-3012 Bern, Switzerland
{varga,bunke}@iam.unibe.ch

Summary. In this chapter, the use of synthetic training data for handwriting recognition is studied. After an overview of the previous works related to the field, the authors' main results regarding this research area are presented and discussed, including a perturbation model for the generation of synthetic text lines from existing cursively handwritten lines of text produced by human writers. The goal of synthetic text line generation is to improve the performance of an off-line cursive handwriting recognition system by providing it with additional training data. It can be expected that by adding synthetic training data the variability of the training set improves, which leads to a higher recognition rate. On the other hand, synthetic training data may bias a recognizer towards unnatural handwriting styles, which could lead to a deterioration of the recognition rate. The proposed perturbation model is evaluated under several experimental conditions, and it is shown that significant improvement of the recognition performance is possible even when the original training set is large and the text lines are provided by a large number of different writers.

1 Introduction

The problem of automatic recognition of scanned handwritten documents is of great significance in numerous scientific, business, industrial, and personal applications that require the reading and processing of human written texts. The ultimate goal is that computers approach, or even surpass, the text recognition performance of humans. Despite the enormous amount of research activities that already have been carried out in the past decades to study this problem, it is considered very difficult and still not satisfactorily solved [1, 2]. Today's commercial systems work in areas where strict task specific knowledge and constraints are available, such as postal address reading [3], and the processing of bank checks [4] and forms [5, 6]. On the other hand, the more challenging task of recognizing unconstrained handwriting has also many potential applications, for example, office automation, digital libraries, and personal digital

T. Varga and H. Bunke: *Perturbation Models for Generating Synthetic Training Data in Handwriting Recognition*, Studies in Computational Intelligence (SCI) **90**, 333–360 (2008)
www.springerlink.com

assisting devices. In this chapter the problem of unconstrained recognition is addressed.

Despite the existence of the numerous elaborated and mature handwriting recognition techniques [7, 8, 9, 10, 11, 12], machines' reading performance is still considerably lower than that of humans. This inspired researchers to focus not only on the development of novel recognition algorithms, but also on the improvement of other aspects of handwriting recognition systems. These efforts include multiple classifier combination [13, 14, 15], the better utilization of the available a-priori, e.g. linguistic knowledge [16, 17], as well as the collection of large, publicly available datasets of human written texts [18, 19, 20], which enables better training of the recognizers and also an objective comparison of their performances.

As an alternative, to overcome the difficulties and inherent limitations of collecting a large number of human written samples, the present chapter investigates the generation and use of synthetic training data for off-line cursive handwriting recognition. It has been shown in many works before that the size and quality of the training data has a great impact on the performance of handwriting recognition systems. A general observation is that the more texts are used for training, the better recognition performance can be achieved [21, 22, 23, 24].

In this work it is examined whether this observation holds if the training set is augmented by synthetically generated texts. The motivation is that augmenting the training set by computer generated text samples is much faster and cheaper than collecting additional human written samples. To achieve our goal, a perturbation model is presented to generate synthetic text lines from existing cursively handwritten lines of text produced by human writers. Our purpose is to add synthetic data to the natural training data, rendered by human writers, so as to enlarge the training set. The basic idea of the approach is to use continuous nonlinear functions that control a class of geometrical transformations applied on the existing handwritten texts. The functions ensure that the distortions performed are not reversed by standard preprocessing operations of handwriting recognition systems. Besides the geometrical distortions, thinning and thickening operations are also part of the model.

A closer examination reveals, however, that the use of synthetic training data does not necessarily lead to an improvement of the recognition rate, because of two adversarial effects. First, it can be expected that the variability of the training set improves, which potentially leads to a higher recognition rate. On the other hand, synthetic training data may bias a recognizer towards unnatural handwriting styles, which can lead to a deterioration of the recognition rate, particularly if natural handwriting is used for testing.

The aim in this chapter is to find configurations of our recognizer and the synthetic handwriting generation process, by which the recognition performance can be significantly improved. The parameters examined include the number of Gaussian mixture components in the recognizer used for distribution estimation, distortion strength, training set size, and the number of writers in the training set. It is shown that significant improvement of the recognition performance is possible even when the original training set is large and the text lines are provided by many different writers. But to really achieve an improvement in this case, one has also to consider the capacity of the recognition system, which needs to be appropriately adjusted when expanding the training set with synthetic text lines. Parts of this work have been published in [25, 26]. The current chapter provides a synoptic presentation and overview of the authors' previous work on synthetic text line generation for the training of handwriting recognition systems.

The chapter is organized as follows. In Section 2, an overview of the related previous works on synthetic text generation is given. Section 3 introduces our perturbation model, while in Section 4 a concise description of the off-line handwriting recognition system used for the experiments is given. Experimental results are presented in Section 5. Finally, Section 6 provides some conclusions and suggestions for future work.

2 Synthetically Generated Text

The concept of synthetic text relates to both machine printed and handwritten documents. Synthesizing text means that real-world processes that affect the final appearance of a text are simulated by a computer program. For example, in the case of machine printed documents the printing and scanning defects, while in the case of handwriting the different writing instruments or the whole writing process can be modeled and simulated by computer.

Synthetic texts can be generated in numerous ways, and they have widespread use in the field of document analysis and recognition. In the following, a brief overview is given. Approaches for both machine printed and handwritten synthetic text generation are presented, since they often have similar aims, and thus the findings and developments of one field can also affect and stimulate the other one and vice versa.

2.1 Improving and Evaluating Recognition Systems

The two main difficulties that contemporary text recognizers have to face are the degraded quality of document images as well as the great variation of the possible text styles [27, 28, 29]. The quality of document images usually degrades to various extent during printing, scanning, photocopying, and faxing. Style variation means that either different fonts might be used (machine printed text), or many individual writing styles can occur (handwritten text).

One way to alleviate the above mentioned problems is to train the recognizers using sets of text samples that are more representative to the specific recognition task under consideration. This idea is supported by two facts. First of all, every recognizer needs to be trained, i.e. it has to learn how the different characters and/or words may look like. Furthermore, in the past decade researchers in the field of image pattern recognition realized that any further improvement of recognition performance depends as much on the size and quality of the training data as on the underlying features and classification algorithms used [30]. As a rule of thumb says, the classifier that is trained on the most data wins.

A straightforward way to improve the training set is to collect more real-world text samples [18, 19, 20]. The effectiveness of this approach has been experimentally justified by numerous works in the literature, yielding higher recognition performance for increased training set sizes [21, 22, 23, 24]. Unfortunately, collecting real-world samples is a rather expensive and time consuming procedure, and truthing the collected data is error-prone [31, 32]. A possible solution to these drawbacks is to create text image databases automatically by generating synthetic data, which is cheap, fast, and far less error-prone. Furthermore, it enables the generation of much larger databases than those acquired by the conventional method. The main weakness of the synthetic approach is that the generated data may not be as representative as real-world data.

In machine printed OCR (Optical Character Recognition), especially when the possible fonts are a-priori known, the concept of representativeness of the training set can be approached from the side of document degradation. In [33, 34, 35], defects caused by the use of printing and imaging devices are explicitly modeled and applied to ideal input images (e.g. Postscript document) to generate realistic image populations. Such synthetic data can then be used to build huge and more representative training sets for document image recognition systems [36, 37, 38]. The ability of controlling the degree of degradation makes it also possible to carry out systematic design and evaluation of OCR systems [36, 39, 40, 41].

For handwriting recognition, no parameterized model of real-world image populations is available, due to the lack of mathematical models accounting for the enormous variations present in human handwriting. Nevertheless, several attempts to generate synthetic data for handwriting recognition systems are reported.

In [42], human written character tuples are used to build up synthetic text pages. Other approaches apply random perturbations on human written characters [21, 43, 44, 45, 46], or words [47, 48]. In [49], realistic off-line characters are generated from on-line patterns using different painting modes.

Generating synthetic handwriting does not necessarily require to use human written texts as a basis. In [50] and [51], characters are generated by perturbation of the structural description of character prototypes.

Those works where the application of synthetic training data yielded improved recognition performance over natural training data are mainly related to the field of isolated character recognition [21, 43, 45, 46]. The natural training set was augmented by perturbed versions of human written samples, and the larger training set enabled better training of the recognizer. However, to the knowledge of the authors, for the problem of general, cursive handwritten word and text line recognition, no similar results besides those of the authors (see e.g. [25, 26]) involving synthetically generated text images have been reported.

Finally, perturbation approaches can also be applied in the recognition phase, making the recognizer insensitive to small transformations or distortions of the image to be recognized [44, 47, 52].

2.2 Handwritten Notes and Communications

The use of handwriting has the ability to make a message or a letter look more natural and personal. One way to facilitate the input of such messages for electronic communication is to design methods that are able to generate handwriting-style texts, particularly in the style of a specific person.

Such methods have several possible applications. For example, using a word processor, editable handwritten messages could be inputted much faster directly from the keyboard. For pen-based computers, errors made by the user could be corrected automatically by substituting the erroneous part of text by its corrected version, using the same writing style.

In [53], texts showing a person's handwriting style are synthesized from a set of tuples of letters, collected previously from that person, by simply concatenating an appropriate series of static images of tuples together.

Learning-based approaches are presented in [54], [55], and [56], to generate Hangul characters, handwritten numerals, and cursive text, respectively, of a specific person's handwriting style. These methods need temporal (on-line) information to create a stochastic model of an individual style.

A method that is based on character prototypes instead of human written samples is presented in [57]. Korean characters are synthesized using templates of ideal characters, and a motor model of handwriting generation (see [58]) adapted to the characteristics of Korean script. The templates consist of strokes of predefined writing order. After the geometrical perturbation of a template, beta curvilinear velocity and pen-lifting profiles are generated for the strokes, which are overlapped in time. Finally, the character is drawn using the generated velocity and pen-lifting profiles.

One possible application of the method is to build handwriting-style fonts for word processors. On the other hand, the method can provide training data for handwriting recognizers. Although the generated characters look natural and represent various styles, they were not used for training purposes.

2.3 Reading-Based CAPTCHAs

At present, there is a clear gap between the reading abilities of humans and machines. Particularly, humans are remarkably good at reading seriously degraded (e.g. deformed, occluded, or noisy) images of text, while modern OCR systems usually fail when facing such an image [59].

This observation can be used to design so-called CAPTCHAs (Completely Automatic Public Turing test to tell Computers and Humans Apart), to distinguish humans from computers [60, 61, 62]. The main application of CAPTCHAs is to prevent computer programs from automatic registration to publicly available services offered on the Internet. For example, this way spammers can be prevented from registering automatically thousands of free e-mail accounts for their fraudulent activities.

Several reading-based CAPTCHAs were proposed in the literature. All of them synthesize a degraded text image that is used to challenge the applicant to read it. The approval for the access to the required resource is then based on the correctness of the answer the applicant types in. The challenges may contain machine printed texts [60, 59, 63, 64, 65, 66, 67], or handwriting [68]. Reading-based CAPTCHAs that are already in industrial use include [60], [66], and [67].

3 Perturbation Model

Variation in human handwriting is due to many sources, including letter shape variation, variety of writing instruments, and others. In this section, a perturbation model for the distortion of cursive handwritten text lines is presented, where these sources of variation are modeled by geometrical transformations as well as thinning and thickening operations.

3.1 Previous Work and Design Goals

In the field of handwritten character recognition, numerous methods are reported to perturb character images. Among other geometrical transformations, translation, scaling, rotation, shearing, shrinking, interpolation between character samples, and also nonlinear deformations were tried [21, 43, 45, 46]. Other types of perturbations include erosion and dilation [21], and pixel inversion noise [45].

Although they seem to be very different approaches, surprisingly almost all of the transformations mentioned in the previous paragraph have been applied successfully to generate additional training samples for character recognition systems, yielding improvements in the recognition performance.[1] Thus the character recognition experiments suggest that most of the perturbations might improve the recognition rate. Furthermore, there is no comparative study showing that one or more of these approaches are superior to the others.

With this background from character recognition research in mind, the design of our perturbation model was motivated by two important aspects: simplicity and nonlinearity. Simplicity is achieved by applying the same concept

[1] The only exception is shrinking, which deteriorated the system performance in [21]

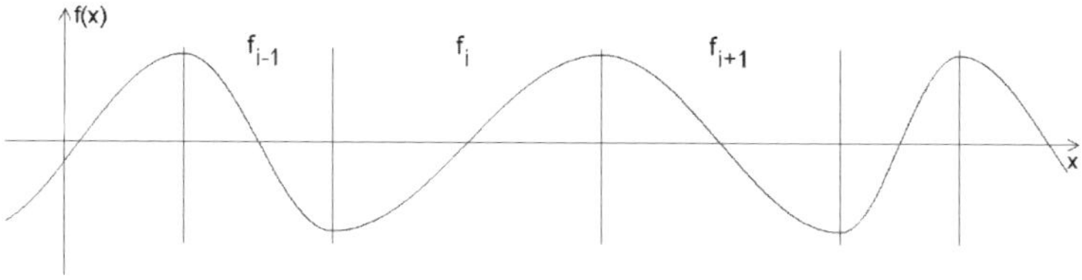

Fig. 1. Example of a CosineWave function

(underlying function, see Subsection 3.2) to each type of geometrical transformation, and considering only some basic types of distortions (shearing, scaling and shifting along one of the main axes). Nonlinearity is needed so that the distortions applied on the handwriting cannot be reversed by standard linear preprocessing operations of a state-of-the-art handwriting recognition system (see Section 4).

The perturbation model incorporates some parameters with a range of possible values, from which a random value is picked each time before distorting a text line. There is a constraint on the text lines to be distorted: they have to be skew and slant corrected, because of the nature of the applied geometrical transformations. This constraint is not severe, because skew and slant correction are very common preprocessing steps found in almost any handwriting recognition system. In the following subsections the perturbation model is described in greater detail.

3.2 Underlying Functions

Each geometrical transformation in the model is controlled by a continuous nonlinear function, which determines the strength of the considered transformation. These functions will be called *underlying functions.*

The underlying functions are synthesized from a simple function, called *CosineWave.* A CosineWave is the concatenation of n functions, $f_1, f_2, \ldots, f_n$, where $f_i : [0, l_i] \to \mathbb{R}$, $f_i(x) = (-1)^i \cdot a \cdot \cos(\frac{\pi}{l_i} \cdot x)$, $l_i > 0$. An example is shown in Fig. 1. The functions f_i (separated by vertical line segments in Fig. 1) are called *components.* The *length* of component f_i is l_i and its *amplitude* is $|a|$. The amplitude does not depend on i, i.e. it is the same for all components.

To randomly generate a CosineWave instance, three ranges of parameter values need to be defined:

- $[a_{min}, a_{max}]$ for the amplitude $|a|$,
- $[l_{min}, l_{max}]$ for the component length,
- $[x_{min}, x_{max}]$ for the interval to be covered by the concatenation of all components.

The generation of a CosineWave is based on the following steps. First the amplitude is selected by picking a value $\alpha \in [a_{min}, a_{max}]$ randomly and letting $a = \alpha$ or $a = -\alpha$ with a 50% probability each. Then l_1 is decided by randomly

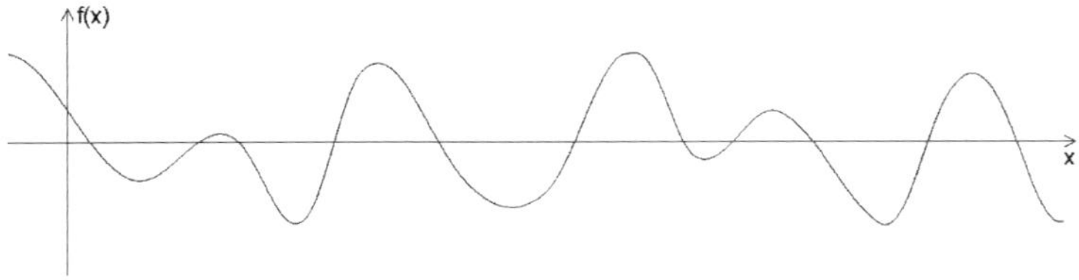

Fig. 2. Example of a sum of two CosineWave functions

picking a value from $[l_{min}, l_{max}]$. Finally the beginning of the first component (i.e. f_1) is chosen randomly from the $[x_{min} - l_1, x_{min}]$ interval. From this point on we only have to add additional components, one after the other, with randomly chosen lengths, until x_{max} is reached. For randomly picking a value from an interval, always the uniform distribution over that interval is used.

An underlying function is obtained by summing up a number, m, of such CosineWave functions. Fig. 2 depicts an example of such an underlying function with $m = 2$.

3.3 Geometrical Transformations

The underlying functions control several geometrical transformations, which are divided into two groups: the *line level* transformations applied on whole lines of text, and the *connected component level* transformations applied on the individual connected components of the considered line of text. The underlying function of each transformation is randomly generated, as described in Subsection 3.2. The parameters x_{min} and x_{max} are always defined by the actual size of the image to be distorted. In the following the geometrical transformations will be defined and illustrated by figures. Note that the figures are only for illustration purposes, and weaker instances of the distortions are actually used in the experiments described later on.

There are four classes of geometrical transformations on the line level. Their purpose is to change properties, such as slant, horizontal and vertical size, and the position of characters with respect to the baseline. The line level transformations are these:

- **Shearing:** The underlying function, denoted by $f(x)$, of this transformation defines the tangent of the shearing angle for each x coordinate. Shearing is performed with respect to the lower baseline. An example is shown in Fig. 3. In this example and the following ones, the original text line is shown at the bottom, the underlying function in the middle, and the result of the distortion on top.
- **Horizontal scaling:** Here the underlying function determines the horizontal scaling factor, $1 + f(x)$, for each x coordinate. This transformation

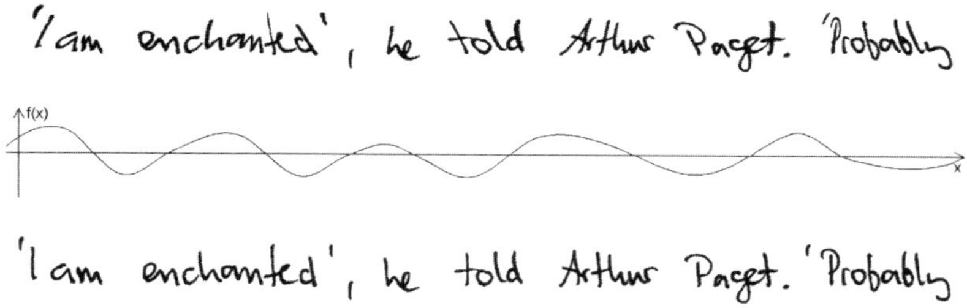

Fig. 3. Illustration of shearing. The original text line is at the bottom, the underlying function is in the middle, and result of the distortion is on top

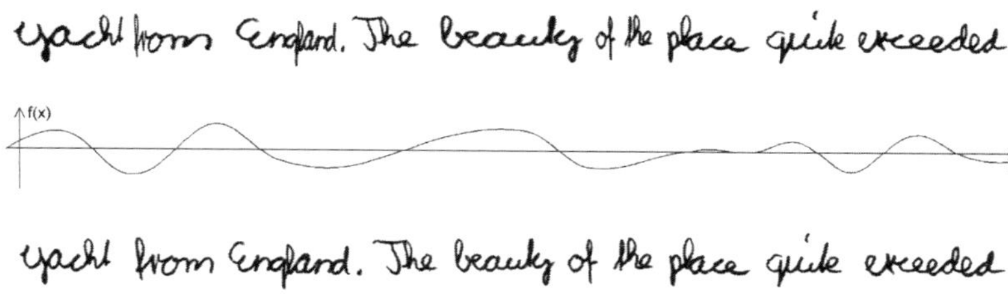

Fig. 4. Illustration of horizontal scaling

is performed through horizontal shifting of the pixel columns.[2] An example of this operation is shown in Fig. 4.

- **Vertical scaling:** The underlying function determines the vertical scaling factor, $1 + f(x)$, for each x coordinate. Scaling is performed with respect to the lower baseline. An example can be seen in Fig. 5.
- **Baseline bending:** This operation shifts the pixel columns in vertical direction, by the amount of $h \cdot f(x)$ for each x coordinate, where h is the height of the body of the text (i.e. the distance between the upper and lower baselines). An example is given in Fig. 6.[3]

The perturbation model also includes transformations, similar to the ones described above, on the level of connected components. These transformations change the structure of the writing in a local context, i.e. within each connected component. After the application of these transformations, the resulting connected components are scaled in both horizontal and vertical direction so that their bounding boxes regain their original sizes, and then they are placed in the image exactly at their original locations. For each connected component, individual underlying functions are generated. There are three classes of such transformations:

[2] The appropriate shifting value at x is given by $\int_0^x (1 + f(x))dx = x + \int_0^x f(x)dx$

[3] It can be observed that the baseline is usually not a straight line, but rather of a wavy shape

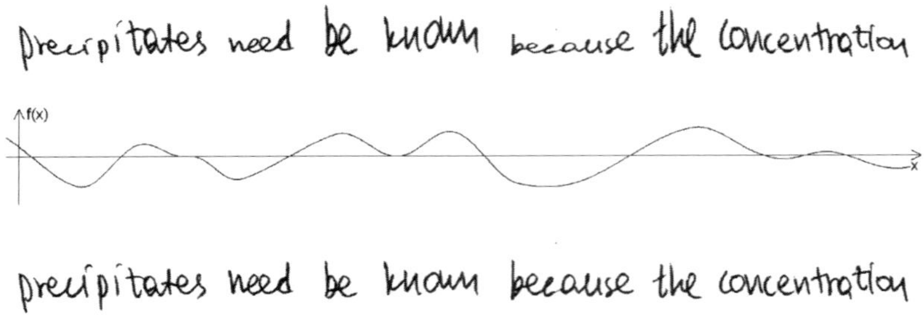

Fig. 5. Illustration of vertical scaling

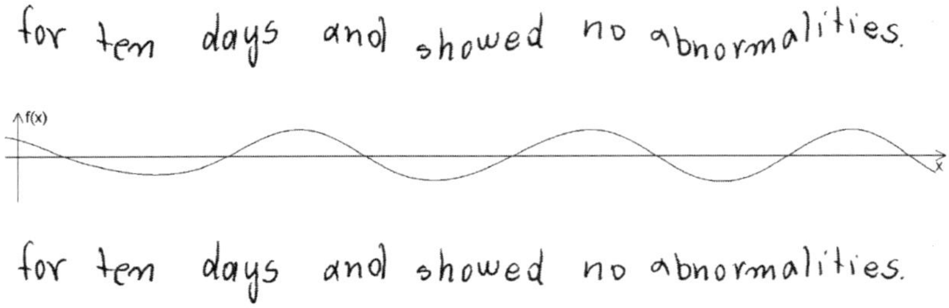

Fig. 6. Illustration of baseline bending

- **Horizontal scaling:** This transformation is identical to the line level horizontal scaling as described before, but it is applied to individual connected components rather than whole lines of text.
- **Vertical scaling 1:** This is the counterpart of horizontal scaling in the vertical direction.
- **Vertical scaling 2:** This transformation is identical to the line level vertical scaling, except that scaling is performed with respect to the horizontal middle-line of the bounding box.

The effect of all three transformations applied one after the other is shown in Fig. 7. In this figure, the lower text line is the original one, and above its distorted version is displayed. One can observe that in spite of the distortions the connected components underwent, their bounding boxes have remained the same.

3.4 Thinning and Thickening Operations

The appearance of a text line can also be changed by varying the thickness of its strokes. In the present perturbation model this is done by applying thinning or thickening steps iteratively. The method is based on a grayscale variant of the MB2 thinning algorithm [69]. (A general way to get the grayscale version

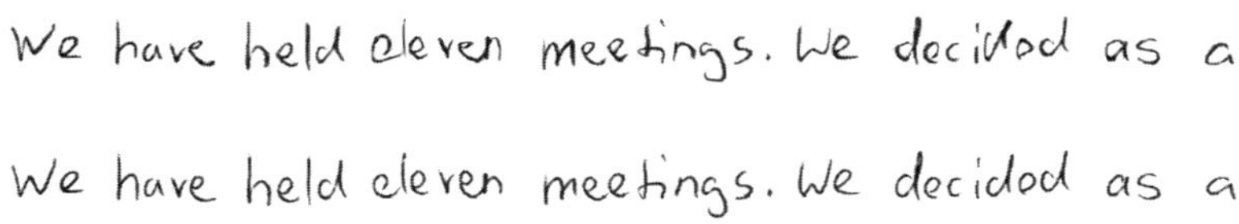

Fig. 7. Illustration of connected component level distortions. The original text line is below, and the result of the distortions is above

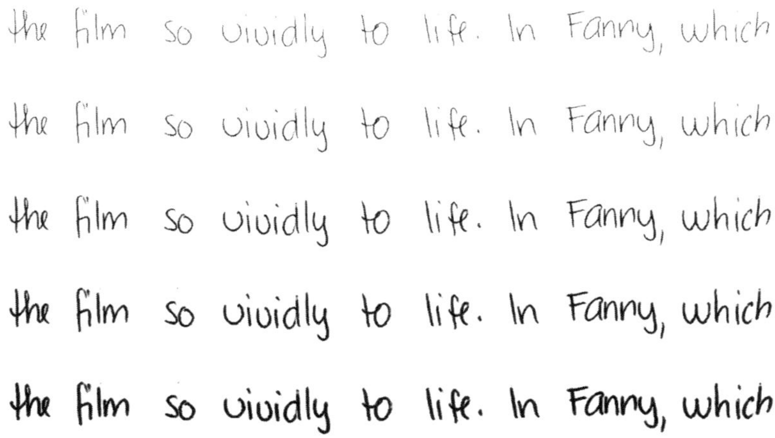

Fig. 8. Illustration of thinning (above) and thickening (below) operations. The original text line is in the middle

of a specific type of thinning algorithm operating on binary images can be found in [70]). Thinning and thickening could also be performed using the morphological erosion and dilation operators, respectively, but this would not be safe when applied iteratively, because part of the original writing might be lost after too many steps of erosion. An illustration is given in Fig. 8, where the original text line is located in the middle, and above (below) it the results of two successive thinning (thickening) steps can be seen. The choice whether thinning or thickening is applied, as well as the number of steps (including zero) is randomly made.

3.5 Distorted Text Line Generation

Now that the main constituents of the perturbation model have been introduced, a simple scheme for the distortion of whole text lines can be designed. The steps of the perturbation method for distorting a given skew and slant corrected text line are the following:

1. Apply each of the line level transformations to the text line, one after the other, in the order given in Subsection 3.3.

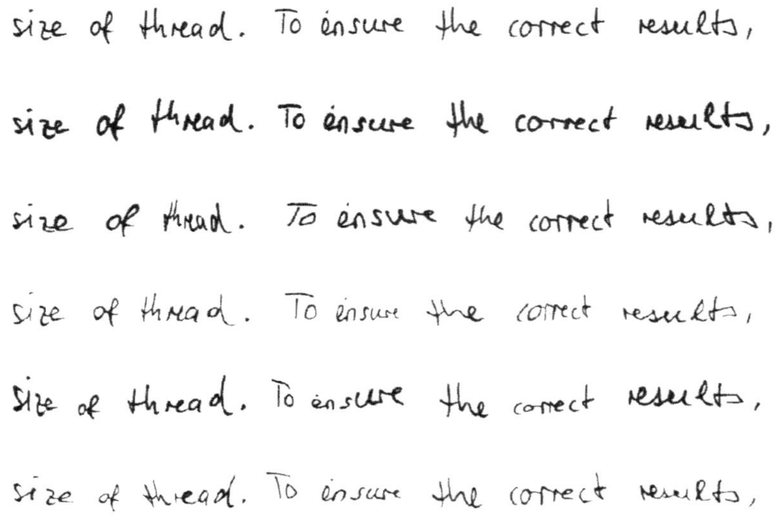

Fig. 9. Demonstration of the perturbation method. The original human written text line is on top, and below it five distorted versions can be seen

2. For each individual connected component, apply the connected component level transformations, and make sure that the bounding boxes remain the same with respect to both size and location.
3. Apply thinning or thickening operations.

Of course, these steps are not required to be always rigorously followed. In particular, one can omit one or several of the transformations. The method is demonstrated in Fig. 9. The original human written text line is on top, and below there are five synthetically generated versions of that line. It can be seen that all of the characters have somewhat changed in each generated line. Note that due to the random nature of the perturbation method, virtually all generated text lines are different. Other examples are given in Section 5.

4 Handwriting Recognition System

The application considered in this chapter is the off-line recognition of cursively handwritten text lines. The recognizer used is the Hidden Markov Model (HMM) based cursive handwritten text line recognizer described in [12]. The recognizer takes, as a basic input unit, a complete line of text, which is first normalized with respect to skew, slant, baseline location and writing width.[4]

[4] Text line normalization is also applied in the training phase. Since the text lines to be distorted have to be skew and slant corrected, synthetic training text line

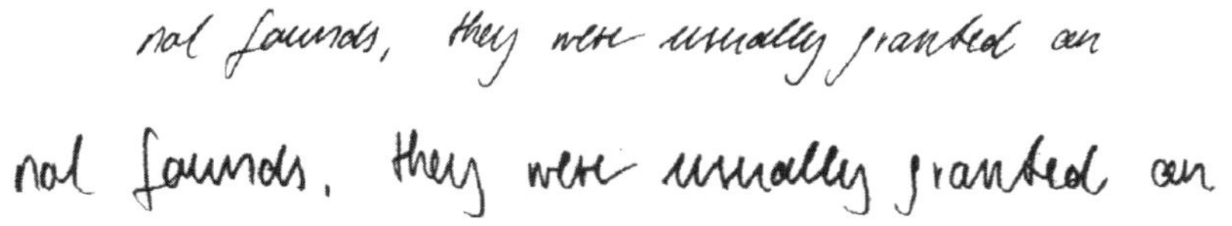

Fig. 10. Example of an input text line, before (above) and after (below) normalization

An example is shown in Fig. 10. Normalization with respect to baseline location means that the body of the text line (the part which is located between the upper and lower baselines), the ascender part (above the upper baseline), and the descender part (below the lower baseline) will be vertically scaled to a predefined height. Writing width normalization is performed by a horizontal scaling operation, and its purpose is to scale the characters so that they have a predefined average width value.

For feature extraction, a sliding window of one pixel width is moved from left to right over the input text line, and nine geometrical features are extracted at each window position. Thus an input text line is converted into a sequence of feature vectors in a 9-dimensional feature space. The nine features used in the system are the average gray value of the window, the center of gravity, the second order moment of the window, the position and the gradient of the upper and lower contours, the number of black-white transitions in vertical direction, and the average gray value between the upper and lower contour [12].

For each character, an HMM is built. In all HMMs the linear topology is used, i.e. there are only two transitions per state, one to itself and one to the next state. In the emitting states, the observation probability distributions are estimated by mixtures of Gaussian components. In other words, continuous HMMs are used. The character models are concatenated to represent words and sequences of words. For training, the Baum-Welch algorithm [71] is applied. In the recognition phase, the Viterbi algorithm [71] with bigram language modeling [17] is used to find the most probable word sequence. As a consequence, the difficult task of explicitly segmenting a line of text into isolated words is avoided, and the segmentation is obtained as a byproduct of the Viterbi decoding applied in the recognition phase. The output of the recognizer is a sequence of words. In the experiments described in the following, the recognition rate will always be measured on the word level.

---

generation takes place right after the skew and the slant of the text line have been normalized

5 Experimental Evaluation

The purpose of the experiments is to investigate whether the performance of the off-line handwritten text recognizer described in Section 4 can be improved by adding synthetically generated text lines to the training set. Two configurations with respect to training set size and number of writers are examined: small training set with only a few writers, and large training set with many writers.

For the experiments, subsets of the IAM-Database [20] were used. This database includes over 1,500 scanned forms of handwritten text from more than 600 different writers. In the database, the individual text lines of the scanned forms are extracted already, allowing us to perform off-line handwritten text line recognition experiments directly without any further segmentation steps.[5]

All the experiments presented in this section are writer-independent, i.e. the population of writers who contributed to the training set is disjoint from those who produced the test set. This makes the task of the recognizer very hard, because the writing styles found in the training set can be totally different from those in the test set, especially if the training set was provided by only a few writers. However, when a given training set is less representative of the test set, greater benefit can be expected from the additional synthetic training data.

If not mentioned otherwise, all the three steps described in Subsection 3.5 are applied to distort a natural text line. Underlying functions are obtained by summing up two randomly generated *CosineWave* functions (two is the minimum number to achieve peaks with different amplitudes, see Figs. 1 and 2). Concerning thinning and thickening operations, there are only three possible events allowed: one step of thinning, one step of thickening, or zero steps (i.e. nothing happens), with zero steps having the maximal probability of the three alternatives, while the two other events are equally probable.

5.1 Small Training Set with a Small Number of Writers

The experiments described in this subsection are conducted in order to test the potential of the proposed method in relatively simple scenarios, i.e. the case of a small training set and only of few writers. For the experiments, 541 text lines from 6 different writers, were considered.[6] The underlying lexicon consisted of 412 different words. The six writers who produced the data used in the experiments will be denoted by a, b, c, d, e and f in the following. Subsets of writers will be represented by sequences of these letters. For example, *abc* stands for writers a, b, and c.

Three groups of experiments were conducted, in which the text lines of the training sets were distorted by applying three different subsets of the

[5] See also: http://www.iam.unibe.ch/~fki/iamDB

[6] Each writer produced approximately 90 text lines

Table 1. Results of the experiments described in Subsection 5.1 (in %)

	original	all dist.	line level	cc. level
a	33.14	48.98	47.06	38.69
b	38.68	43.07	40.41	42.61
c	39.16	49.31	46.80	44.41
d	30.56	53.14	48.62	43.02
e	54.40	59.61	58.88	54.24
f	18.83	31.98	26.90	27.76
ab	60.69	73.46	75.79	54.92
cd	56.84	61.30	62.44	59.66
ef	63.84	68.46	67.54	67.51
abc	75.19	74.11	75.78	74.83
def	65.35	68.87	67.04	68.74

distortions described in Section 3. The three subsets were the set of *all distortions*, the set of geometrical transformations on the *line level*, and the set of *connected component level* geometrical transformations. In each case, five distorted text lines per given training text line were generated and added to the training set. So the extended training set was six times larger than the original one.

Fig. 11 shows examples of natural and synthetically generated pairs of text lines used in the experiments where all the distortions were applied. For each pair of text lines the natural one is shown below, while the synthetic one is above it. The first pair belongs to writer *a*, the second to writer *b*, and so on.

The recognition results of the three experiments are shown in Table 1, where the rows correspond to the different training modalities. The test set is always the complement of the training set, and consists of natural text only. For example, the test set corresponding to the first row consists of all natural text lines written by writers *bcdef*, while the training set is given by all natural text lines produced by writer *a* plus five distorted instances of each natural text line. In the first column, the results achieved by the original system that uses only natural training data are given for the purpose of reference. The other columns contain the results of the three groups of experiments using expanded training sets, i.e. the results for all, line level, and connected component level distortions, respectively. In those three columns each number corresponds to the median recognition rate of three independent experimental runs. In each run a different recognition rate is usually obtained because of the random nature of the distortion procedure.

In Table 1 it can be observed that adding synthetic training data leads to an improvement of the recognition rate in 29 out of 33 cases. Some of the improvements are quite substantial, for example, the improvement from 33.14% to 48.98% in row *a*.

Augmenting the training set of a handwriting recognition system by synthetic data as proposed in this chapter may have two adversarial effects on

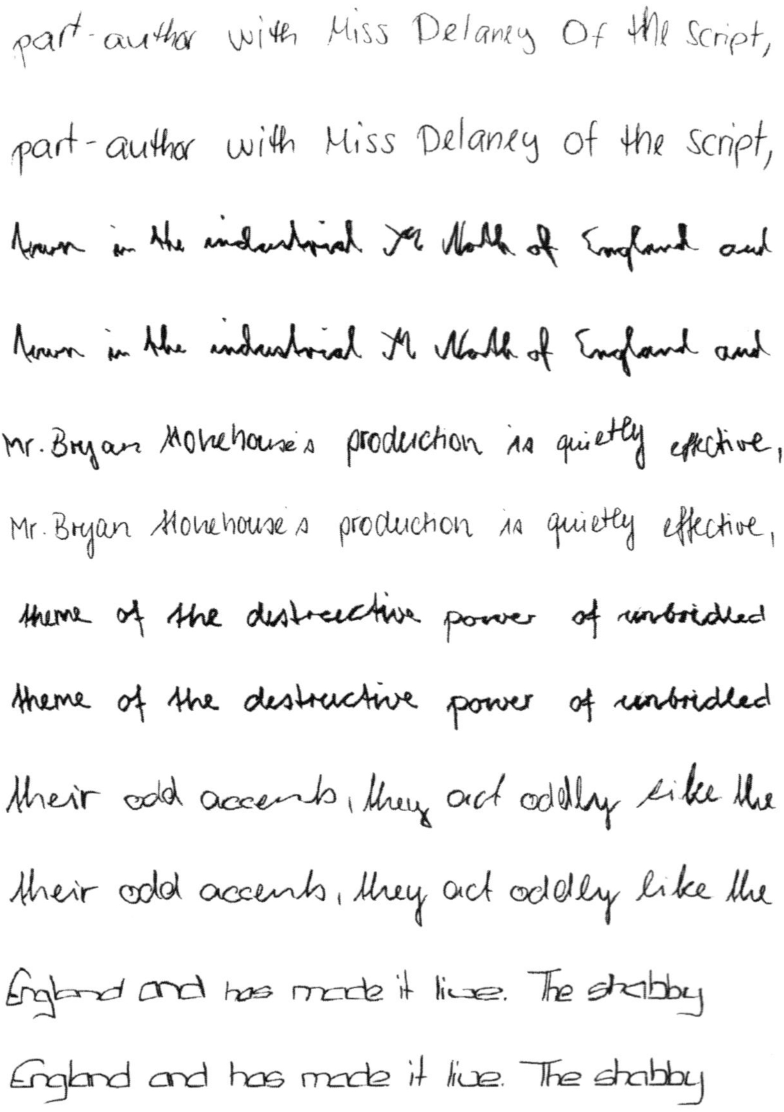

Fig. 11. Natural (below) and synthetic (above) text lines for writers *a*-*f*

the recognition rate. First, adding synthetic data increases the variability of the training set, which may be beneficial when the original training set has a low variability, i.e. when it was produced by only one or a few writers. On the other hand, the distortions may produce unnatural looking words and characters, which may bias the recognizer in an undesired way, because the test set includes only natural handwriting.

The greatest increase in recognition performance can be observed in Table 1 for those cases when there is only one writer in the training set. Then the variability of the training set is low and the addition of synthetic data leads to a better modeling of the test set. In this case, the application of all distortions outperforms the use of only line level or connected component level distortions. Where multiple writers are used for training, the variability of the training set is larger and the increase in recognition performance becomes smaller when synthetic training data is added. Also, in this case using all distortions does not always result in higher recognition rate than applying just line level or connected component level distortions.

Since in the majority of the experimental runs, an improvement of the recognition rate was observed, it can be concluded that the use of synthetic training data can potentially lead to improved handwriting recognition systems, in case of only a few writers in the training set.

In all experiments described in this subsection, single Gaussians were used in the HMMs' states to estimate observation probability distributions (see also Section 4). As we will see in the following, the number of Gaussians should be increased if the training set contains handwriting samples from many writers.

5.2 Large Training Set with Many Writers

In the following, the case where there are many writers and a large training set is considered. For the experiments, a subset of the IAM-Database different from that used in the previous subsection was considered, consisting of 1,993 text lines produced by 400 different writers, and the underlying lexicon contained 6,012 words. This set of text lines was randomly divided into *training*, *validation* and *test set*, such that their sets of writers were pairwise disjoint. The training and validation set contained 1,433 lines from 288 writers, and 160 text lines from 32 writers, respectively. The test set contained 400 text lines from 80 writers.

First, the training and the validation set were used to find the optimal parameters for the system that uses natural training data only, and for the system that uses a mixture of natural and synthetic training data. In the following, these two optimized systems will be referred to as *Original System* and *Expanded System*, respectively.

The optimization was performed in terms of *capacity* and *distortion strength*. The capacity of the recognition system is defined as the number of free parameters to be estimated from the training set. It determines how much information the recognizer can store to express its knowledge about the handwriting represented by the training set. A capacity too high may cause overfitting on the training data. On the other hand, a capacity too low may lead to a poor handwriting model. Since the synthetically expanded training set contains increased variability (both natural and unnatural), its optimal capacity is expected to be higher than the recognizer's optimal capacity for the original training set. That is, if the capacity of the system is not increased after the expansion of the training set, there is the danger that the capacity may be too low, such that the system is biased towards the unnatural variability introduced by the additional synthetic text lines, to an extent which may cause the recognition performance to drop. In the experiments, the capacity was varied through changing the number of Gaussian mixture components used for estimating the feature value distributions in the states of the Hidden Markov Models (see Section 4). The number of Gaussian mixtures, Ga, is the same in all HMMs. If this parameter, Ga, is increased, then it enables the system to model the distributions of the features extracted from the handwriting more accurately. Thus the capacity of the system is increased.

The second parameter to optimize was the distortion strength, which can be controlled by changing the interval of the possible amplitude values for the underlying functions described in Section 3. Four levels of strength were defined based on a subjective assessment: *very weak*, *weak*, *middle* and *strong*. Note that these terms indicate only the relative order of the four levels, rather than absolute categories.[7] In Fig. 12, two examples are shown, where the text lines on top were distorted using all four different distortion strengths. For the distorted text line generation, all of the distortions were applied, in the way described in Subsection 3.5. A trade-off between quality and variability of the generated text lines can be observed, which is governed by the distortion strength. That is, stronger distortions usually introduce more variability, but on the other hand, the generated text lines tend to look less natural. Thus tuning the distortion strength is expected to be beneficial.

Detailed results of the optimization stage are reported in Table 2. In the HMM training procedure, the training set, consisting of natural and synthetic training data, was used, while the recognition rates were measured on the validation set, which consisted of natural text lines only. Column *original* corresponds to the system using exclusively natural training data. According to the best result, the system with $Ga = 15$ is chosen as the *Original System*, which achieved a recognition rate of 70.48%. The other four columns, namely *very weak*, *weak*, *middle* and *strong*, show the recognition rates of the system using a mixture of natural and synthetic training data. For each text line in the training set, always five distorted text lines were generated, thus the expanded training set was always six times larger than the original one. Those results which correspond to statistically significant improvements with respect to the *Original System* (with a significance level higher than 90%), are highlighted using boldface.[8]

It can be seen that increasing the capacity is beneficial for expanded training sets. Rows $Ga = 6$ and $Ga = 12$ show the effects of low capacity after

[7] The strength was increased by *jointly* increasing the amplitude parameters for all the transformations, sampling in equal steps in terms of the parameter values of the perturbation model. For thinning/thickening, the probability of zero steps was decreased

[8] The significance level of an improvement was calculated from the writer level recognition rates, by applying a statistical z-test for matched samples

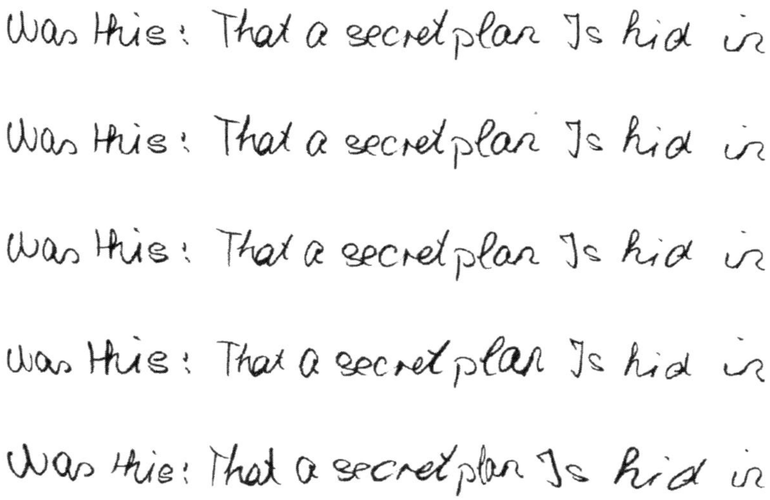

a)

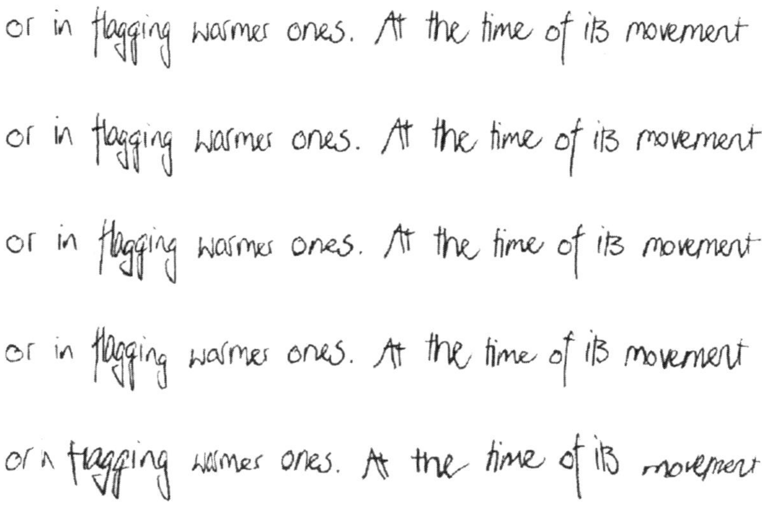

b)

Fig. 12. Illustration of levels of distortion strength used in the experiments of Subsection 5.2. From top to bottom, for both a) and b) parts: original, very weak, weak, middle and strong

Table 2. Results of the optimization stage of the experiments of Subsection 5.2 (in %). Statistically significant improvements are highlighted using boldface

	original	very weak	weak	middle	strong
Ga=6	67.04	65.45	66.12	65.52	62.81
Ga=12	69.95	69.69	71.41	69.76	70.09
Ga=15	70.48	70.88	**72.27**	**71.54**	70.48
Ga=18	70.15	**72.20**	**72.47**	**72.40**	71.01
Ga=21	69.62	71.61	**72.40**	**72.01**	71.54
Ga=24	70.48	71.34	**73.00**	**73.33**	71.21
Ga=27	70.22	71.48	**72.87**	**73.86**	**71.67**
Ga=30	69.49	**71.67**	**72.14**	**73.20**	**71.74**

training set expansion with synthetic data, resulting in lower recognition rates in the majority of the cases. With an increasing strength of the distortions, the optimal capacities become higher: from column *original* to column *strong* the optimal Ga's were 15, 18, 24, 27 and 30, respectively. This can be explained by the increasing variability of the training set. (Note that for strength *strong*, the optimal capacity is possibly above $Ga = 30$.) The most significant improvements came at strengths *weak* and *middle*. All significant improvements in these columns have a significance level greater than 95%. The most significant area is at strength *middle*, from $Ga = 24$ to $Ga = 30$. Here the significance level is greater than 99%. Thus the *Expanded System* was chosen among these, namely the one with $Ga = 27$, where the recognition rate was 73.86%.

After the optimization stage, the *Original System* was trained on the union of the training and validation set, and the *Expanded System* on the union of the expanded training and expanded validation set. For each natural text line in the validation set, five synthetic text lines were generated at strength *middle* to get the expanded validation set. Then, using the test set for testing on previously unseen examples, the recognition results of the *Original System* and the *Expanded System* were 76.85% and 79.54%, respectively, as shown in Table 3. This shows that using synthetic text lines, the recognition performance could be improved by more than 2.5%. The significance level of this improvement is greater than 99%. (The recognition rates on the test set differ a lot from those measured on the validation set. This can be explained by the relatively small size of the validation set. The magnitude of the validation set is limited by the amount of text lines in the training set, so that the training set has approximately the same optimal capacity as its union with the validation set. This way the negative effects of too low capacity can be avoided at the testing phase. But the choice of the training set size is also constrained by the computational complexity of the training process, since the training of HMMs using a large number of Gaussian mixtures is a rather time consuming procedure.)

Table 3. Results on the test set of the experiments of Subsection 5.2

	Ga	strength	recognition rate
Original System	15	–	76.85%
Expanded System	27	middle	79.54%

We also note that the synthetic training set expansion methodology presented above consists of such optimizations that must always be done, independently of the underlying datasets:

- The most appropriate distortion strength between zero and extremely strong can only be found empirically, because it may depend on the details of the recognizer under consideration, as well as on the concrete dataset.
- Finding the optimal number of Gaussians (or more generally, the optimal capacity) is a must in a multi-Gaussian system, because it is dependent on the characteristics of the training set. The same optimization is needed for the synthetically expanded training set, in order to have a fair comparison with the original system.[9]

Thus, the experiments show that expansion of the available set of text lines by synthetically generated instances makes it possible to significantly improve the recognition performance of a handwritten text line recognizer, even when the original training set is large and contains handwriting from many writers.

5.3 Capacity and Saturation

The main goal of synthetic training set expansion was to improve the recognition performance, by adding synthetic text lines to the original, i.e. human written, training set. With respect to this goal, an important observation of the experiments was that the number of Gaussians needed to be appropriately increased so that the synthetic training set expansion can improve the recognition rate.

To further examine this phenomenon, an experiment was conducted using gradually increasing training sets of an increasing number of writers, while keeping the test set as well as the number of Gaussian components (i.e. the capacity) fixed. The natural training and validation set defined in Subsection 5.2 was used for training and testing, respectively. The numbers of Gaussians considered were 1 and 6. The two corresponding curves of recognition rates are shown in Fig. 13, where different proportions of the training set were used for training, while the test set was always the same. The percentages on the horizontal axis are to be understood with respect to the union of the training set and the validation set (the union consists of $1433+160 = 1,593$ text lines).

Based on these curves, two statements can be made:

[9] It was also demonstrated in this subsection why the optimization of the capacity should not be overlooked, see Table 2

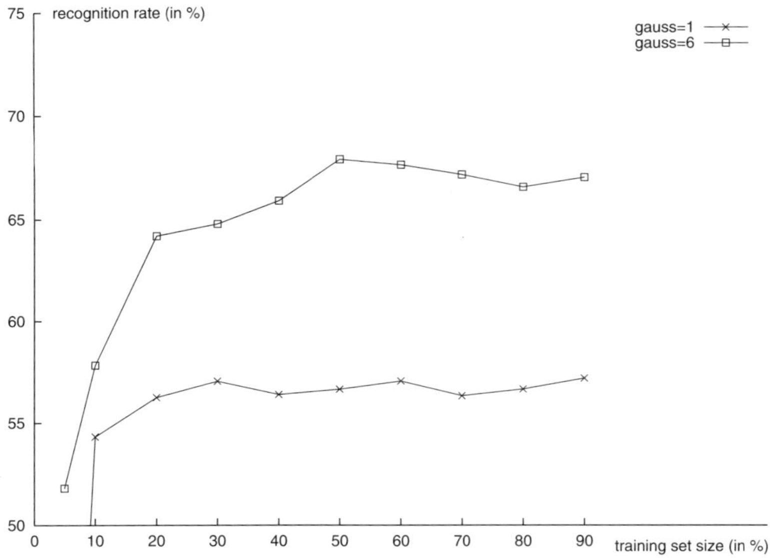

Fig. 13. Recognition rates on the test set using increasing training set sizes and fixed capacity of the recognizer

- For 1 Gaussian, we cannot expect further improvements above approximately 20% of $1,593 \approx 320$ training text lines.
- For 6 Gaussians, we cannot expect further improvements above approximately 50% of $1,593 \approx 800$ training text lines.

This leads to the intuitive notion of *saturation*, which means that given a fixed capacity of the handwriting recognition system, from a certain amount of natural training data no further improvements in the recognition rate can be expected. In other words, it cannot be predicted whether increasing the training set size yields (slightly) improved or deteriorated recognition performance. Furthermore, in Fig. 13 it also can be seen that in case of a higher capacity of 6 Gaussians, the recognizer needs more natural training data to get saturated.

Apparently, if the amount of natural training data already causes the system to be saturated, we cannot expect any positive change in the recognition rate through the expansion with synthetic data either, since even additional natural data does not help.[10] To the contrary, the negative effect of unnaturality inherent in the synthetic data can become dominant, causing the recognition rate to drop.

As an example for 6 Gaussians, in Table 2 the recognition rate dropped because the system was already saturated (note that the same data was used here

[10] Assuming that natural data is more appropriate than synthetic data for the estimation of details of natural handwriting

to illustrate saturation). In other words, the too low capacity of the system after synthetic training set expansion manifested itself through saturation.

To overcome the problem of saturation, in Subsection 5.2 the capacity of the recognizer had to be increased, in order to make room for further improvement when synthetic training set expansion is applied.

6 Conclusions and Future Work

In this chapter, the generation and use of synthetic training data in handwriting recognition was discussed. First, an overview of the related works of the field was given, including both machine printed and handwritten synthetic text generation.

The most important results of the authors' research in the field of synthetic handwriting generation for training purposes were also presented. A method for training set expansion by generating randomly perturbed versions of natural text lines rendered by human writers was presented and evaluated under several experimental conditions in writer-independent experiments. It was demonstrated that using such expanded training sets, improvements in the recognition rate can be achieved rather easily when the original training set is small and contains handwriting from only a limited number of writers. In the second experiment, it was shown that significant improvement in the recognition rate is possible to achieve even in the case of a large training set provided by many writers. In this case, the applied distortion strength needs to be adjusted, and the capacity of the recognizer (i.e. the number of Gaussians used for distribution estimations) plays an important role. The capacity has to be optimized after training set expansion, because the optimal capacity of the recognition system trained on the expanded training set is expected to be higher than the optimal capacity of the system trained on the original training set. If the capacity is not properly adjusted when using the synthetically expanded training set, there is the danger that the capacity may become too low, such that the system is biased towards unnatural handwriting styles in an undesired way, causing the recognition performance to drop.

Finally, based on the empirical observations of the experiments, the intuitive concept of saturation was introduced. The most important point is that the saturation has to be taken into account, because neither synthetic nor natural training set expansion can improve the recognition rate when the recognition system is already saturated by the available amount of natural training data. To cope with this problem, in the experiments the capacity of the recognizer was increased to open up room for further improvement.

As for possible future work, we plan to use not only one but several distortion strengths when expanding the training set. This may produce smoother training data than, for example, having only natural and strongly distorted text lines, but nothing between these two levels. Another idea is not to add all the generated texts to the training set, but perform a kind of pre-selection of the most appropriate ones, by using an rejection mechanism. Style dependent distortions as well as distortion strengths may also facilitate the creation of expanded training sets of better quality.

Since the problem of synthetic training data was addressed from a rather general point of view in the experiments, many questions mostly related to the enhancement of the baseline perturbation method are still open, e.g. considering other types of distortions as well as underlying functions, or examining the suitability of the individual distortions.

Our current work makes use of HMM for handwritten text line recognition. However, similar effects can be expected when dealing with other types of recognizers, for example, nearest neighbor classifier [21, 46] and neural networks [43, 45].

References

1. Bunke, H.: Recognition of Cursive Roman Handwriting – Past, Present and Future. In: Proc. 7th Int. Conf. on Document Analysis and Recognition, Edinburgh, Scotland (2003) 448–459
2. Plamondon, R., Srihari, S.: On-line and Off-line Handwriting Recognition: A Comprehensive Survey. IEEE Trans. on Pattern Analysis and Machine Intelligence **22**(1) (2000) 63–84
3. Srihari, S.: Handwritten Address Interpretation: a Task of Many Pattern Recognition Problems. Int. Journal of Pattern Recognition and Artificial Intelligence **14**(5) (2000) 663–674
4. Impedovo, S., Wang, P., Bunke, H., eds.: Automatic Bankcheck Processing. World Scientific (1997)
5. Gopisetty, S., Lorie, R., Mao, J., Mohiuddin, M., Sorin, A., Yair, E.: Automated Forms-processing Software and Services. IBM Journal of Research and Development **40**(2) (1996) 211–230
6. Ye, X., Cheriet, M., Suen, C.: A Generic Method of Cleaning and Enhancing Handwritten Data from Business Forms. Int. Journal on Document Analysis and Recognition **4**(2) (2001) 84–96
7. Arica, N., Yarman-Vural, F.: An Overview of Character Recognition Focused on Off-line Handwriting. IEEE Trans. on Systems, Man, and Cybernetics – Part C: Applications and Reviews **31**(2) (2001) 216–233
8. Mori, S., Suen, C., Yamamoto, K.: Historical Review of OCR Research and Development. In O'Gorman, L., Kasturi, R., eds.: Document Image Analysis. IEEE Computer Society Press (1995) 244–273
9. Simon, J.C.: Off-line Cursive Word Recognition. Proceedings of the IEEE **80**(7) (1992) 1150–1161
10. Steinherz, T., Rivlin, E., Intrator, N.: Offline Cursive Script Word Recognition – a Survey. Int. Journal on Document Analysis and Recognition **2**(2) (1999) 90–110
11. Vinciarelli, A.: A Survey on Off-line Cursive Word Recognition. Pattern Recognition **35**(7) (2002) 1433–1446
12. Marti, U.V., Bunke, H.: Using a Statistical Language Model to Improve the Performance of an HMM-based Cursive Handwriting Recognition System. Int. Journal of Pattern Recognition and Artificial Intelligence **15**(1) (2001) 65–90

13. Kittler, J., Hatef, M., Duin, R., Matas, J.: On Combining Classifiers. IEEE Trans. on Pattern Analysis and Machine Intelligence **20**(3) (1998) 226–239
14. Roli, F., Kittler, J., Windeatt, T., eds.: Proc. 5th Int. Workshop on Multiple Classifier Systems, Cagliari, Italy, Springer (2004)
15. Kuncheva, L.: Combining Pattern Classifiers: Methods and Algorithms. Wiley-Interscience (2004)
16. Lorette, G.: Handwriting Recognition or Reading? – What is the Situation at the Dawn of the 3rd Millenium? Int. Journal on Document Analysis and Recognition **2**(1) (1999) 2–12
17. Rosenfeld, R.: Two Decades of Statistical Language Modeling: Where do We Go from Here? Proc. of the IEEE **88**(8) (2000) 1270–1278
18. Elliman, D., Sherkat, N.: A Truthing Tool for Generating a Database of Cursive Words. In: Proc. 6th Int. Conf. on Document Analysis and Recognition, Seattle, WA, USA (2001) 1255–1262
19. Guyon, I., Haralick, R., Hull, J., Phillips, I.: Data Sets for OCR and Document Image Understanding Research. In Bunke, H., Wang, P., eds.: Handbook of Character Recognition and Document Image Analysis. World Scientific (1997) 779–799
20. Marti, U.V., Bunke, H.: The IAM-Database: an English Sentence Database for Off-line Handwriting Recognition. Int. Journal on Document Analysis and Recognition **5**(1) (2002) 39–46
21. Cano, J., Pérez-Cortes, J., Arlandis, J., Llobet, R.: Training Set Expansion in Handwritten Character Recognition. In: Proc. 9th SSPR / 4th SPR, Windsor, Ontario, Canada (2002) 548–556
22. Günter, S., Bunke, H.: Multiple Classifier Systems in Offline Handwritten Word Recognition – On the Influence of Training Set and Vocabulary Size. Int. Journal of Pattern Recognition and Artificial Intelligence **18**(7) (2004) 1302–1320
23. Rowley, H., Goyal, M., Bennett, J.: The Effect of Large Training Set Sizes on Online Japanese Kanji and English Cursive Recognizers. In: Proc. 8th Int. Workshop on Frontiers in Handwriting Recognition, Niagara-on-the-Lake, Ontario, Canada (2002) 36–40
24. Velek, O., Nakagawa, M.: The Impact of Large Training Sets on the Recognition Rate of Off-line Japanese Kanji Character Classifiers. In: Proc. 5th IAPR Workshop on Document Analysis Systems, Princeton, New Jersey, USA (2002) 106–109
25. Varga, T., Bunke, H.: Generation of Synthetic Training Data for an HMM-based Handwriting Recognition System. In: Proc. 7th Int. Conf. on Document Analysis and Recognition, Edinburgh, Scotland (2003) 618–622
26. Varga, T., Bunke, H.: Off-line Handwritten Textline Recognition Using a Mixture of Natural and Synthetic Training Data. In: Proc. 17th Int. Conf. on Pattern Recognition, Cambridge, United Kingdom (2004) 545–549
27. Kasturi, R., O'Gorman, L., Govindaraju, V.: Document Image Analysis: A Primer. Sādhanā **27**(1) (2002) 3–22
28. Nagy, G., Nartker, T., Rice, S.: Optical Character Recognition: An Illustrated Guide to the Frontier. In: Proc. IS&T/SPIE Symposium on Electronic Imaging: Science and Technology. Volume 3967., San Jose, CA, USA (2000) 58–69
29. Rice, S., Jenkins, F., Nartker, T.: The Fifth Annual Test of OCR Accuracy. Technical Report ISRI-TR-96-01, University of Nevada, Las Vegas, Nevada, USA (1996)

30. Baird, H.: The State of the Art of Document Image Degradation Modeling. In: Proc. 4th IAPR Workshop on Document Analysis Systems, Rio de Janeiro, Brasil (2000) 1–13
31. Stork, D.: Toward a Computational Theory of Data Acquisition and Truthing. In Helmbold, D., Williamson, B., eds.: Computational Learning Theory. Volume 2111 of Lecture Notes in Computer Science. Springer (2001) 194–207
32. Vuurpijl, L., Niels, R., Erp, M., Schomaker, L., Ratzlaff, E.: Verifying the UNIPEN devset. In: Proc. 9th Int. Workshop on Frontiers in Handwriting Recognition, Kokubunji, Tokyo, Japan (2004) 586–591
33. Baird, H.: Document Image Defect Models. In Baird, H., Bunke, H., Yamamoto, K., eds.: Structured Document Image Analysis. Springer (1992) 546–556
34. Doermann, D., Yao, S.: Generating Synthetic Data for Text Analysis Systems. In: Proc. 4th Annual Symposium on Document Analysis and Information Retrieval, Las Vegas, Nevada, USA (1995) 449–467
35. Kanungo, T., Haralick, R., Phillips, I.: Global and Local Document Degradation Models. In: Proc. 2nd Int. Conf. on Document Analysis and Recognition, Tsukuba Science City, Japan (1993) 730–734
36. Baird, H.: Document Image Defect Models and their Uses. In: Proc. 2nd Int. Conf. on Document Analysis and Recognition, Tsukuba Science City, Japan (1993) 62–67
37. Ho, T., Baird, H.: Large-Scale Simulation Studies in Image Pattern Recognition. IEEE Trans. on Pattern Analysis and Machine Intelligence **19**(10) (1997) 1067–1079
38. Märgner, V., Pechwitz, M.: Synthetic Data for Arabic OCR System Development. In: Proc. 6th Int. Conf. on Document Analysis and Recognition, Seattle, WA, USA (2001) 1159–1163
39. Baird, H., Fossey, R.: A 100-Font Classifier. In: Proc. 1st Int. Conf. on Document Analysis and Recognition, St.-Malo, France (1991) 332–340
40. Baird, H., Nagy, G.: A Self-Correcting 100-Font Classifier. In: Proc. IS&T/SPIE Symposium on Electronic Imaging: Science and Technology. Volume 2181., San Jose, California, USA (1994) 106–115
41. Ho, T., Baird, H.: Evaluation of OCR Accuracy Using Synthetic Data. In: Proc. 4th Annual Symposium on Document Analysis and Information Retrieval, Las Vegas, Nevada, USA (1995) 413–422
42. Helmers, M., Bunke, H.: Generation and Use of Synthetic Training Data in Cursive Handwriting Recognition. In: Proc. 1st Iberian Conf. on Pattern Recognition and Image Analysis, Puerto de Andratx, Mallorca, Spain (2003) 336–345
43. Drucker, H., Schapire, R., Simard, P.: Improving Performance in Neural Networks Using a Boosting Algorithm. In Hanson et. al., S., ed.: Advances in Neural Information Processing Systems 5. Morgan Kaufmann (1993) 42–49
44. Ha, T., Bunke, H.: Off-line Handwritten Numeral Recognition by Perturbation Method. IEEE Trans. on Pattern Analysis and Machine Intelligence **19**(5) (1997) 535–539
45. Mao, J., Mohiuddin, K.: Improving OCR Performance Using Character Degradation Models and Boosting Algorithm. Pattern Recognition Letters **18** (1997) 1415–1419
46. Mori, M., Suzuki, A., Shio, A., Ohtsuka, S.: Generating New Samples from Handwritten Numerals based on Point Correspondence. In: Proc. 7th Int. Workshop on Frontiers in Handwriting Recognition, Amsterdam, The Netherlands (2000) 281–290

47. Kaufmann, G., Bunke, H., Ha, T.: Recognition of Cursively Handwritten Words Using a Combined Normalization/Perturbation Approach. In Downton, A., Impedovo, S., eds.: Progress in Handwriting Recognition. World Scientific (1997) 21–28
48. Setlur, S., Govindaraju, V.: Generating Manifold Samples from a Handwritten Word. Pattern Recognition Letters **15** (1994) 901–905
49. Velek, O., Liu, C.L., Nakagawa, M.: Generating Realistic Kanji Character Images from On-line Patterns. In: Proc. 6th Int. Conf. on Document Analysis and Recognition, Seattle, WA, USA (2001) 556–560
50. Govindan, V., Shivaprasad, A.: Artificial Database for Character Recognition Research. Pattern Recognition Letters **12**(10) (1991) 645–648
51. Tung, C., Lee, H.: Performance Analysis of an OCR System via a Handwritten Character Generator. Pattern Recognition **27**(2) (1994) 221–232
52. Simard, P., Cun, Y., Denker, J.: Efficient Pattern Recognition Using a New Transformation Distance. In et al., S.H., ed.: Advances in Neural Information Processing Systems 5. Morgan Kaufmann (1993) 50–58
53. Guyon, I.: Handwriting Synthesis from Handwritten Glyphs. In: Proc. 5th Int. Workshop Frontiers in Handwriting Recognition, Essex, United Kingdom (1996) 309–312
54. Choi, H., Cho, S., Kim, J.: Generation of Handwritten Characters with Bayesian Network based On-line Handwriting Recognizers. In: Proc. 7th Int. Conf. on Document Analysis and Recognition, Edinburgh, Scotland (2003) 995–1001
55. Choi, H., Cho, S., Kim, J.: Writer-Dependent Online Handwriting Generation with Bayesian Networks. In: Proc. 9th Int. Workshop on Frontiers in Handwriting Recognition, Kokubunji, Tokyo, Japan (2004) 130–135
56. Wang, J., Wu, C., Xu, Y.Q., Shum, H.Y., Ji, L.: Learning based Cursive Handwriting Synthesis. In: Proc. 8th Int. Workshop on Frontiers in Handwriting Recognition, Niagara-on-the-Lake, Ontario, Canada (2002) 157–162
57. Lee, D.H., Cho, H.G.: A New Synthesizing Method for Handwriting Korean Scripts. Int. Journal of Pattern Recognition and Artificial Intelligence **12**(1) (1998) 46–61
58. Plamondon, R., Guerfali, W.: The Generation of Handwriting with Delta-lognormal Synergies. Biological Cybernetics **78** (1998) 119–132
59. Baird, H., Coates, A., Fateman, R.: PessimalPrint: a Reverse Turing Test. Int. Journal on Document Analysis and Recognition **5**(2-3) (2003) 158–163
60. Ahn, L., Blum, M., Hopper, N., Langford, J.: The CAPTCHA Web Page. `http://www.captcha.net` (2000)
61. Ahn, L., Blum, M., Hopper, N., Langford, J.: CAPTCHA: Telling Humans and Computers Apart. In Biham, E., ed.: Advances in Cryptology. Volume 2656 of Lecture Notes in Computer Science. Springer (2003) 294–311
62. Ahn, L., Blum, M., Langford, J.: Telling Humans and Computers Apart Automatically – How Lazy Cryptographers Do AI. Communications of the ACM **47**(2) (2004) 57–60
63. Baird, H., Luk, M.: Protecting Websites with Reading-based CAPTCHAs. In: Proc. 2nd Int. Web Document Analysis Workshop, Edinburgh, Scotland (2003) 53–56
64. Baird, H., Riopka, T.: ScatterType: a Reading CAPTCHA Resistant to Segmentation Attack. In: Proc. 12th SPIE/IS&T Document Recognition and Retrieval Conference. Volume 5676., San Jose, California, USA (2005)

65. Chew, M., Baird, H.: BaffleText: a Human Interactive Proof. In: Proc. 10th SPIE/IS&T Document Recognition and Retrieval Conference. Volume 5010., Santa Clara, California, USA (2003) 305–316
66. Lillibridge, M., Abadi, M., Bharat, K., Broder, A.: Method for Selectively Restricting Access to Computer Systems. U.S. Patent No. 6,195,698 (2000)
67. Simard, P., Szeliski, R., Couvreur, J., Calinov, I.: Using Character Recognition and Segmentation to Tell Computer from Humans. In: Proc. 7th Int. Conf. on Document Analysis and Recognition, Edinburgh, Scotland (2003) 418–423
68. Rusu, A., Govindaraju, V.: Handwritten CAPTCHA: Using the Difference in the Abilities of Humans and Machines in Reading Handwritten Words. In: 9th Int. Workshop on Frontiers in Handwriting Recognition, Kokubunji, Tokyo, Japan (2004) 226–231
69. Manzanera, A., Bernard, T.: Improved Low Complexity Fully Parallel Thinning Algorithm. In: Proc. 10th Int. Conf. on Image Analysis and Processing, Venice, Italy (1999) 215–220
70. Soille, P.: Morphological Image Analysis. Springer (1999)
71. Rabiner, L.: A Tutorial on Hidden Markov Models and Selected Applications in Speech Recognition. Proceedings of the IEEE **77**(2) (1989) 257–286

Review of Classifier Combination Methods

Sergey Tulyakov[1], Stefan Jaeger[2], and Venu Govindaraju[1], and David Doermann[2]

[1] Center for Unified Biometrics and Sensors, University at Buffalo, Amherst, NY 14228, USA tulyakov@cedar.buffalo.edu,venu@cubs.buffalo.edu
[2] Institute for Advanced Computer Studies, University of Maryland, College Park, MD 20742, USA {jaeger,doermann}@umiacs.umd.edu

Summary. Classifier combination methods have proved to be an effective tool to increase the performance of pattern recognition applications. In this chapter we review and categorize major advancements in this field. Despite a significant number of publications describing successful classifier combination implementations, the theoretical basis is still missing and achieved improvements are inconsistent. By introducing different categories of classifier combinations in this review we attempt to put forward more specific directions for future theoretical research. We also introduce a retraining effect and effects of locality based training as important properties of classifier combinations. Such effects have significant influence on the performance of combinations, and their study is necessary for complete theoretical understanding of combination algorithms.

1 Introduction

The efforts to automate the combination of expert opinions have been studied extensively in the second half of the twentieth century [1]. These studies have covered diverse application areas: economic and military decisions, natural phenomena forecasts, technology applications. The combinations presented in these studies can be separated into mathematical and behavioral approaches [2]. The mathematical combinations try to construct models and derive combination rules using logic and statistics. The behavioral methods assume discussions between experts, and direct human involvement in the combination process. The mathematical approaches gained more attention with the development of computer expert systems. Expert opinions could be of different nature dependent on the considered applications: numbers, functions, etc. For example, the work of R. Clemen contains combinations of multiple types of data, and, in particular, considers combinations of experts' estimations of probability density functions [2].

The pattern classification field developed around the end of the twentieth century deals with the more specific problem of assigning input signals to two

S. Tulyakov et al.: *Review of Classifier Combination Methods*, Studies in Computational Intelligence (SCI) **90**, 361–386 (2008)
www.springerlink.com

or more classes. The combined experts are classifiers and the result of the combination is also a classifier. The outputs of classifiers can be represented as vectors of numbers where the dimension of vectors is equal to the number of classes. As a result, the combination problem can be defined as a problem of finding the combination function accepting N-dimensional score vectors from M classifiers and outputting N final classification scores (Figure 1), where the function is optimal in some sense, e.g. minimizing the misclassification cost. In this chapter, we will deal exclusively with the mathematical methods for classifier combination from a pattern classification perspective.

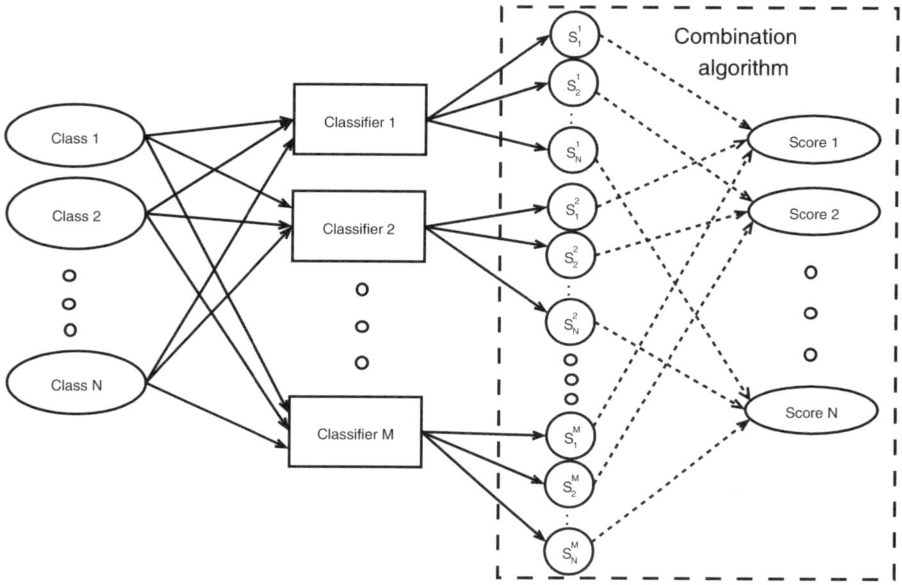

Fig. 1. Classifier combination takes a set of s_i^j - score for class i by classifier j and produces combination scores S_i for each class i

In the last ten years, we have seen a major boost of publications in optical character recognition and biometrics. Pattern classification applications of these two fields include image classification, e.g character recognition and word recognition, speech recognition, person authentication by voice, face image, fingerprints or other biometric characteristics. As a result, these two fields are the most popular application targets for multiple classifier systems so far [3, 4, 5, 6].

In this chapter, we will present different categorizations of classifier combination methods. Our goal is to give a better understanding of the current problems in this field. We will also provide descriptions of major classifier combination methods and their applications in document analysis. We organized this chapter as follows: Section 2 introduces different categorizations of classifier combination methods. Section 3 discusses ensemble techniques, while Section 4 focuses on non-ensemble techniques. In Section 5, we address additional issues important to classifier combination, such as retraining.

2 Defining the Classifier Combination Problem

In order to provide a more complete description of the classifier combination field we attempt to categorize different types of classifier combinations in this section.

2.1 Score Combination Functions and Combination Decisions

Classifier combination techniques operate on the outputs of individual classifiers and usually fall into one of two categories. In the first approach the outputs are treated as inputs to a generic classifier, and the combination algorithm is created by training this, sometimes called 'secondary', classifier. For example, Dar-Shyang Lee [7] used a neural network to operate on the outputs of the individual classifiers and to produce the combined matching score. The advantage of using such a generic combinator is that it can learn the combination algorithm and can automatically account for the strengths and score ranges of the individual classifiers. In the second approach, a function or a rule combines the classifier scores in a predetermined manner.

The final goal of classifier combination is to create a classifier which operates on the same type of input as the base classifiers and separates the same types of classes. Using combination rules implies some final step of classification decision. If we denote the score assigned to class i by base classifier j as s_i^j, then the typical combination rule is some function f and the final combined score for class i is $S_i = f(\{s_i^j\}_{j=1,\dots,M})$. The sample is classified as belonging to class $\arg\max_i S_i$. Thus the combination rules can be viewed as a classifier operating on base classifiers' scores, involving some combination function f and the $\arg\max$ decision, showing that there is no real conceptual difference between the categories mentioned above.

Generic classifiers used for combinations do not have to be necessarily constructed following the above described scheme, but in practice we see this theme commonly used. For example, in multilayer perceptron classifiers the last layer has each node containing a final score for one class. These scores are then compared and the maximum is chosen. Similarly, k-nearest neighbor classifier can produce scores for all classes as ratios of the number of representatives of a particular class in a neighborhood to k. The class with the highest ratio is then assigned to a sample.

In summary, combination rules can be considered as a special type of classifier of particular $\arg\max f$ form. Combination functions f are usually simple functions, such as sum, weighted sum, max, etc. Generic classifiers such as neural networks and k-nearest neighbor, on the other hand, imply more complicated functions.

2.2 Combinations of Fixed Classifiers and Ensembles of Classifiers

One main categorization is based on whether the combination uses a fixed (usually less than 10) set of classifiers, as opposed to a large pool of classifiers (potentially infinite) from which one selects or generates new classifiers. The first type of combinations assumes classifiers are trained on different features or different sensor inputs. The advantage comes from the diversity of the classifiers' strengths on different input patterns. Each classifier might be an expert on certain types of input patterns. The second type of combinations assumes large number of classifiers, or ability to generate classifiers. In the second type of combination the large number of classifiers are usually obtained by selecting different subsets of training samples from one large training set, or by selecting different subsets of features from the set of all available features, and by training the classifiers with respect to selected training subset or subset of features.

2.3 Operating Level of Classifiers

Combination methods can also be grouped based on the level at which they operate. Combinations of the first type operate at the *feature* level. The features of each classifier are combined to form a joint feature vector and classification is subsequently performed in the new feature space. The advantage of this approach is that using the features from two sets at the same time can potentially provide additional information about the classes. For example, if two digit recognizers are combined in such a fashion, and one recognizer uses a feature indicating the enclosed area, and the other recognizer has a feature indicating the number of contours, then the combination of these two features in a single recognizer will allow class '0' to be easily separated from the other classes. Note that individually, the first recognizer might have difficulty separating '0' from '8', and the second recognizer might have difficulty separating '0' from '6' or '9'. However, the disadvantage of this approach is that the increased number of feature vectors will require a large training set and complex classification schemes. If the features used in the different classifiers are not related, then there is no reason for combination at the feature level.

Combinations can also operate at the decision or score level, that is they use outputs of the classifiers for combination. This is a popular approach because the knowledge of the internal structure of classifiers and their feature vectors is not needed. Though there is a possibility that representational information is lost during such combinations, this is usually compensated by the lower complexity of the combination method and superior training of the final system. In the subsequent sections we will only consider classifier combinations at the decision level.

The following is an example of the application where feature level combination can improve a classifier's performance. Bertolami and Bunke [8] compared the combinations at the feature and the decision levels for handwriting recognition. Their handwriting recognizer uses a sliding window to extract pixel and geometrical features for HMM matching. The combination at the feature level has a single HMM trained on the composite vector of these features. The combination at the decision level has two HMMs trained on separate pixel and geometrical feature vectors, and the recognition results, word matching scores, are combined together with the language model. The combination at the feature level seems to achieve better results, and authors explain its effectiveness by improved alignment of the HMM recognizer.

Note that combination on the feature level is not conceptually different from trying to incorporate different types of information into a single feature vector. For example, Favata [9] constructed handwritten word recognizers which utilized a character recognizer based on gradient, concavity and structural features. Thus feature level combination rather delegates the combination task to the base classifier (HMM in above example) instead of solving it.

2.4 Output Types of Combined Classifiers

Another way to categorize classifier combination is by the outputs of the classifiers used in the combination. Three types of classifier outputs are usually considered [3]:

- Type I (abstract level): This is the lowest level since a classifier provides the least amount of information on this level. Classifier output is merely a single class label or an unordered set of candidate classes.
- Type II (rank level): Classifier output on the rank level is an ordered sequence of candidate classes, the so-called n-best list. The candidate class at the first position is the most likely class, while the class positioned at the end of the list is the most unlikely. Note that there are no confidence values attached to the class labels on rank level. Only their position in the n-best list indicates their relative likelihood.
- Type III (measurement level): In addition to the ordered n-best lists of candidate classes on the rank level, classifier output on the measurement level has confidence values assigned to each entry of the n-best list. These confidences, or scores, can be arbitrary real numbers, depending on the classification architecture used. The measurement level contains therefore the most information among all three output levels.

In principle, a combination method can operate on any of these levels. For combinations based solely on label sets or rankings of class labels, i.e. output on abstract and rank level, several voting techniques have been proposed and experimentally investigated [10, 11, 12]. The advantage of classifier output on abstract and rank level is that different confidence characteristics have no negative impact on the final outcome, simply because confidence plays no role in the decision process. Nevertheless, the confidence of a classifier in a particular candidate class usually provides useful information that a simple class ranking cannot reflect. This suggests the use of combination methods that operate on the measurement level, and which can exploit the confidence assigned to each candidate class. Nowadays most classifiers do provide information on measurement level, so that applying combination schemes on the measurement level should be possible for most practical applications. On measurement level, however, we have to take into account that each classifier in a multiple classifier system may output quite different confidence values, with different ranges, scales, means etc. This may be a minor problem for classifier ensembles generated with Bagging and Boosting (see Section 3 since all classifiers in the ensemble are based on the same classification architecture, only their training sets differ. Each classifier will therefore provide similar output. However, for classifiers based on different classification architectures, this output will in general be different. Since different architectures lead more likely to complementary classifiers, which are especially promising for combination purposes, we need effective methods for making outputs of different classifiers comparable.

If the combination involves classifiers with different output types, the output is usually converted to any one of the above: to type I [3], to type II [10], or to type III [7]. Most existing classifier combination research deals with classifier outputs of type III (measurement level), among which we can find combinations with fixed structure, e.g. sum of scores [3, 13], or combinations that can be trained using available training samples (weighted sum, logistic regression [10], Dempster-Shafer rules [3], neural network [7], etc.).

2.5 Complexity Types of Classifier Combinations

In [14] we proposed a new framework for combining classifiers based on the structure of combination functions. Though we applied it only to biometric person authentication applications there, it can provide a useful insight into other applications as well.

As Figure 1 shows, the combination algorithm is a map

$$\{s_k^j\}_{j=1,\dots,M;k=1,\dots,N} \Rightarrow \{S_i\}_{i=1,\dots,N} \qquad (1)$$

of NM classifiers' scores into the N-dimensional final score space, where N is the number of classes. The complexity types define how such maps are constructed. Specifically, the combination type is determined by whether all classifiers' scores participate in the derivation of each final combined score, and whether only a single combination function is trained for all classes or there is an individual combination function for each class.

We separate combination algorithms into four different types depending on the number of classifier's scores they take into account and the number of combination functions required to be trained:

1. Low complexity combinations: $S_i = f(\{s_i^j\}_{j=1,\dots,M})$. Combinations of this type require only one combination function to be trained, and the combination function takes as input scores for one particular class.
2. Medium complexity I combinations: $S_i = f_i(\{s_i^j\}_{j=1,\dots,M})$. Combinations of this type have separate score combining functions for each class and each such function takes as input parameters only the scores related to its class.
3. Medium complexity II combinations:

$$S_i = f(\{s_i^j\}_{j=1,\dots,M}, \{s_k^j\}_{j=1,\dots,M;k=1,\dots,N,k\neq i}) \tag{2}$$

   This function takes as parameters not only the scores related to one class, but all output scores of classifiers. Combination scores for each class are calculated using the same function, but scores for class i are given a special place as parameters. Applying function f for different classes effectively means permutation of the function's parameters.
4. High complexity combinations: $S_i = f_i(\{s_k^j\}_{j=1,\dots,M;k=1,\dots,N})$. Functions calculating final scores are different for all classes, and they take as parameters all output base classifier scores.

In order to illustrate the different combination types we can use a matrix representation as shown in Figure 2. Each row corresponds to a set of scores output by a particular classifier, and each column has scores assigned by classifiers to a particular class.

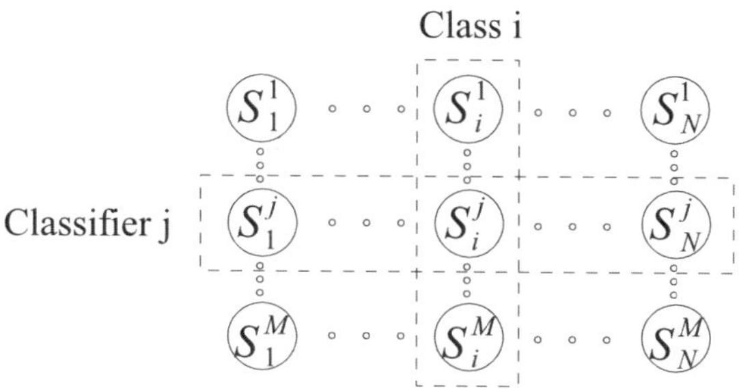

Fig. 2. Output classifier scores arranged in a matrix; s_i^j - score for class i by classifier j

Figure 3 shows four complexity types of combinations using the score matrix. The combinations of low and medium I complexity types use only scores of one particular class to derive a final combined score for this class. Medium II and high complexity combinations use scores related to all classes to derive a final combined score of any single class. Low and medium II complexity combinations have a single combination function f used for all classes, and

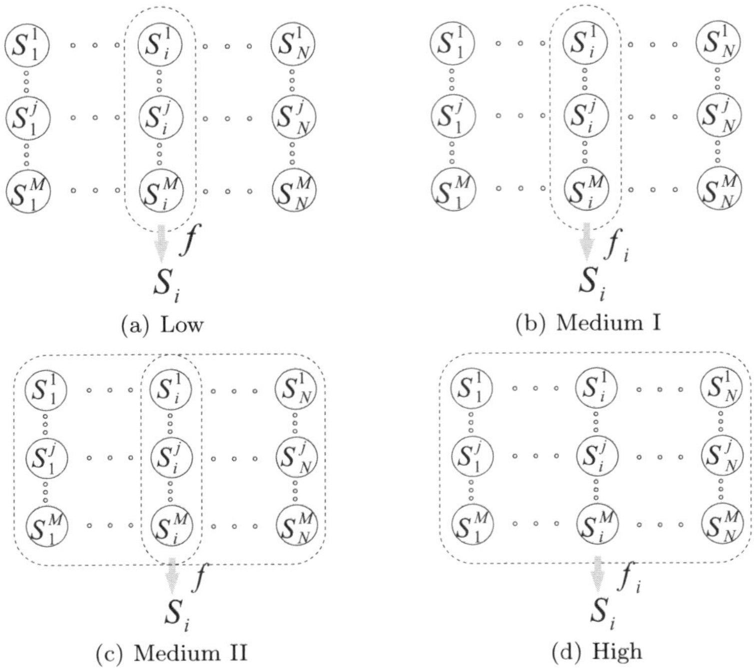

Fig. 3. The range of scores considered by each combination type and combination functions

medium I and high complexity combinations might have different combination functions f_i for different classes i.

As an example, simple combination rules (sum, weighted sum, product, etc.) typically produce combinations of low complexity type. Combinations which try to find the separate combination function for each class [15, 16] are of medium I complexity type. The rank-based combination methods (e.g. Borda count in Section 4) represent the combinations of medium II complexity type, since calculating rank requires comparing the original score with other scores produced during the same identification trial. Behavior-knowledge spaces (BKS, see Section 4) are an example of high complexity combination type, since they are both rank-based and can have user-specific combination functions. One way to obtain combinations of different types is to use different score normalizations before combining normalized scores by a simple combination rule of low complexity. For example, by using class-specific Z-normalization or identification trial specific T-normalization [17], we are dealing respectively with medium I or medium II complexity combination types.

Higher complexity combinations can potentially produce better classification results since more information is used. On the other hand the availability of training samples will limit the types of possible combinations. Thus the choice of combination type in any particular application is a trade-off between classifying capabilities of combination functions and the availability of sufficient training samples. When the complexity is lowered it is important to see if any useful information is lost. If such loss happens, the combination algorithm should be modified to compensate for it.

Different generic classifiers such as neural networks, decision trees, etc., can be used for classifier combinations within each complexity class. However, the choice of the generic classifiers or combination functions is less important than the choice of the complexity type.

2.6 Classification and Combination

From Figure 1 and the discussion in Section 2.1 we can view the problem of combining classifiers as a classification problem in the score space $\{s_i^j\}_{j=1,\dots,M;i=1,\dots,N}$. Any generic pattern classification algorithm trained in this score space can act as a combination algorithm. Does it make sense to search for other, more specialized methods of combination? In other words, does classifier combination field has anything new to offer with respect to traditional pattern classification research?

One difference between the combination problem and the general pattern classification problem is that in the combination problem features (scores) have a specific meaning of being related to a particular class or being produced by a particular classifier. In the general pattern classification problem we do not assign such meaning to features. Thus intuitively we tend to construct combination algorithms which take such meaning of scores into consideration. The four combination complexity types presented in the previous section are based on this intuition, as they pay special attention to the scores s_i of class i while deriving a combined score S_i for this class.

The meaning of the scores, though, does not provide any theoretical basis for choosing a particular combination method, and in fact can lead to constructing suboptimal combination algorithms. For example, by constructing combinations of low and medium I complexity types we effectively disregard any interdependencies between scores related to different classes. As we showed in [14] and [18] such dependencies can provide useful information for the combination algorithm.

The following is a list of cases which might not be solved optimally in traditional pattern classification algorithms. The task of classifier combination can be defined as developing specialized combination algorithms for these cases.

1. The situation of having to deal with a large number of classes arises frequently in the pattern recognition field. For example, biometric person identification, speech and handwriting recognition are applications with very large number of classes. The number of samples of each class available for training can be small, such as in biometric applications where

a single person template is enrolled into the database, or even zero for speech and handwriting recognition when the class is determined by the lexicon word.

2. The number of classifiers M is large. For example, taking multiple training sets in bagging and boosting techniques yields arbitrarily large number of classifiers. The usual method of combination in these cases is to use some a priori rule, e.g. sum rule.
3. Additional information about classifiers is available. For example, in the case of multimodal biometrics combination it is safe to assume that classifiers act independently. This might be used to better estimate the joint score density of M classifiers as a product of M separately estimated score densities of each classifier.
4. Additional information about classes is available. Consider the problem of classifying word images into classes represented by a lexicon: The relation between classes can be expressed through classifier independent methods, for example, by using the string edit distance. Potentially classifier combination methods could benefit from such additional information.

The cases listed above present situations where generic pattern classification methods in score space are not sufficient or suboptimal. The first two cases describe scenarios where the feature space has very large dimensions. By adopting a combination of reduced complexity we are able to train combination algorithm and achieve performance improvement. If neither the number of classifiers nor the number of classes is large, the generic pattern classification algorithm operating in the score space can solve the combination problem.

When additional information besides training score vectors is available as in scenarios 3 and 4 it should be possible to improve on the generic classification algorithms which use only a sample of available score vectors for training, but no other information.

3 Classifier Ensembles

The focus of this chapter is to explore the combinations on a fixed set of classifiers. We assume that there are only few classifiers and we can collect some statistical data about these classifiers using a training set. The purpose of the combination algorithm is to learn the behavior of these classifiers and produce an efficient combination function.

In this section, however, we shortly address another approach to combination that includes methods trying not only to find the best combination algorithm, but also trying to find the best set of classifiers for the combination. This type of combination usually requires a method for generating a large number of classifiers. Few methods for generating classifiers for such combinations exist. One of the methods is based on bootstrapping the training set in order to obtain a multitude of subsets and train a classifier on each of these subsets. Another method is based on the random selection of the subsets of features from one large feature set and training classifiers on these feature subsets [19]. A third method applies different training conditions, e.g. choosing random initial weights for neural network training or choosing dimensions for decision trees [20]. The ultimate method for generating classifiers is a random separation of feature space into the regions related to particular classes [21].

Simplest methods of combination apply some fixed functions to the outputs of all the generated classifiers (majority voting, bagging [22]). More complex methods, such as boosting [23, 24], stack generalization [25], attempt to select only those classifiers which will contribute to the combination.

Although there is substantial research on the classifier ensembles, very few theoretical results exist. Most explanations use bias and variance framework which is presented below. But such approaches can only give asymptotic explanations of observed performance improvements. Ideally, the theoretical foundation for classifier ensembles should use statistical learning theory [26, 27]. But it seems that such work will be quite difficult. For example, it is noted in [28] that an unrestricted ensemble of classifiers has a higher complexity than individual combined classifiers. The same paper presents an interesting explanation of the performance improvements based on the classifier's margin - the statistical measure of the difference between scores given to correct and incorrect classification attempts. Another theoretical approach to the classifier ensemble problem was developed by Kleinberg in the theory of stochastic discrimination [29, 21]. This approach considers very general type of classifiers (which are determined by the regions in the feature space) and outlines criteria on how these classifiers should participate in the combination.

3.1 Reductions of Trained Classifier Variances

One way to explain the improvements observed in ensemble combination methods (bagging, boosting) is to decompose the added error of the classifiers into bias and variance components [30, 31, 22]. There are few definitions of such decompositions [24]. Bias generally shows the difference between optimal Bayesian classification and average of trained classifiers, where average means real averaging of scores or voting and average is taken over all possible trained classifiers. The variance shows the difference between a typical trained classifier and an average one.

The framework of Tumer and Ghosh [32] associates trained classifiers with the approximated feature vector densities of each class. This framework has been used in many papers on classifier combination recently [33, 34, 35, 36]. In this framework, trained classifiers provide approximations to the true posterior class probabilities or to the true class densities:

$$f_i^m(x) = p_i(x) + \epsilon_i^m(x) \tag{3}$$

where i is the class index and m is the index of a trained classifier. For a fixed point x the error term can be represented as a random variable where randomness is determined by the random choice of the classifier or used training set. By representing it as a sum of mean β and zero-mean random variable η we get

$$\epsilon_i^m(x) = \beta_i(x) + \eta_i^m(x) \tag{4}$$

For simplicity, assume that the considered classifiers are unbiased, that is $\beta_i(x) = 0$ for any x, i. If point x is located on the decision boundary between classes i and j then the added error of the classifier is proportional to the sum of the variances of η_i and η_j:

$$E_{add}^m \sim \sigma_{\eta_i^m}^2 + \sigma_{\eta_j^m}^2 \tag{5}$$

If we average M such trained classifiers and if the error random variables η_i^m are independent and identically distributed as η_i, then we would expect the added error to be reduced M times:

$$E_{add}^{ave} \sim \sigma_{\eta_i^{ave}}^2 + \sigma_{\eta_j^{ave}}^2 = \frac{\sigma_{\eta_i}^2 + \sigma_{\eta_j}^2}{M} \tag{6}$$

The application of the described theory is very limited in practice since too many assumptions about classifiers are required. Kuncheva [34] even compiles a list of used assumptions. Besides independence assumption of errors, we need to hypothesize about error distributions, that is the distributions of the random variable η_i. The tricky part is that η_i is the difference between true distribution $p_i(x)$ and our best guess about this distribution. If we knew what the difference is, we would have been able to improve our guess in the first place. Although there is some research [33, 37] into trying to make assumptions about these estimation error distributions and seeing which combination rule is better for a particular hypothesized distribution, the results are not proven in practice.

3.2 Bagging

Researchers have very often concentrated on improving single-classifier systems mainly because of their lack in sufficient resources for simultaneously developing several different classifiers. A simple method for generating multiple classifiers in those cases is to run several training sessions with the same single-classifier system and different subsets of the training set, or slightly modified classifier parameters. Each training session then creates an individual classifier. The first more systematic approach to this idea was proposed in [22] and became popular under the name "Bagging." This method draws the training sets with replacement from the original training set, each set resulting in a slightly different classifier after training. The technique used for generating the individual training sets is also known as bootstrap technique and aims at reducing the error of statistical estimators. In practice, bagging has shown good results. However, the performance gains are usually small when bagging is applied to weak classifiers. In these cases, another intensively investigated technique for generating multiple classifiers is more suitable: Boosting.

3.3 Boosting

Boosting has its root in a theoretical framework for studying machine learning, and deals with the question whether an almost randomly guessing classifier can be boosted into an arbitrarily accurate learning algorithm. Boosting attaches a weight to each instance in the training set [38, 24, 39]. The weights are updated after each training cycle according to the performance of the classifier on the corresponding training samples. Initially, all weights are set equally, but on each round, the weights of incorrectly classified samples are increased so that the classifier is forced to focus on the hard examples in the training set [38].

A very popular type of boosting is AdaBoost (Adaptive Boosting), which was introduced by Freund and Schapire in 1995 to expand the boosting approach introduced by Schapire. The AdaBoost algorithm generates a set of classifiers and votes them. It changes the weights of the training samples based on classifiers previously built (trials). The goal is to force the final classifiers to minimize expected error over different input distributions. The final classifier is formed using a weighted voting scheme. Details of AdaBoost, in particular the AdaBoost variant called AdaBoost.M1, can be found in [24].

Boosting has been successfully applied to a wide range of applications. Nevertheless, we will not go more into the details of boosting and other ensemble combinations in this chapter. The reason is that the focus of classifier ensembles techniques lies more on the generation of classifiers and less on their actual combination.

4 Non-Ensemble Combinations

Non-ensemble combinations typically use a smaller number of classifiers than ensemble-based classifier systems. Instead of combining a large number of automatically generated homogeneous classifiers, non-ensemble classifiers try to combine heterogeneous classifiers complementing each other. The advantage of complementary classifiers is that each classifier can concentrate on its own small subproblem instead of trying to cope with the classification problem as a whole, which may be too hard for a single classifier. Ideally, the expertise of the specialized classifiers do not overlap. There are several ways to generate heterogeneous classifiers. The perhaps easiest method is to train the same classifier with different feature sets and/or different training parameters. Another possibility is to use multiple classification architectures, which produce different decision boundaries for the same feature set. However, this is not only more expensive in the sense that it requires the development of independent classifiers, but it also raises the question of how to combine the output provided by multiple classifiers.

Many combination schemes have been proposed in the literature. As we have already discussed, they range from simple schemes to relatively complex combination strategies. This large number of proposed techniques shows the uncertainty researchers still have in this field. Up till now, researchers have not been able to show the general superiority of a particular combination scheme, neither theoretically nor empirically. Though several researchers have come up with theoretical explanations supporting one or more of the proposed schemes, a commonly accepted theoretical framework for classifier combination is still missing.

4.1 Elementary Combination Schemes on Rank Level

The probably simplest way of combining classifiers are voting techniques on rank level. Voting techniques do not consider the confidence values that may have been attached to each class by the individual classifiers. This simplification allows easy integration of all different kinds of classifier architectures.

4.1.1 Majority Voting

A straightforward voting technique is majority voting. It considers only the most likely class provided by each classifier and chooses the most frequent class label among this crisp output set. In order to alleviate the problem of ties, the number of classifiers used for voting is usually odd. A trainable variant of majority voting is weighted majority voting, which multiplies each vote by a weight before the actual voting. The weight for each classifier can be obtained; e.g., by estimating the classifiers' accuracies on a validation set. Another voting technique that takes the entire n-best list of a classifier into account, and not only the crisp 1-best candidate class, is Borda count.

4.1.2 Borda Count

Borda count is a voting technique on rank level [11]. For every class, Borda count adds the ranks in the n-best lists of each classifier, with the first entry in the n-best list; i.e., the most likely class label, contributing the highest rank number and the last entry having the lowest rank number. The final output label for a given test pattern X is the class with highest overall rank sum. In mathematical terms, this reads as follows: Let N be the number of classifiers, and r_i^j the rank of class i in the n-best list of the j-th classifier. The overall rank r_i of class i is thus given by

$$r_i = \sum_{j=1}^{N} r_i^j \tag{7}$$

The test pattern X is assigned the class i with the maximum overall rank count r_i.

Borda count is very simple to compute and requires no training. Similar to majority vote, there is a trainable variant that associates weights to the ranks of individual classifiers. The overall rank count for class i then computes as

$$r_i = \sum_{j=1}^{N} w_j r_i^j \tag{8}$$

Again, the weights can be the performance of each individual classifier measured on a training or validation set.

While voting techniques provide a fast and easy method for improving classification rates, they nevertheless leave the impression of not realizing the full potential of classifier combination by throwing away valuable information on measurement level. This has lead scientists to experiment with elementary combination schemes on this level as well.

4.2 Elementary Combination Schemes on Measurement Level

Elementary combination schemes on measurement level apply simple rules for combination, such as sum-rule, product-rule, and max-rule. Sum-rule simply adds the score provided by each classifier of a classifier ensemble for every class, and assigns the class label with the maximum score to the given input pattern. Analogously, product-rule multiplies the score for every class and then outputs the class with the maximum score. Interesting theoretical results, including error estimations, have been derived for those simple combination schemes. For instance, Kittler et al. showed that sum-rule is less sensitive to noise than other rules [13]. Despite their simplicity, simple combination schemes have resulted in high recognition rates, and it is by no means obvious that more complex methods are superior to simpler ones, such as sum-rule.

The main problem with elementary combination schemes is the incompatibility of confidence values. As discussed in Subsection 2.4, the output of classifiers can be of different type. Even for classifier output on measurement level (Type III), the output can be of different nature; e.g., similarity measures, likelihoods in the statistical sense, or distances to hyperplanes. In fact, this is why many researchers prefer using the name "score," instead of the names "confidence" or "likelihood," for the values assigned to each class on measurement level. This general name should emphasize the fact that those values are not the correct a posteriori class probabilities and have, in general, neither the same range and scale nor the same distribution. In other words, scores need to be normalized before they can be combined in a meaningful way with elementary combination schemes.

Similar to the situation for combination schemes, there is no commonly accepted method for normalizing scores of different classifiers. A couple of normalization techniques have been proposed. According to Jain et al., a good normalization scheme must be robust and efficient [40, 41]. In this context, robustness refers to insensitivity to score outliers and efficiency refers to the proximity of the normalized values to the optimal values when the distribution of scores is known.

The perhaps easiest normalization technique is the min-max normalization. For a given set of matching scores $\{s_k\}, k = 1, 2, \ldots, n$, the min-max normalized scores $\{s'_k\}$ are given by

$$s'_k = \frac{s_k - min}{max - min}, \tag{9}$$

where max and min are the maximum and minimum estimated from the given set of matching scores $\{s_k\}$, respectively. This simple type of normalization retains the original distribution of scores except for a scaling factor, transforming all the scores into a common range $[0\,;1]$. The obvious disadvantage of min-max normalization is its sensitivity to outliers in the data used for estimation of min and max. Another simple normalization method, which is, however, sensitive to outliers as well, is the so-called z-score. The z-score computes the arithmetic mean μ and the standard deviation σ on the set of scores $\{s_k\}$, and normalizes each score with

$$s'_k = \frac{s_k - \mu}{\sigma} \tag{10}$$

It is biased towards Gaussian distributions and does not guarantee a common numerical range for the normalized scores. A normalization scheme that is insensitive to outliers, but also does not guarantee a common numerical range, is MAD. This stands for "median absolute deviation," and is defined as follows:

$$s'_k = \frac{s_k - median}{MAD}, \tag{11}$$

with $MAD = median(|s_k - median|)$. Note that the median makes this normalization robust against extreme points. However, MAD normalization does a poor job in retaining the distribution of the original scores. In [41], Jain et al. list two more normalization methods, namely a technique based on a double sigmoid function suggested by Cappelli et al. [42], and a technique called "tanh normalization" proposed by Hampel et al. [43]. The latter technique is both robust and efficient.

Instead of going into the details of these two normalization methods, we suggest an alternative normalization technique that we have successfully applied to document processing applications, in particular handwriting recognition and script identification. This technique was first proposed in [44] and [45]. Its basic idea is to perform a warping on the set of scores, aligning the nominal progress of score values with the progress in recognition rate. In mathematical terms, we can state this as follows:

$$s'_k = \frac{\sum_{i=0}^{k} n_{correct}(s_i)}{N} \tag{12}$$

The help function $n_{correct}(s_i)$ computes the number of patterns that were correctly classified with the original score s'_k on an evaluation set with N patterns. The new normalized scores s'_k thus describe a monotonously increasing partial sum, with the increments depending on the progress in recognition rate. We can easily see that the normalized scores fall all into the same numerical range $[0\,;1]$. In addition, the normalized scores also show robustness against outliers because the partial sums are computed over a range of original scores, thus averaging the effect of outliers. Using the normalization scheme in (12), we were able to clearly improve the recognition rate of a combined on-line/off-line handwriting recognizer in [44, 45]. Combination of off-line and on-line handwriting recognition is an especially fruitful application domain of classifier combination. It allows combination of the advantages of off-line recognition with the benefits of on-line recognition, namely the independence from stroke order and stroke number in off-line data, such as scanned handwritten documents, and the useful dynamic information contained in on-line data, such as data captured by a Tablet PC or graphic tablet. Especially on-line handwriting recognition can benefit from a combined recognition approach because off-line images can easily be generated from on-line data.

In a later work, we elaborated the idea into an information-theoretical approach to sensor fusion, identifying the partial sum in (12) with an exponential distribution [46, 47, 48]. In this information-theoretical context, the normalized scores read as follows:

$$s'_k = -E * \ln\left(1 - p(s_k)\right) \tag{13}$$

The function $p(s_k)$ is an exponential distribution with an expectation value E that also appears as a scaler upfront the logarithmic expression. The function $p(s_k)$ thus describes an exponential distribution defining the partial sums in (12). Note that the new normalized scores, which we refer to as "informational confidence," are information defined in the Shannon sense as the negative logarithm of a probability [49]. With the normalized scores being information, sum-rule now becomes the natural combination scheme. For more details on this information-theoretical technique, including practical experiments, we refer readers to the references [46, 47, 48] and to another chapter in this book.

4.3 Dempster-Shafer Theory of Evidence

Among the first more complex approaches for classifier combination was the Dempster-Shafer theory of evidence [50, 51]. As its name already suggests, this theory was developed by Arthur P. Dempster and Glenn Shafer in the sixties and seventies. It was first adopted by researchers in Artificial Intelligence in order to process probabilities in expert systems, but has soon been adopted for other application areas, such as sensor fusion and classifier combination. Dempster-Shafer theory is a generalization of the Bayesian theory of probability and differs in several aspects: First, Dempster-Shafer's theory introduces degrees of belief that do not necessarily meet the mathematical properties of probabilities. Second, it assigns probabilities to sets of possible outcomes rather than single events only. Third, it considers probability intervals that contain the precise probability for sets of possible outcomes. The two main ideas of Dempster-Shafer's theory, as they are usually presented in textbooks, are to obtain degrees of belief for one question from subjective probabilities for a related question, and Dempster's rule for combining such degrees of belief when they are based on independent items of evidence [Glenn Shafer]. Dempster's rule of combination is a generalization of Bayes' rule. It operates on masses assigning probabilities to sets of outcomes. For two sets of masses m_1 and m_2, Dempster's rule defines the joint mass $m_{1,2}(X)$ for an outcome set X as follows:

$$m_{1,2}(X) = \begin{cases} 0 & \text{if} \quad X = \emptyset \\ \frac{1}{1-K} \sum_{A_i \cap B_j = X} m_1(A_i) m_2(B_j) & \text{if} \quad X \neq \emptyset \end{cases} \tag{14}$$

where

$$K = \sum_{A_i \cap B_j = \emptyset} m_1(A_i) m_2(B_j) \tag{15}$$

The Dempster-Shafer approach has produced good results in document processing and its still used today, notwithstanding its higher complexity compared to other approaches [52].

4.4 Behavior Knowledge Space

An equally complex, but perhaps more popular approach is the Behavior-Knowledge Space (BKS) method introduced in [53]. The BKS method is a trainable combination scheme on abstract level, requiring neither measurements nor ordered sets of candidate classes. It tries to estimate the a posteriori probabilities by computing the frequency of each class for every possible set of classifier decisions, based on a given training set. The result is a lookup table that associates the final classification result with each combination of classifier outputs; i.e., each combination of outputs in the lookup table is represented by its most often encountered class label. Given a specific classifier decision $S_1, \ldots, S_N$ from N individual classifiers, the a posteriori probability $\hat{P}(c_i|S_1, \ldots, S_N)$ of class c_i is estimated as follows

$$\hat{P}(c_i|S_1, \ldots, S_N) = \frac{N(c_i|S_1, \ldots, S_N)}{\sum_i N(c_i|S_1, \ldots, S_N)} \tag{16}$$

where $N(c_i|S_1, \ldots, S_N)$ counts the frequency of class c_i for each possible combination of crisp classifier outputs.

In order to provide reasonable performance, the BKS method needs to be trained with large datasets so that meaningful statistics can be computed for each combination in the lookup table. If k is the number of classes and N is the number of combined classifiers, then BKS requires estimates of k^{N+1} a posteriori probabilities. This can pose problems when this number is high but the available set of training patterns is only small.

In addition to the methods presented in this section, several other techniques have been proposed in the recent past [20, 54]. They have gained some importance, but are out of the scope of this chapter.

5 Additional Issues in Classifier Combinations

5.1 The Retraining Effect

The combined classifiers usually produce complementary results and combination algorithms utilize this property. But there is another reason why combinations might produce the performance improvement. The improvement might be the result of the combination algorithm retraining the imperfect matching score outputs of combined classifiers on new training data.

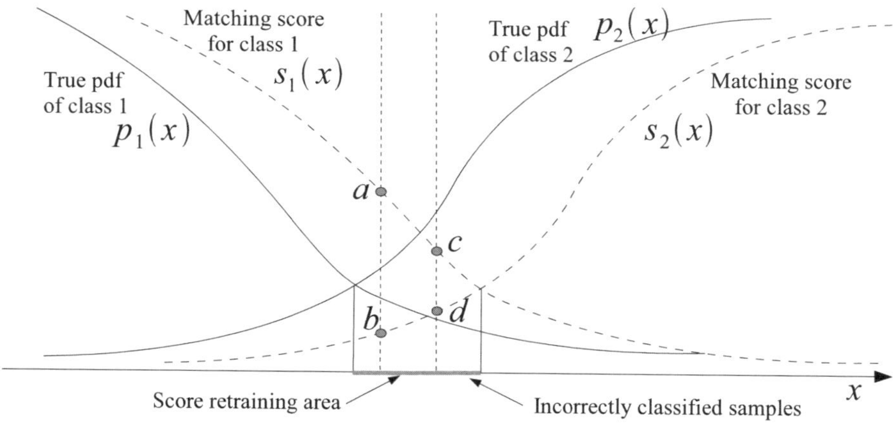

Fig. 4. Using additional training samples in the region of incorrectly classified samples might result in the correction of classification errors

Suppose we have two classes with equal prior probability and equal misclassification cost as in Figure 4. The optimal classification algorithm should make a decision based on whether the probability density function (pdf) of one class is bigger or less than that of the other class: $p_1(x) <> p_2(x)$. Suppose we have one classifier and it outputs a matching scores $s_1(x)$ and $s_2(x)$ for two classes approximating the pdfs of the corresponding classes. The decision of this classifier will be based on its scores: $s_1(x) <> s_2(x)$. The difference between these two decisions produces an area in which classification decisions are different from ideal optimal decisions ("Incorrectly classified samples" area of Figure 4). The classification performance can be improved if the decisions in this area are reversed.

The range of matching scores can be separated into regions of the type $c \leq s_1(x) \leq a$, or $b \leq s_2(x) \leq d$ or the intersections of those ("Score retraining area" of Figure 4). By using an additional set of training samples, the postprocessing algorithm can calculate the frequencies of training samples of each class belonging to these regions, say F_1 and F_2. If the number of training samples falling in these regions is large, then the frequencies will approximate the original class pdfs: $F_1 \sim p_1(x)$ and $F_2 \sim p_2(x)$. Consequently, the scores $s_1(x)$ and $s_2(x)$ could be replaced by new scores corresponding to frequencies F_1 and F_2, respectively. It is quite possible that the classification based on new scores will be correct in previously incorrectly classified regions.

This technique was implemented in [55] for the problem of classifying handwritten digits. The algorithm was able to improve the correctness of a classifier with 97% recognition rate by around 0.5% and of a classifier with 90% recognition rate by around 1.5%. Approximately 30,000 digit images were used for retraining and no information about the original training sets was provided.

This type of score postprocessing can be considered as a classifier combination function f which works only with one classifier and transforms the scores of this classifier s_1, s_2 into the combined scores S_1, S_2:

$$c \leq s_1 \leq a \quad \& \quad b \leq s_2 \leq d \quad \Longrightarrow \quad (S_1, S_2) = f(s_1, s_2) = (F_1, F_2)$$

This combination function is trained on newly supplied training samples and achieves improvement only due to the extraneous training data. Similarly, if we consider traditional combinations utilizing two or more classifiers, then some of the performance improvement might be attributed to using additional training data.

As far as we know, the retraining effect on classifier combinations has not been investigated so far. In particular, it would be interesting to know for each practical application, what part of the improvement is due to the retraining.

5.2 Locality Based Combination Methods

The retraining algorithm of the previous section implied the ability to separate training samples based on their class matching scores. The score for the test sample can be corrected if we are able to find a reference subset of training samples having similar performance. There exist few attempts to explore similar ideas in classifier combinations.

Behavior-knowledge spaces [53] can be considered as one of such locality based methods. The reference set is the subset of the training samples having same ranks assigned by all classifiers as a test sample. The choice of such reference sets might seem not very efficient - their number is large, and even single rank change places training sample in a different reference set. But experimental results seem to confirm the validity of such reference sets. Though the full technique is applicable only in cases with small number of classes and classifiers (two classifiers of ten digit classes in [53]), it can be extended to other cases by grouping together relevant subsets, e.g. subsets having same ranks for first few classes. Note, that rank information implicitly accounts for the dependence between scores assigned to different classes. By incorporating this information into the combination algorithm we effectively perform medium II or high complexity combinations.

Another type of forming reference sets is to use some additional information about classes. In the method of local accuracy estimates [56] the reference set for each test sample is found by applying the k-nearest neighbor algorithm to the original feature space. Subsequently, the strength of each classifier is determined on this reference set, and the classification decision is made by the classifier with highest performance. Clearly, the method could be extended to not only selecting the best performing classifier for the current test sample, but to combine the scores of classifiers according to their strengths. In any case, the important information about the test sample and its relationships with training samples is supplied by the k-nearest neighbor method. In a sense, the k-nearest neighbor method can be viewed as an additional classifier, which is combined with other classifiers. Similar approaches are also investigated in [57, 58].

Opposite to the presentation of the retraining effect in the previous subsection, the methods of finding reference sets discussed here do not rely exclusively on the matching scores of classifiers. Thus, by using additional information these methods can potentially perform better than the generic classifier operating in the score space. But, due to the similarity with retraining methods, it might be that a significant part of the performance improvement is caused by the retraining effect.

5.3 Locality Based Methods with Large Number of Classes

The interesting problem of finding the reference set for the test sample arises in the situations with large or variable number of classes. As we discussed in Subsection 2.6, such problems are not easily solvable with traditional pattern classification algorithms. For example, the recognition of handwritten words might include a large word lexicon, and the training data might simply not include all these words. In biometric person authentication only a small number of training samples, e.g. a single sample, per person is available. How can a reference set of training data be found in these cases?

The string edit distance might provide a valuable neighborhood information for the handwritten word application in order to find a reference set. In a more complicated approach we introduced the concept of local lexicon density in [59]. The lexicon density is determined not only by the lexicon itself, but also by the strength of the word recognizer using this lexicon. In addition to finding a reference set, the presented algorithm could be used to estimate the strength of each recognizer in the neighborhood of the test sample. Though the method of estimating lexicon density seems to have good performance, it has not been applied to problem of combining word recognizers yet.

The problem of finding a reference set for biometric person authentication has been investigated before in speaker identification research [60, 61]. The set of persons having similar biometrics to the queried person (cohort) is found using training samples of all enrolled persons. During testing, a matching score for the query person is judged against the matching scores of the cohort persons. These research was mostly used for making decisions on the matching scores, and not in combining different matchers. Some research has appeared recently [15, 16] trying to find user specific combinations of biometric matchers. But these methods use locality information only implicitly by estimating performance on the whole set of enrolled persons, and not on the cohort set. As far as we know, the cohort method has not been applied to classifier combinations so far.

The local neighborhood in cohorts is determined by the matching distances among enrolled templates. These distances can be used in constructing so called 'background models' [61]. In our research we used the set of matching scores between input template and enrolled templates to construct so called 'identification models' [18]. Both models can be used for finding reference sets in classifier combination algorithms. In addition, both models can be user-generic or user-specific, and both models can be used in a single combination method.

6 Conclusion

This chapter presented an overview of classifier combination methods. We categorized these methods according to complexity type, output type, ensemble vs non-ensemble combinations, etc. We also tried to define the classifier combination field by specifying cases, e.g. the presence of a large number of classes, which can not be readily solved by traditional pattern classification algorithms. We briefly presented main research directions in ensemble classifier combinations (Section 3) and non-ensemble classifier combinations (Section 4). In the last section, we discussed the retraining effect and issues of locality in classifier combinations. This section also presented some potential new research topics.

Overall, our goal was to show the current state of the art in classifier combination. Though significant progress has been made so far in devising new combination methods, the research is still mostly limited to the low complexity type of combinations. Exploring other complexity types of combinations and understanding locality properties of classifiers can be a fertile ground for future research.

References

1. Cooke, R.: Experts in Uncertainty: Opinion and Subjective Probability in Science. Oxford University Press (1991)
2. Clemen, R., Winkler, R.: Combining probability distributions from experts in risk analysis. Risk Analysis **19** (1999) 187–203
3. Xu, L., Krzyzak, A., Suen, C.Y.: Methods for combining multiple classifiers and their applications to handwriting recognition. IEEE transactions on System, Man, and Cybernetics **23**(3) (1992) 418–435
4. Gader, P., Mohamed, M., Keller, J.: Fusion of Handwritten Word Classifiers. Pattern Recognition Letters **17** (1996) 577–584
5. Sirlantzis, K., Hoque, S., Fairhurst, M.C.: Trainable Multiple Classifier Schemes for Handwritten Character Recognition. In: 3rd International Workshop on Multiple Classifier Systems (MCS), Cagliari, Italy, Lecture Notes in Computer Science, Springer-Verlag (2002) 169–178
6. Wang, W., Brakensiek, A., Rigoll, G.: Combination of Multiple Classifiers for Handwritten Word Recognition. In: Proc. of the 8th International Workshop on Frontiers in Handwriting Recognition (IWFHR-8), Niagara-on-the-Lake, Canada (2002) 117–122
7. Lee, D.S.: Theory of Classifier Combination: The Neural Network Approach. Ph.D Thesis, SUNY at Buffalo (1995)
8. Bertolami, R., Bunke, H.: Early feature stream integration versus decision level combination in a multiple classifier system for text line recognition. In: Pattern Recognition, 2006. ICPR 2006. 18th International Conference on. Volume 2. (2006) 845–848
9. Favata, J.: Character model word recognition. In: Fifth International Workshop on Frontiers in Handwriting Recognition, Essex, England (1996) 437–440
10. Ho, T.K., Hull, J.J., Srihari, S.N.: Decision combination in multiple classifier systems. IEEE Trans. on Pattern Analysis and Machine Intelligence **16**(1) (1994) 66–75
11. Erp, M.V., Vuurpijl, L.G., Schomaker, L.: An Overview and Comparison of Voting Methods for Pattern Recognition. In: Proc. of the 8th International Workshop on Frontiers in Handwriting Recognition (IWFHR-8), Niagara-on-the-Lake, Canada (2002) 195–200
12. Kang, H.J., Kim, J.: A Probabilistic Framework for Combining Multiple Classifiers at Abstract Level. In: Fourth International Conference on Document Analysis and Recognition (ICDAR), Ulm, Germany (1997) 870–874
13. Kittler, J., Hatef, M., Duin, R., Matas, J.: On combining classifiers. IEEE Trans. on Pattern Analysis and Machine Intelligence (1998) 226–239
14. Tulyakov, S., Govindaraju, V.: Classifier combination types for biometric applications. In: 2006 IEEE Computer Society Conference on Computer Vision and Pattern Recognition (CVPR 2006), Workshop on Biometrics, New York, USA (2006)
15. Jain, A., Ross, A.: Learning user-specific parameters in a multibiometric system. In: International Conference on Image Processing. 2002. Volume 1. (2002) I–57–I–60 vol.1
16. Fierrez-Aguilar, J., Garcia-Romero, D., Ortega-Garcia, J., Gonzalez-Rodriguez, J.: Bayesian adaptation for user-dependent multimodal biometric authentication. Pattern Recognition **38**(8) (2005) 1317–1319

17. Auckenthaler, R., Carey, M., Lloyd-Thomas, H.: Score normalization for text-independent speaker verification systems. Digital Signal Processing **10**(1-3) (2000) 42–54
18. Tulyakov, S., Govindaraju, V.: Identification model for classifier combinations. In: Biometrics Consortium Conference, Baltimore, MD (2006)
19. Last, M., Bunke, H., Kandel, A.: A feature-based serial approach to classifier combination. Pattern Analysis and Applications **5**(4) (2002) 385–398
20. Ho, T.K.: The random subspace method for constructing decision forests. Pattern Analysis and Machine Intelligence, IEEE Transactions on **20**(8) (1998) 832–844
21. Kleinberg, E.M.: On the algorithmic implementation of stochastic discrimination. IEEE Transactions on Pattern Analysis and Machine Intelligence **22**(5) (2000) 473–490
22. Breiman, L.: Bagging predictors. Machine Learning **24**(2) (1996) 123–140
23. Schapire, R.: The strength of weak learnability. Machine Learning **5** (1990) 197–227
24. Freund, Y., Schapire, R.: Experiments with a New Boosting Algorithm. In: Proc. of 13th Int. Conf. on Machine Learning, Bari, Italy (1996) 148–156
25. Wolpert, D.H.: Stacked generalization. Neural Networks **5**(2) (1992) 241–260
26. Vapnik, V.: The Nature of Statistical Learning Theory. Springer-Verlag, New York (1995)
27. Vapnik, V.: Statistical Learning Theory. Wiley, New York (1998)
28. Schapire, R., Freund, Y., Bartlett, P., Lee, W.S.: Boosting the margin: A new explanation for the effectiveness of voting methods. The Annals of Statistics **26**(5) (1998) 1651–1686
29. Kleinberg, E.M.: Stochastic discrimination. Annals of Mathematics and Artificial Intelligence **1** (1990)
30. Kong, E., Dietterich, T.: Error-correcting output coding corrects bias and variance. In: 12th International Conference on Machine Learning. (1995) 313–321
31. Tibshirani, R.: Bias, variance and prediction error for classification rules. Technical Report 41, Department of Statistics, University of Toronto (1996)
32. Tumer, K., Ghosh, J.: Linear and order statistics combiners for pattern classification. In Sharkey, A.J., ed.: Combining Artificial Neural Nets: Ensembles and Multi-Net Systems. Springer-Verlag, London (1999) 127–162
33. Kuncheva, L.: A theoretical study on six classifier fusion strategies. Pattern Analysis and Machine Intelligence, IEEE Transactions on **24**(2) (2002) 281–286
34. Kuncheva, L.I.: Combining Pattern Classifiers: Methods and Algorithms. Wiley InterScience (2004)
35. Fumera, G., Roli, F.: Performance analysis and comparison of linear combiners for classifier fusion. In: SSPR/SPR. (2002) 424–432
36. Fumera, G., Roli, F.: Analysis of error-reject trade-off in linearly combined multiple classifiers. Pattern Recognition **37**(6) (2004) 1245–1265
37. Alkoot, F.M., Kittler, J.: Experimental evaluation of expert fusion strategies. Pattern Recognition Letters **20**(11-13) (1999) 1361–1369
38. Freund, Y., Schapire, R.: A Short Introduction to Boosting. Journal of Japanese Society for Artificial Intelligence **14**(5) (1999) 771–780
39. Guenter, S., Bunke, H.: New Boosting Algorithms for Classification Problems with Large Number of Classes Applied to a Handwritten Word Recognition Task. In: 4th International Workshop on Multiple Classifier Systems (MCS),

Guildford, UK, Lecture Notes in Computer Science, Springer-Verlag (2003) 326–335
40. Huber, P.J.: Robust Statistics. Wiley, New York (1981)
41. Jain, A., Nandakumar, K., Ross, A.: Score Normalization in Multimodal Biometric Systems. Pattern Recognition **38**(12) (2005) 2270–2285
42. Cappelli, R., Maio, D., Maltoni, D.: Combining Fingerprint Classifiers. In: First Int. Workshop on Multiple Classifier Systems. (2000) 351–361
43. Hampel, F.R., Rousseeuw, P.J., Ronchetti, E., Stahel, W.: Robust Statistics: The Approach Based on Influence Functions. Wiley, New York (1986)
44. Velek, O., Jaeger, S., Nakagawa, M.: A New Warping Technique for Normalizing Likelihood of Multiple Classifiers and its Effectiveness in Combined On-Line/Off-Line Japanese Character Recognition. In: 8th International Workshop on Frontiers in Handwriting Recognition (IWFHR), Niagara-on-the-Lake, Canada (2002) 177–182
45. Velek, O., Jaeger, S., Nakagawa, M.: Accumulated-Recognition-Rate Normalization for Combining Multiple On/Off-line Japanese Character Classifiers Tested on a Large Database. In: 4th International Workshop on Multiple Classifier Systems (MCS), Guildford, UK, Lecture Notes in Computer Science, Springer-Verlag (2003) 196–205
46. Jaeger, S.: Informational Classifier Fusion. In: Proc. of the 17th Int. Conf. on Pattern Recognition, Cambridge, UK (2004) 216–219
47. Jaeger, S.: Using Informational Confidence Values for Classifier Combination: An Experiment with Combined On-Line/Off-Line Japanese Character Recognition. In: Proc. of the 9th Int. Workshop on Frontiers in Handwriting Recognition, Tokyo, Japan (2004) 87–92
48. Jaeger, S., Ma, H., Doermann, D.: Identifying Script on Word-Level with Informational Confidence. In: Int. Conf. on Document Analysis and Recognition (ICDAR), Seoul, Korea (2005) 416–420
49. Shannon, C.E.: A Mathematical Theory of Communication. Bell System Tech. J. **27**(623-656) (1948) 379–423
50. Dempster, A.P.: A Generalization of Bayesian Inference. Journal of the Royal Statistical Society **30** (1968) 205–247
51. Shafer, G.: A Mathematical Theory of Evidence. Princeton University Press (1976)
52. Mandler, E., Schuermann, J.: Combining the Classification Results of Independent Classifiers Based on the Dempster/Shafer Theory of Evidence. In E.S. Gelsema, L.K., ed.: Pattern Recognition and Artificial Intelligence. (1988) 381–393
53. Huang, Y., Suen, C.: A Method of Combining Multiple Experts for Recognition of Unconstrained Handwritten Numerals. IEEE Trans. on Pattern Analysis and Machine Intelligence (TPAMI) **17**(1) (1995) 90–94
54. Ianakiev, K., Govindaraju, V.: Architecture for Classifier Combination Using Entropy Measures. In: 1st International Workshop on Multiple Classifier Systems (MCS), Cagliari, Italy, Lecture Notes in Computer Science, Springer-Verlag (2000) 340–350
55. Ianakiev, K., Govindaraju, V.: Deriving pseudo-probabilities of correctness given scores. In Chen, D., Cheng, X., eds.: Pattern Recognition and String Matching,. Kluwer Publishers (2002)

56. Woods, K., Kegelmeyer, W.P., J., Bowyer, K.: Combination of multiple classifiers using local accuracy estimates. Pattern Analysis and Machine Intelligence, IEEE Transactions on **19**(4) (1997) 405–410
57. Giacinto, G., Roli, F.: Dynamic classifier selection based on multiple classifier behaviour. Pattern Recognition **34**(9) (2001) 1879–1881 TY - JOUR.
58. Kuncheva, L.: Clustering-and-selection model for classifier combination. In: Knowledge-Based Intelligent Engineering Systems and Allied Technologies, 2000. Proceedings. Fourth International Conference on. Volume 1. (2000) 185–188 vol. 1
59. Govindaraju, V., Slavik, P., Xue, H.: Use of lexicon density in evaluating word recognizers. Pattern Analysis and Machine Intelligence, IEEE Transactions on **24**(6) (2002) 789–800
60. Colombi, J., Reider, J., Campbell, J.: Allowing good impostors to test. In: Signals, Systems & Computers, 1997. Conference Record of the Thirty-First Asilomar Conference on. Volume 1. (1997) 296–300 vol. 1
61. Rosenberg, A., Parthasarathy, S.: Speaker background models for connected digit password speaker verification. In: Acoustics, Speech, and Signal Processing, 1996. ICASSP-96. Conference Proceedings., 1996 IEEE International Conference on. Volume 1. (1996) 81–84 vol. 1

Machine Learning for Signature Verification

Sargur N. Srihari[1], Harish Srinivasan[1], Siyuan Chen[1], and Matthew J. Beal[2]

[1] Center of Excellence for Document Analysis and Recognition (CEDAR), Buffalo NY {srihari@cedar,hs32@cedar,schen7@cedar,mbeal@cse}.buffalo.edu
[2] Department of Computer Science and Engineering, University at Buffalo, The State University of New York, Buffalo NY, USA

Summary. *Signature verification is a common task in forensic document analysis. It's aim is to determine whether a questioned signature matches known signature samples. From the viewpoint of automating the task it can be viewed as one that involves machine learning from a population of signatures. There are two types of learning tasks to be accomplished: person-independent (or general) learning and person-dependent (or special) learning. General learning is from a population of genuine and forged signatures of several individuals, where the differences between genuines and forgeries across all individuals are learnt. The general learning model allows a questioned signature to be compared to a single genuine signature. In special learning, a person's signature is learnt from multiple samples of only that person's signature - where within-person similarities are learnt. When a sufficient number of samples are available, special learning performs better than general learning (5% higher accuracy). With special learning, verification accuracy increases with the number of samples. An interactive software implementation of signature verification involving both the learning and performance phases is described.*

1 Introduction

The most common task in the field of forensic document analysis [1, 2, 3, 4, 5] is that of authenticating signatures. The problem most frequently brought to a document examiner is the question relating to the authenticity of a signature: *Does this questioned signature (Q) match the known, true signatures (K) of this subject?* [6] A forensic document examiner - also known as a questioned document (QD) examiner - uses years of training in examining signatures in making a decision in forensic case work.

The training of a document examiner involves years of learning from signatures that are both genuine and forged. In case-work, exemplars are usually only available for genuine signatures of a particular individual, from which the characteristics of the genuine signature are learnt.

S.N. Srihari et al.: *Machine Learning for Signature Verification*, Studies in Computational Intelligence (SCI) **90**, 387–408 (2008)
www.springerlink.com

Automating the task of signature verification is the subject of this chapter. It is naturally formulated as a machine learning task. A program is said to exhibit machine learning capability in performing a task if it is able to learn from exemplars, improve as the number of exemplars increase, etc. [7]. The performance task of signature verification is one of determining whether a questioned signature is genuine or not. An example of a case where a questioned signature sample is to be matched against multiple known samples of a particular writer is shown in Fig. 1.

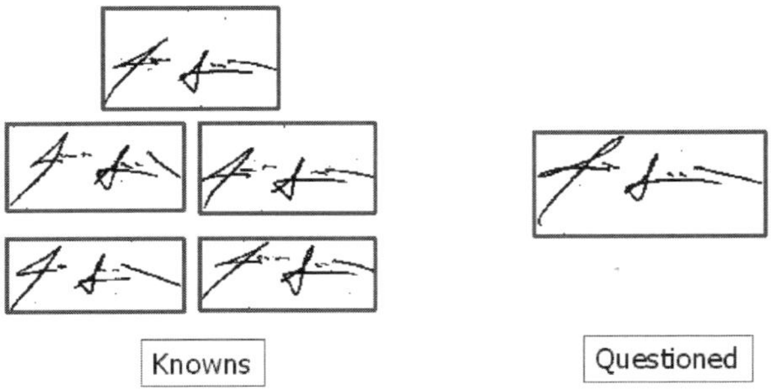

Fig. 1. Signature verification where a questioned signature (right) is matched against four knowns

Paralleling the learning tasks of the human questioned document examiner, the machine learning tasks can be stated as *general learning* (which is person-independent) or *special learning* (which is person-dependent) [8]. In the case of general learning the goal is to learn from a large population of genuine and forged signature samples. The focus is on differentiating between genuine-genuine differences and genuine-forgery differences. The learning problem is stated as learning a two-class classification problem where the input consists of the difference between a pair of signatures. The verification task is performed by comparing the questioned signature against each known signature. The general learning problem can be viewed as one where learning takes place with near misses as counter-examples [9].

Special learning focuses on learning from genuine samples of a particular person. The focus is on learning the differences between members of the class of genuines. The verification task is essentially a one-class problem of determining whether the questioned signature belongs to that class or not.

There is scattered literature on automatic methods of signature verification [10, 11, 12, 13, 14]. Automatic methods of writer verification - which is the task of determining whether a sample of handwriting, not necessarily a signature, was written by a given individual - are also relevant [15]. Identification is the task of determining as to who among a given set of individuals might have written the questioned writing. The handwriting verification and identification tasks parallel those of biometric verification and identification for which there is a large literature. The use of a machine learning paradigm for biometrics has been proposed recently [16].

The rest of this chapter is organized as follows. Section 2 describes the processes of computing the features of a signature and matching the features of two signatures. Section 3 describes the two methods of learning. Section 4 deals with how the learnt knowledge is used in evaluating a questioned signature (called the performance task). A comparison of the accuracies of the two strategies on a database of genuines and forgeries is described in Section 5. Section 6 describes an interactive software implementation of the methods described. Section 7 is a chapter summary.

2 Feature Extraction and Similarity Computation

Signatures are relied upon for identification due to the fact that each person develops unique habits of pen movement which serve to represent his or her signature. Thus at the heart of any automatic signature verification system are two algorithms: one for extracting features and the other for determining the similarities of two signatures based on the features. Features are elements that capture the uniqueness. In the QD literature such elements are termed *discriminating elements* or *elements of comparison*. A given person's samples can have a (possibly variable) number of elements and the combination of elements have greater discriminating power.

A human document examiner uses a chart of elemental characteristics [6]. Such elements are ticks, smoothness of curves, smoothness of pressure changes, placement, expansion and spacing, top of writing, base of writing, angulation/slant, overall pressure, pressure change patterns, gross forms, variations, connective forms and micro-forms. The elemental characteristics such as speed, proportion, pressure and design are used to determine higher level characteristics such as rhythm, form and balance.

Automatic signature verification methods described in the literature use an entirely different set of features. Some are based on image texture such as wavelets while others focus on geometry and topology of the signature image. Types of features used for signature verification are wavelet descriptors [17], projection distribution functions [18, 14, 19], extended shadow code [18] and geometric features [20].

2.1 GSC Features

A quasi-multiresolution approach for features are the Gradient, Structural and Concavity, or GSC, features [21, 22]. Gradient features measure the local scale characteristics of image, structural features measure the intermediate scale ones, and concavity can measure the characteristics over the scale of whole image. Following this philosophy, three types of feature maps are drawn and the corresponding local histograms of each cell is quantized into binary features. Fig. 2(a) shows an example of a signature, which has a 4x8 grid imposed on it for extracting GSC features; rows and columns of the grid are drawn based on the black pixel distributions along the horizontal and vertical directions. A large number of binary features have been extracted from these grids, as shown in Fig. 2(b), which are *global word shape features* [23]; there are 1024 bits which are obtained by concatenating 384 gradient bits, 384 structural bits and 256 concavity bits.

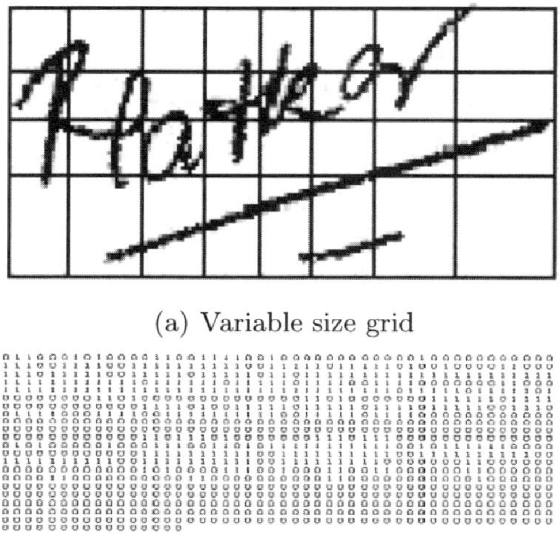

(a) Variable size grid

(b) 1024-bit binary feature vector

Fig. 2. Signature feature computation using a grid: (a) variable size 4x8 grid, and (b) binary feature vector representing gradient, structural and concavity features

A similarity or distance measure is used to compute a score that signifies the strength of match between two signatures. The similarity measure converts the pairwise data from feature space to *distance* space.

Several similarity measures can be used with binary vectors, including the well-known Hamming distance. Much experimentation with binary-valued GSC features, has led to the *correlation* measure of distance as yielding the best accuracy in matching handwriting shapes [24]. It is defined as follows. Let S_{ij} $(i,j \in \{0,1\})$ be the number of occurrences of matches with i in the first vector and j in the second vector at the corresponding positions, the dissimilarity D between the two feature vectors X and Y is given by the formula:

$$D(X,Y) = \frac{1}{2} - \frac{S_{11}S_{00} - S_{10}S_{01}}{2\sqrt{(S_{10}+S_{11})(S_{01}+S_{00})(S_{11}+S_{01})(S_{00}+S_{10})}}$$

It can be observed that the range of $D(X,Y)$ has been normalized to $[0,1]$. That is, when $X = Y$, $D(X,Y) = 0$, and when they are completely different, $D(X,Y) = 1$.

A refined method to compute the features and obtain the distance values is discussed next.

2.2 GSC Features using Flexible Templates

The GSC features which are based on non-overlapping rectangular cells placed on the signature can be improved upon by considering the specific pair of signatures being compared.

2.2.1 Image Grid for Signature Matching

The GSC binary feature vector depends upon the grid used to partition the image– one of which is shown in Fig. 2(a). The simplest one with equal cell size can be used for character recognition. A more complex linear grid is obtained from two one-dimensional projection histograms obtained along horizontal and vertical directions so that the foreground pixel masses of cells are equal both column- and row- wise. To achieve good alignment between signature images a non-linear sampling grid should be applied. The problem of searching a specific geometric transformation or mapping function to match two closely related images is a classic image registration problem. Thus the generation of non-linear grid consists of two stages: point mapping and transformation searching, which is also the general solution of matching problem without prior knowledge of registration. In the first stage, the extrema of strokes are labeled as the landmarks of signatures, then an appropriate graph matching algorithm is applied to match these two point sets. Based on the registration information obtained from the first stage a geometric transformation which minimizes a specific cost funtion can be found. The non-linear grid is naturally the projection of the reference grid by the mapping function.

2.2.2 Landmark Mapping

The extrema along the strokes where the curvature of contours has local maximum record representative information of personal writing. After labeling the extremas along the contours as landmarks of images the Scott and Longuet-Higgins algorithm [25] can be used to match the landmark sets of two images. In this algorithm the two point sets are placed at the same surface and a proximity matrix G is constructed to indicate their spatial relations, i.e., $G = exp(-r_{ij}^2/\sigma^2)$, where r_{ij} is the Euclidean distance between points i and j. The pairing matrix P is then constructed to approximate a permutation matrix which corresponds to a matching function. Using Scott and Longuet-Higgins algorithm point-to-point matching can be constructed between two point sets while igoring some bad matches (Fig. 3).

Fig. 3. Extremas matching between two signature contour images

2.3 Cell Alignment by Spline Mapping

With registration information from the previous stage, a geometric transformation can be constructed to map the landmarks of the reference image– also called control points– to correspondence in the test image. Thin-plate spline warping is a powerful spatial transformation to achieve this goal. An imaginary infinite thin steel plate is constrained to lie over the displacement points and the spline is the superposition of eigenvectors of the bending energy matrix. The spline algebra tries to express the deformation by minimizing the physical bending energy over the flat surface. The resulting plate surface is differentiable everywhere. The spline function

$$f(x,y) = a_1 + a_x x + a_y y + \sum_{i=1}^{n} W_i U\left(|P_i - (x,y)|\right)$$

along x and y coordinates can map every point in the reference image into one in the test image. Please see [26] for details.

Thus from the reference grid we can generate the corresponding grid using the mapping function obtained. The aligned cell is defined as the region whose boundaries are the generated grids (Fig. 4).

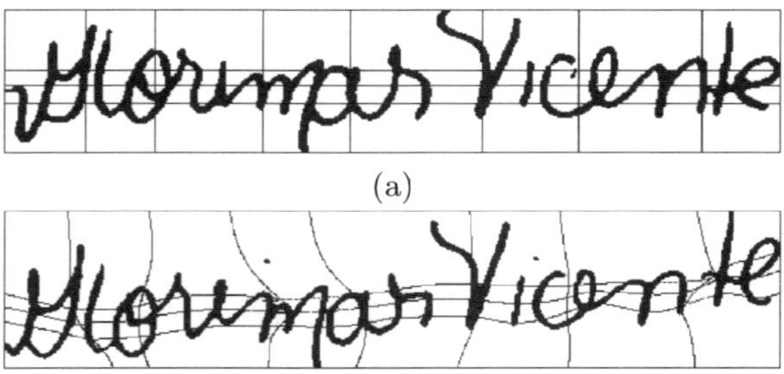

(a)

(b)

Fig. 4. Signature grids: (a) using horizontal and vertical pixel histograms, the image is divided by a 4×8 grid, (b) mapping grid obtained by spline warping

3 Learning Strategies

Person-independent or general learning is a one-step approach that learns from a large population of genuine and forged samples. On the other hand person-dependent(person specific) learning focuses on learning from the genuine samples of a specific individual.

3.1 Person-Independent (General) Learning

The general learning approach uses two sets of signature pairs: genuine-genuine and genuine-forgery. Features are extracted for these samples and a similarity measure is used to compute the distance between each pair. Let $\mathbf{D_S}$ denote the vector of distances between all pairs in set *one*, which represents the distribution of distances when samples truly came from the same person. Similarly let $\mathbf{D_D}$ denote the vector of distances between all pairs in set *two*, which represents the distribution of distances when samples truly came from different persons. These distributions can be modeled using known distributions such as Gaussian or gamma. Fig. 5 shows typical distribution curves obtained when distances are modeled using a gamma distribution. The Gaussian assigns non-zero probabilities to negative values of distance although such values are never encountered. Since this problem in not there with the Gamma, it is to be preferred. The probability density function of the gamma distributions is as follows:

$$Gamma(x) = \frac{x^{\alpha-1}\exp^{(-x/\beta)}}{(\Gamma(\alpha))\beta^{\alpha}}$$

Here α and β are gamma parameters which can be evaluated from the mean and variance as follows $\alpha = \mu^2/\sigma^2$ and $\beta = \sigma^2/\mu$. 'α' is called the shape parameter and 'β' is the scale parameter. The parameters that need to be learnt for such a model are typically derived from the sufficient statistics of the distribution, and are namely μ (mean) and σ (variance) for a Gaussian, or α (shape) and β (width) for a gamma. These distributions are referred to as genuine-genuine and genuine-impostor distributions in the domain of biometrics.

3.2 Person-Dependent Learning (Person Specific Learning)

In questioned document case work there are typically multiple genuine signatures available. They can be used to learn the variation across them - so as to determine whether the questioned signature is within the range of variation. First, pairs of known samples are compared using a similarity measure to obtain a distribution over distances between features of samples - this represents the distribution of the variation/similarities amongst samples - for the individual. The corresponding classification method involves comparing the

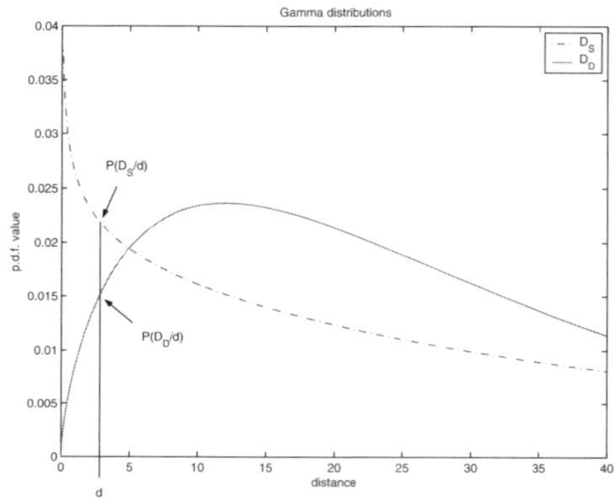

Fig. 5. Parametric models of two distributions: genuine-genuine distances (D_S), and genuine-forgery distances (D_D). In both cases the best fits with gamma distributions are illustrated. Values of the gamma probability density functions at d are shown

questioned sample against all available known samples to obtain another distribution in distance space. The Kolmogorov-Smirnov test, KL-divergence and other information-theoretic methods can be used to obtain a probability of similarity of the two distributions, which is the probability of the questioned sample belonging to the ensemble of knowns. These methods are discussed below.

3.2.1 Within-Person Distribution

If a given person has N samples, $\binom{N}{2}$ defined as $\frac{N!}{N!(N-r)!}$ pairs of samples can be compared as shown in Figure 6. In each comparison, the distance between the features is computed. This calculation maps feature space to distance space. The result of all $\binom{N}{2}$ comparisons is a $\{\binom{N}{2}\times 1\}$ distance vector. This vector is the distribution in distance space for a given person. For example, in the signature verification problem this vector is the distribution in distance space for the ensemble of genuine known signatures for that writer. A key advantage of mapping from feature space to distance space is that the number of data points in the distribution is $\binom{N}{2}$ as compared to N for a distribution in feature space alone. Also the calculation of the distance between every pair of samples gives a measure of the variation in samples for that writer. In essence the distribution in distance space for a given known person captures the similarities and variation amongst the samples for that person. Let N be the total number of samples and $N_{WD} = \binom{N}{2}$ be the total number of comparisons that can be made which also equals the length of the within-person distribution vector. The within-person distribution can be written as

$$\mathbf{D_W} = (d_1, d_2, \ldots, d_{N_{WD}})^\top \tag{1}$$

where $\top$ denotes the transpose operation and d_j is the distance between the pair of samples taken at the jth comparison, $j \in \{1, \ldots, N_{WD}\}$.

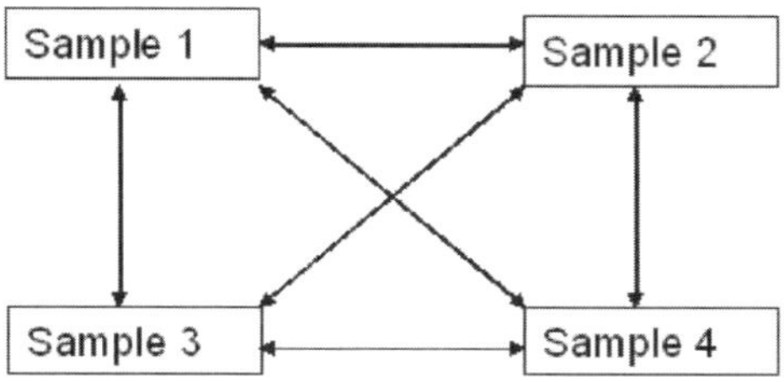

Fig. 6. Person-dependent (special) learning involves comparing all possible genuine-genuine pairs, as shown for four genuine samples, to form the vector $\mathbf{D_W}$, which in this example is of length $N_{WD} = \binom{4}{2} = 6$ comparisons

4 Performance Task

The performance task of signature verification is to answer the question whether or not a questioned signature belongs to the genuine signature set. The person-independent method uses knowledge from a general population to determine whether two samples, one a questioned and the other a genuine, belong to the same person. This task is called 1:1 verification. Person-dependent classification tasks involves matching one questioned sample against multiple known samples from the person. Details of the two performance algorithms are given below.

4.1 Person-Independent Classification

The process of 1:1 verification(one input sample compared with one known sample) starts with feature extraction and then computing the distance d between the features using a similarity measure. From the learning described in Section 3.1, the likelihood ratio defined as $\frac{P(D_S|d)}{P(D_D|d)}$ can be calculated, where $P(D_S|d)$ is the probability density function value under the D_S distribution at the distance d and $P(D_D|d)$ is the probability density function value under the D_D distribution at the distance d. If the likelihood ratio is greater than 1, then the classification answer is that the two samples do belong the same person and if the ratio is less than 1, they belong to different persons. Figure 5 shows how the likelihood ratio is obtained. If there are a total of N known samples from a person, then for one questioned sample N, $1:1$ verifications can be performed and the likelihood ratios are multiplied. In these circumstances it is convenient to deal with log-likelihood-ratios rather than with just likelihood ratios. The log-likelihood-ratio (LLR) is given by $\log P(D_S|d) - \log P(D_D|d)$. The decision of same-person is favored if $\log P(D_S|d) - \log P(D_D|d) > 0$, and the decision of different-person chosen if $\log P(D_S|d) - \log P(D_D|d) < 0$. When N of these $1:1$ verifications are performed these LLR's are summed and then the decision is taken.

4.2 Person-Dependent Classification

When multiple genuines are available then the within-person distribution is obtained in accordance with equation 1. A questioned can be compared against the ensemble of knowns for verification. The classification process consists of two steps.

(i) obtaining questioned *vs* known distribution; and

(ii) comparison of two distributions: questioned *vs* known distribution and within-person distribution.

Questioned vs *Known Distribution*

Using Equation (1) the within-person distribution is obtained by comparing every possible pair of samples from within the given person's samples. Analogous to this, the questioned sample can be compared with every one of the N knowns in a similar way to obtain the questioned vs known distribution. The questioned vs known distribution is given by

$$D_{QK} = (d_1, d_2, \ldots, d_N)^\top , \tag{2}$$

where d_j is the distance between the questioned sample and the jth known sample, $j \in \{1, \ldots, N\}$.

Comparing Distributions

Once the two distributions are obtained, namely the within-person distribution, denoted D_w (Equation (1)), and the Questioned vs Known distribution, D_{QK} (Equation (2)), the task now is to compare the two distributions to obtain a probability of similarity. The intuition is that if the questioned sample did indeed belong to the ensemble of the knowns, then the two distributions must be the same (to within some sampling noise). There are various ways of comparing two distributions and these are described in the following sections.

Kolmogorov-Smirnov Test

The Kolmogorov-Smirnov (KS) test can be applied to obtain a probability of similarity between two distributions. The KS test is applicable to unbinned distributions that are functions of a single independent variable, that is, to data sets where each data point can be associated with a single number [27].

The test first obtains the cumulative distribution function of each of the two distributions to be compared, and then computes the statistic, D, which is a particularly simple measure: it is defined as the maximum value of the absolute difference between the two cumulative distribution functions. Therefore, if comparing two different cumulative distribution functions $S_{N1}(x)$ and $S_{N2}(x)$, the KS statistic D is given by $D = \max_{-\infty<x<\infty} |S_{N1}(x) - S_{N2}(x)|$. The statistic D is then mapped to a probability of similarity, P, according to equation 3

$$P_{KS} = Q_{KS}\left(\sqrt{N_e} + 0.12 + (0.11/\sqrt{N_e})D\right), \tag{3}$$

where the $Q_{KS}(\cdot)$ function is given by (see [27] for details):

$$Q_{KS}(\lambda) = 2\sum_{j=1}^{\infty}(-1)^{j-1}e^{-2j^2\lambda^2}, \qquad \text{such that}: \; Q_{KS}(0) = 1, \; Q_{KS}(\infty) = 0, \tag{4}$$

and N_e is the effective number of data points, $N_e = N_1 N_2 (N_1 + N_2)^{-1}$, where N_1 is the number of data points in the first distribution and N_2 the number in the second. In the rest of this section, other methods of comparing two distributions are discussed.

Kullback-Leibler Divergence and other methods

The Kullback-Leibler (KL) divergence is a measure that can be used to compare two binned distributions. The KL divergence measure between two distributions is measured in bits or nats. An information theoretic interpretation is that it represents the average number of bits that are wasted by encoding events from a distribution P with a code which is optimal for a distribution Q (i.e. using codewords of length $-\log q_i$ instead of $-\log p_i$). Jensen's inequality can be used to show that $D_{KL} = KL(P\|Q) \geq 0$ for all probability distributions P and Q, and $D_{KL} = KL(P\|Q) = 0$ iff $P = Q$. Strictly speaking, the KL measure is a *divergence* between distributions and not a distance, since it is neither symmetric nor satisfies the triangle inequality). The KL divergence so obtained can be converted to represent a probability by $\exp(-\zeta D_{KL})$ (for the sake of simplicity we set $\zeta = 1$ in this article). If the divergence D_{KL} is 0, then the probability is 1 signifying that the two distributions are the same. In order to use this method and other methods discussed in the following sections it is first necessary to convert the two unbinned distributions to binned distributions with a probability associated with each bin. The KL divergence between two distributions is given in Equation 5 below, where B is the total number of bins, P_b and Q_b are the probabilities of the bth bin of two distributions respectively. P_{KL} denotes the probability that the two distributions are the same. Other related measures between distributions P and Q that we will examine are given in Equations (7), (9) and (11)

$$\text{Kullback-Leibler:} D_{KL} = \sum_{b=1}^{B} P_b \log(\frac{P_b}{Q_b}) \tag{5}$$

$$P_{KL} = e^{-\zeta D_{KL}} \tag{6}$$

$$\text{Reverse KL:} D_{RKL} = KL(Q\|P) = \sum_{b=1}^{B} Q_b \log(\frac{Q_b}{P_b}) \tag{7}$$

$$P_{RKL} = e^{-\zeta D_{RKL}} \tag{8}$$

$$\text{Symmetric KL:} D_{HKL} = \frac{1}{2}KL\left(P\|Q\right) + \frac{1}{2}KL\left(Q\|P\right) = \frac{D_{KL} + D_{RKL}}{2} \tag{9}$$

$$P_{HKL} = e^{-\zeta D_{HKL}} \tag{10}$$

$$\text{Jensen-Shannon KL:} D_{JS} = \frac{1}{2}KL\left(P\middle\|\frac{P+Q}{2}\right) + \frac{1}{2}KL\left(Q\middle\|\frac{P+Q}{2}\right) \tag{11}$$

$$P_{JS} = e^{-\zeta D_{JS}} \tag{12}$$

Combined KL and KS measure

A combination of the Kolmogorov-Smirnov and Kullback-Leibler measure, denoted KLKS, has been found to outperform the individual measures as will be analyzed in the performance evaluation section following this. The method to combine is very simple and is obtained by averaging the probabilities defined in equations (3) and (5).

$$P_{KLKS} = \frac{P_{KL} + P_{KS}}{2} \tag{13}$$

5 Performance Evaluation

5.1 Experiments

A database of off-line signatures was prepared as a test-bed [13]. Each of 55 individuals contributed 24 signatures thereby creating 1320 genuine signatures. Some were asked to forge three other writers' signatures, eight times per subject, thus creating 1320 forgeries. Few samples from genuine signatures are shown in Figure 7. Ten examples of genuines of one subject (subject no. 21) and ten forgeries of that subject are shown in Figure 8. Each signature was scanned at 300 dpi gray-scale and binarized using a gray-scale histogram. Salt pepper noise removal and slant normalization were two steps involved in image preprocessing. The database had 24 genuines and 24 forgeries available for each writer as in Figure 8. For each test case a writer was chosen and N genuine samples of that writer's signature were used for learning. The remaining $24 - N$ genuine samples were used for testing. Also 24 forged signatures of this writer were used for testing. Two different error types can be defined

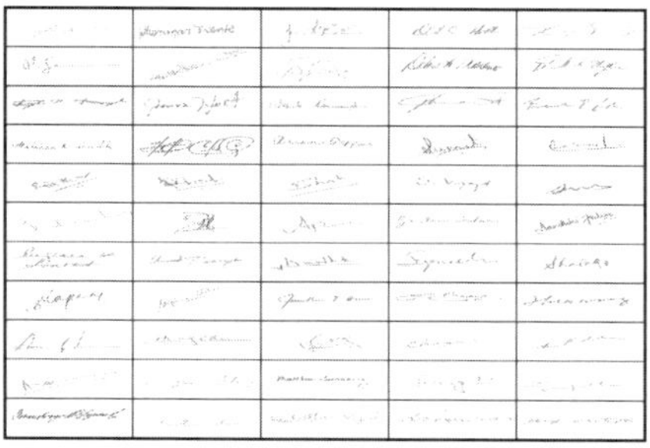

Fig. 7. Genuine signature samples

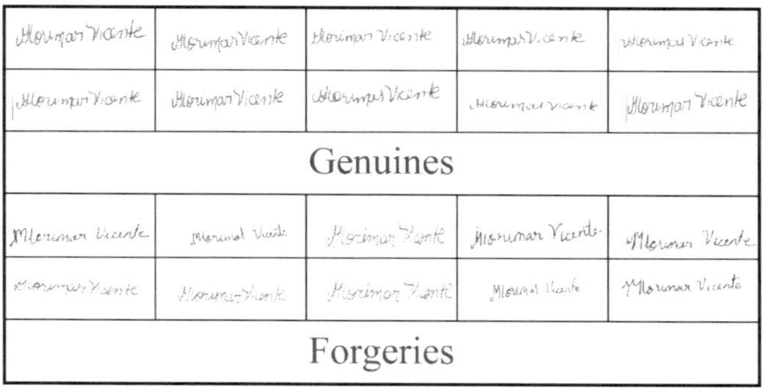

Fig. 8. Samples for one writer: (a) genuines and (b) forgeries

for any biometric person identification problem. False reject rate (Type 1) is the fraction of samples classified as not belonging to the person when truly there were from that person. False acceptance rate (Type 2) is the fraction of samples classified as belonging to the person when truly the samples were not from that person. In the domain of signatures, Type 1 is the fraction of samples classified as forgeries when truly they were genuine and Type 2 the fraction of samples classified as genuine when truly they were forgeries.

5.2 Person-Independent(General) Method

The classification decision boundary discussed in Section 4.1 is given by the sign of the log likelihood-ratio, LLR, $\log P(D_S|d) - \log P(D_D|d)$. A modified decision boundary can be constructed using a threshold α, such that $\log P(D_S|d) - \log P(D_D|d) > \alpha$. When α is varied, we can plot ROC curves as shown in Figure 9. The different subplots in the figure correspond to the ROC curves as the number of known samples is increased from 5 to 20. For each plot, the total error rate defined as (False acceptance+False reject)/2 is minimum at a particular value of α. This is the best setting of α for the specified number of known samples, denoted the *operating point*, and is indicated with an asterix '*'. When 20 samples are used for learning the error rate is approximately 79%.

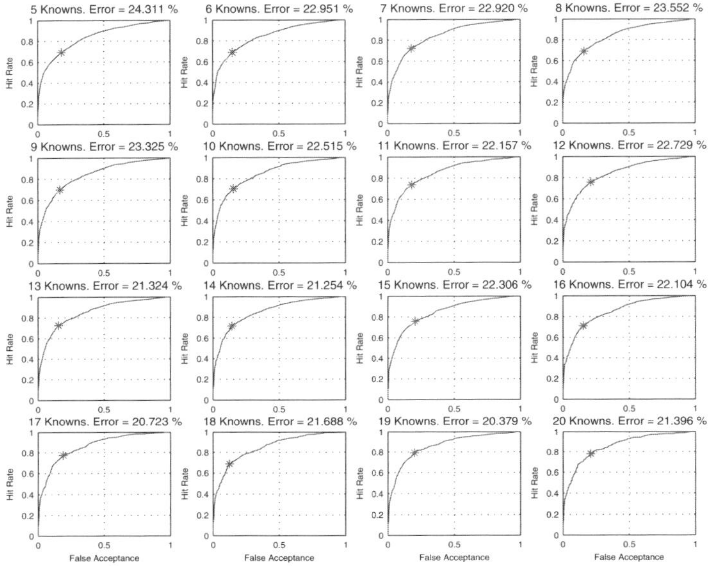

Fig. 9. ROC curves parameterized by α is varied. Each subplot is titled with the number of knowns used for training and the optimum error rate that is possible. The asterix '*' denotes the optimal operating point α for that model.

Figure 10 shows the distribution of LLRs when the questioned samples were genuine and when they were forgeries. A larger region of overlap indicates a higher error rate.

5.3 Person-Dependent Method

The person-dependent classification discussed in Section 4.1 mentioned six different statistics for comparing the two distributions to obtain a probability of match between the questioned sample and the ensemble of knowns. In order to measure error rates for this classificaton technique, once again a decision needs to be made based on the probability of whether or not the questioned sample belongs to the ensemble of knowns. If the probability of match $> \alpha$, then the decision is in favour of the questioned signature to be genuine, and

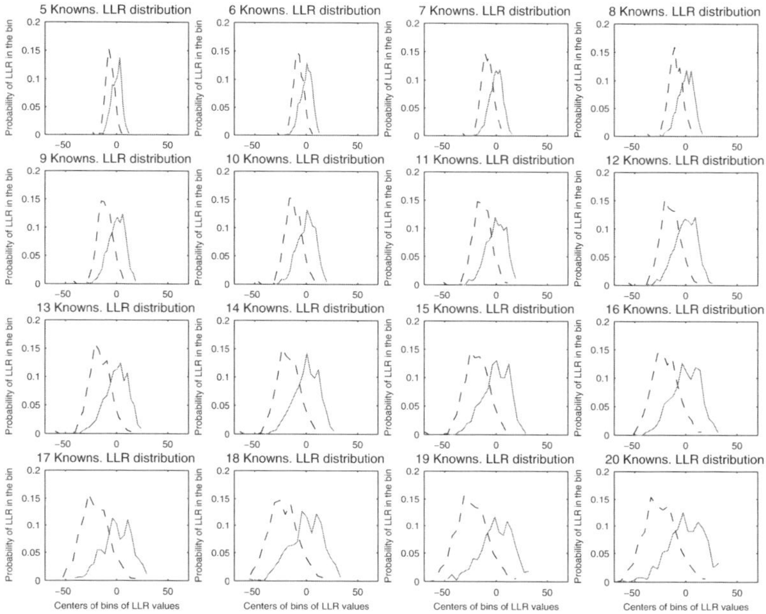

Fig. 10. LLR's obtained from each test case plotted as histograms. The probability (y-axis) that the LLR falls into a range of LLR values (x-axis) is shown for the results of truly genuine (solid) and forgery cases (dotted). Each subplot corresponds to training on a different number of knowns.

if the probability of match $< \alpha$, the decision is in favor of a forgery (this α should not be confused with that used in the person-independent method). By varying the parameter α, once again ROC curves (False Accept vs. False Reject) can be plotted for each of the six measures. The best setting of α is termed as the *operating point.* This setting of α corresponds to the least total error rate possible. Note that the ROC curves are plotted for the test data set and the operating point determined on them. This test data set can be considered as a validation set that helps to determine the operating point. In the curve, the operating point is the point closest to the origin. Table 1 shows the least total error rate possible when different number of known samples were used for training for each of the 6 different measures. Figure 12(a) shows the same table as a graph comparing the different measures and it can be seen that the combined KL and KS measure performs the best. The reason for this can be intutively explained by examining Figure 11(a), where it can be seen that the KS statistic has low false accept rates whereas the KL statistic has low false reject rates. The combination of these two in the KL and KS measure works the best.

Figure 12(b) shows how the operating point (best setting of α) varies with the number of known samples used. It can be seen that in order to obtain the least total error rate, the value of α changes with the number of knowns for

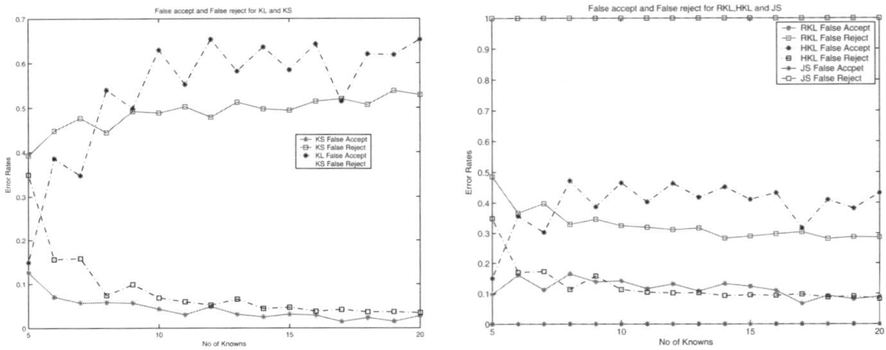

(a) False acceptance and false reject rates are plotted for the KL and the KS statistic as the number of knowns used for training is increased.

(b) False acceptance and false reject rates are plotted as in (a), for the other measures used: RKL, HKL, and JS. The operating point is fixed at 50%.

Fig. 11.

certain measures. The value of α explains a lot about what each statistic learns from the known samples. For example, the high value of α for the KS statistic when large numbers of known samples were used explains that the KS statistic focuses on learning the variation amongst the known samples. Presence of large known samples accounts for greater variation amongst them. Hence if KS focuses on learning the variation, then almost every questioned sample ends up receiving a high probability of match as the majority of questioned samples (genuines and forgeries) invariably fall within the variation. Thus by setting a high value of α the decision that a sample is truly genuine is made only if probability is really high. In simple terms this means that when more samples are used for training the KS statistic will declare a sample as genuine only if the probability of match is really high. In contrast to this measure the KL measure captures the similarities amongst the known samples a lot. This is evident by the low value of α when large number of known signatures are used. Presence of large number of samples accounts for observing more similarities. The KL measure focuses on learning the similarities amongst the samples and it returns a high probability of match very rarely and only when every similarity that is learnt amongst the known samples is present in the questioned sample. Hence the majority of questioned sample receive a low probability of match by the KL measure. To counter this, a low value of α, ensures that the KL measure will declare a sample as forgery only if the probability of match is really low. Similar comments can be made about other measures and it is important to note that those measures for which the operating point does not vary with the number of knowns and those which are around the 50% mark can be a useful property. This basically shows that irrespective of the number of knowns used for training, one can make a decision using the same operating point, and also if the operating point is around the 50% mark there is an equal range of probabilties across which the two different decisions fall. And it is also intuitive that the combined KL KS measure has this fine property. It can be seen that the operating point for the combined KL and KS measure is closest to the 50% mark amongst other measures and is also independent of the number of known samples to some extent. Proceeding with the conclusion that the combined KL KS measure has the above mentioned properties and also outperforms other measures in terms total error rate, we can now consider allowing rejections to reduce the error rates even further. Consider probabilities between $.5 - \beta$ and $.5 + \beta$ for some $\beta > 0$ as the region for reject probabilities. No decision is made if $.5-\beta < Probability < .5+\beta$. This can significantly reduce the total error rate. Figure 13 shows the total error rate as the rejection percentage is changed by changing the value of β. This analysis enables the operator to select a value of β that will induce a certain rejection rate and in turn result in a certain desired error rate. For example, in order to obtain a error rate of 10% with 20 knowns in this data set one should set β to .15 and that accounts for 35% reject rate. Similarly for an error rate of 5% for 20 knowns, β needs be set to .30 which accounts for 62% reject rate.

No. of Knowns	KS	KL	RKL	HKL	JS	KL and KS
5	25.88	24.70	25.61	24.96	25.26	23.87
6	23.54	25.10	25.40	24.57	24.60	22.52
7	22.71	23.35	23.83	23.57	23.31	21.54
8	21.67	23.76	24.60	23.58	23.39	21.20
9	22.17	23.31	24.01	23.03	23.03	20.94
10	21.36	21.93	22.79	21.94	21.63	20.58
11	19.27	20.74	20.96	20.28	20.18	19.02
12	21.13	20.96	21.71	20.42	20.10	19.58
13	20.14	21.73	20.81	21.25	20.78	19.72
14	19.06	20.03	20.84	19.46	19.33	18.41
15	18.28	18.88	19.15	18.10	17.76	17.32
16	19.27	19.08	20.08	18.50	18.38	17.56
17	17.37	17.28	17.36	16.68	16.43	16.07
18	17.79	17.88	18.31	17.58	17.52	17.17
19	17.39	18.09	18.42	17.75	17.37	16.97
20	17.31	17.15	18.58	16.90	17.23	16.40

Table 1. Error rates for signature verification. Measures are Kolmogorov-Smirnov (**KS**), Kullback-Leibler (**KL**), reverse KL (**RKL**), symmetrized KL (**HKL**), Jensen-Shannon (**JS**), and combined KL and KS (**KL and KS**). These are graphed in Figure 12(a)

5.4 Results with Deformed Template

With the construction of a more sophisticated grid, the GSC featuers can be extracted from using these cells in the grid and distances can be computed as before. The refinement method based on flexible template was compared with the original GSC features for performance comparison. In this experiment

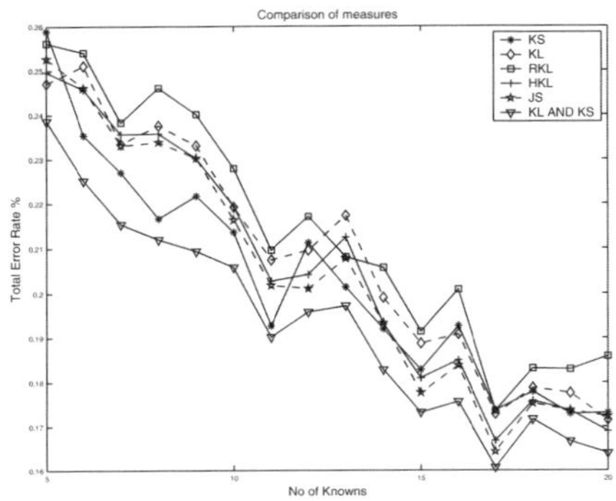

(a) Comparison of measures. The total error rate is plotted for the different measures as the number of knowns used for training is increased. The combined KL and KS measure (lowest trace) outperforms other measures (see text).

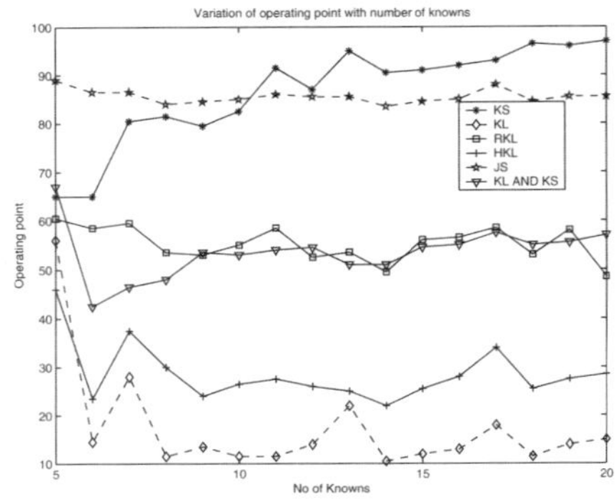

(b) Variation of operating point of the different measures, as a function of the number of knowns used for training. Operating point correponds to the best setting of the decision boundary (probability) for classifying samples as genuine or forgery.

Fig. 12.

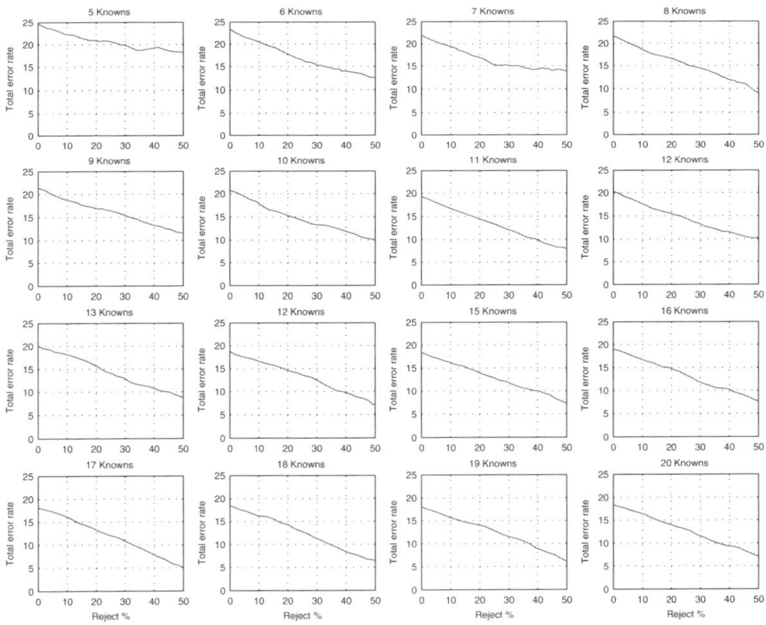

Fig. 13. Error rates as the percentage of allowed rejected (no decision) cases increases. The rejection rate is indirectly controlled by varying the β which assigns the probability region $50 - \beta$ and $50 + \beta$ where no decisions are made and considered as rejects. The different subplots show the plots for different number of knowns used for learning. We have plotted only the trend for the combined KL and KS measure

for each writer we randomly chose 16 signatures from genuine set as known signatures. The testing signature is matched with all the 16 known signatures resulting in 16 distances. Here again, the previously discussed approaches of person-independent and person-dependent can be applied. But for simplicity, a simple averaging of all the 16 distance can be done. The resulting average distance can be compared to a threshold (learnt from a training set), to make a decision of Genuine/Forgery. This resulted in an improvement of performance from 16.07% to as low as 7.9%. It suffices to say that the more sophisticated person-dependent approach using these distances computed from these refined features, will yield even lower error rates for signature verification.

6 Interactive Implementation

Document examination is still a process that must use the expertise of a human examiner since signatures can change with time and there are non-image issues such as speed and pressure of writing, etc. Thus an image-based method described earlier is only one part of the solution. The process of using the algorithms also needs to be interactive so that the user can prepare the inputs and obtain an opinion from the system.

This section describes an interactive software implementation of signature verification where machine learning is a part of the process. CEDAR-FOX [28] is a forensic document examination system. It has graphical interfaces for performing various writer and signature verification/identification tasks.

The learning method chosen for CEDAR-FOX is person-dependent or special learning. This reflects the fact that the questioned document examiner insists on having a set of genuines (size greater than one) in order to perform the task. The combination of KL and KS methods is used.

The learning phase involves the user selecting files corresponding to images of genuine signatures and specifying a name for the compiled learning set. A screen shot of the user interface, where five genuine signatures are used in learning, is shown in Fig 14. CEDAR-FOX allows the user to choose either the standard GSC feature set or the deformed GSC set in feature extraction.

The testing phase involves the user selecting: (i) the file corresponding to the questioned signature, (ii) the name of the compiled learning set and (iii) the choice of the feature extraction algorithm. This is shown in the screen-shot of Fig. 15. The interface allows the user to filter out noise in the signature image, e.g., by automatically erasing extraneous printed text that might accompany the signature image. The result of signature verification is displayed as a probability. In this case the questioned matched the knowns with probability 0.9895.

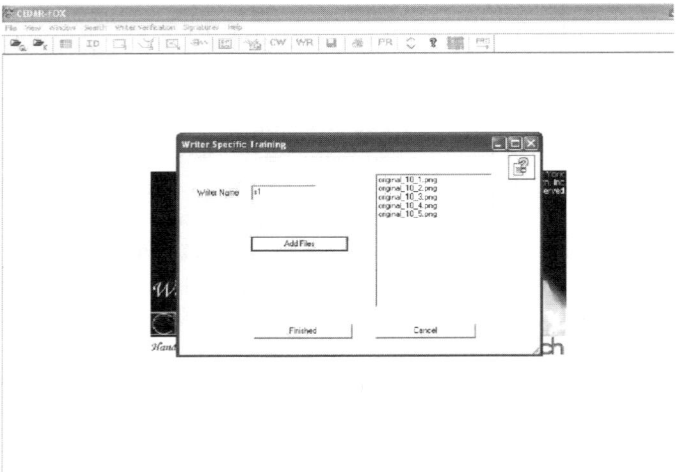

Fig. 14. Signature learning in CEDAR-FOX. The user has interactively slected five genuine image files

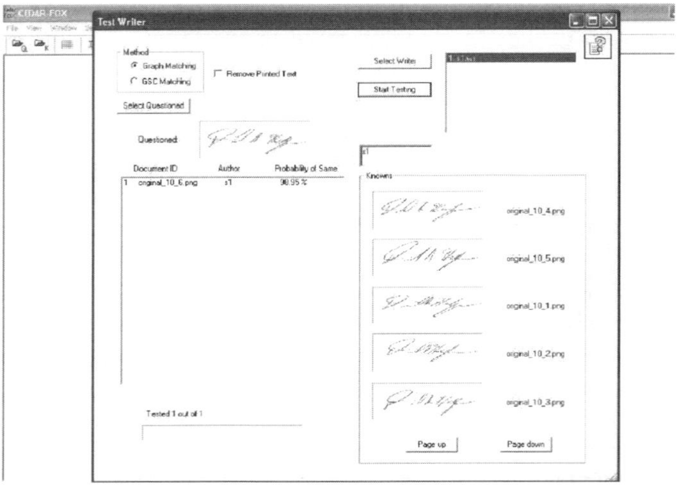

Fig. 15. Signature testing in CEDAR-FOX. The display shows five known signatures, the questioned signature and the probability of the questioned matching the genuines

7 Summary and Discussion

Automatic signature verification is a task where machine learning can be used as a natural part of the process. Two different machine learning approaches, one involving genuines and forgeries in a general set and another involving only genuines for a particular case were described. The first approach is analogous to using counter-examples with near misses in the learning process. Both approaches involve using a similarity measure to compute a distance between features of two signatures. Special learning outperforms general learning particularly as the number of genuines increases. General learning is useful when the number of genuines is very small (less than four). A refined method of extracting features for signatures was also discussed which can further increase verification accuracy. An interactive software implementation of signature verification was described. Future work should consider combining the two types of learning to improve performance.

Acknowledgments: This work was supported in part by the National Institute of Justice grant 2004-IJ-CX-K030

References

1. Osborn, A.: Questioned Documents. Nelson Hall Pub (1929)
2. Robertson, E.W.: Fundamentals of Document Examination. Nelson-Hall (1991)
3. Bradford, R.R., Bradford, R.: Introduction to Handwriting Examination and Identification. Nelson-Hall (1992)
4. Hilton, O.: Scientific Examination of Questioned Documents. CRC Press (1993)
5. Huber, R., Headrick, A.: Handwriting Identification: Facts and Fundamentals. CRC Press (1999)
6. Slyter, S.A.: Forensic Signature Examination. Charles C. Thomas Pub (1995)
7. Mitchell, T.M.: Machine Learning. McGraw-Hill (1997)
8. Srihari, S.N., Xu, A., Kalera, M.K.: Learning strategies and classification methods for off-line signature verification, Proceedings of the Seventh International Workshop on Frontiers in Handwriting Recognition (IWHR), IEEE Computer Society Press (2004) 161–166
9. Winston, P.: Learning structural descriptions from examples. In Winston, P., ed.: The Psychology of Computer Vision. McGraw-Hill (1975) 157–210
10. Leclerc, F., Plamondon, R.: Automatic signature verification: the state of the art, 1989-1993. International Journal of Pattern Recognition and Artificial Intelligence **8**(3) (1994) 643–660
11. Guo, J.K., Doermann, D., Rosenfield, A.: Local correspondences for detecting random forgeries, Proceedings of the International Conference on Document Analysis and Pattern Recognition (1997) 319–323
12. Plamondon, R., Srihari, S.N.: On-line and off-line handwriting recognition: A comprehensive survey. IEEE Transactions on Pattern Analysis and Machine Intelligence **22**(1) (2000) 63–84
13. Kalera, M.K., Zhang, B., Srihari, S.N.: Off-line signature verification and identification using distance statistics. International Journal of Pattern Recognition and Artificial Intelligence **18**(7) (2004) 1339–1360
14. Fang, B., Leung, C.H., Tang, Y.Y., Tse, K.W., Kwok, P.C.K., Wong, Y.K.: Off-line signature verification by the tracking of feature and stroke positions. Pattern Recognition **36** (2003) 91–101
15. Srihari, S.N., Cha, S., Arora, H., Lee, S.: Individuality of handwriting. Journal of Forensic Sciences (2002) 856–872

16. Srinivasan, H., Beal, M., Srihari, S.N.: Machine learning approaches for person verification and identification. Volume 5778., Proc. SPIE: Sensors, and Command, Control, Communications, and Intelligence Technologies for Homeland Security (2005) 574–586
17. Deng, P.S., Liao, H.Y., Ho, C., Tyan, H.R.: Wavelet-base off-line handwritten signature verification. Computer Vision Image Understanding **76**(3) (1999) 173–190
18. Sabourin, R.: Off-line signature verification: Recent advances and perspectives. BSDIA (1997) 84–98
19. Coetzer, J., B.M. Herbst, du Preez, J.: Off-line signature verification using the discrete radon transform and a hidden Markov model. Journal on Applied Signal Processing **4**(2004) 559–571
20. Ferrer, M.A., Alonso, J.B., Travieso, C.M.: Off-line geometric parameters for automatic signature verification using fixed-point arithmetic. IEEE Transactions on Pattern Analysis and Machine Intelligence **27**(6) (2005) 993–997
21. Srikantan, G., Lam, S., Srihari, S.: Gradient based contour encoding for character recognition. Pattern Recognition **7**(1996) 1147–1160
22. Zhang, B., Srihari, S.N.: Analysis of handwriting individuality using handwritten words, Proceedings of the Seventh International Conference on Document Analysis and Recognition, IEEE Computer Society Press (2003) 1142–1146
23. Zhang, B., Srihari, S.N., Huang, C.: Word image retrieval using binary features. In Smith, E.H.B., Hu, J., Allan, J., eds.: SPIE. Volume 5296. (2004) 45–53
24. Zhang, B., Srihari, S.: Properties of binary vector dissimilarity measures. Cary, North Carolina (September, 2003)
25. Scott, G.L., Longuett-Higgins, H.: An algorithm for associating the features of 2 images. Proceedings of the Royal Society of London Series B (Biological) **244**(1991) 21–26
26. Bookstein, F.L.: Principal warps: Thin-plate splines and the decomposition of deformations. IEEE Trans. Pattern Analysis and Machine Intelligence **11**(1989) 567–585
27. Press, W.H., Flannery, B.P., Teukolsky, S.A., Vetterling, W.T.: Numerical Recipes in C: The Art of Scientific Computing. Cambridge University Press (1992)
28. Srihari, S.N., Zhang, B., Tomai, C., Lee, S., Shi, Z., Shin, Y.C.: A system for handwriting matching and recognition, Proc. Symposium on Document Image Understanding Technology (2003) 67–75

Off-line Writer Identification and Verification Using Gaussian Mixture Models

Andreas Schlapbach and Horst Bunke

Institute of Computer Science, University of Bern, Neubrückstrasse 10, CH-3012 Bern, Switzerland
{schlpbch,bunke}@iam.unibe.ch

Summary. This chapter presents an off-line, text-independent system for writer identification and verification. At the core of the system are Gaussian Mixture Models (GMMs). GMMs provide a powerful yet simple means of representing the distribution of features extracted from the text lines of a writer. For each writer, a GMM is built and trained on text lines of that writer. In the identification or verification phase, a text line of unknown origin is presented to each of the models. As a result of the recognition process each model returns a log-likelihood score. These scores are used for both the identification and the verification task. Three types of confidence measures are defined on the scores: simple score based, cohort model based, and world model based confidence measures. Experiments demonstrate a very good performance of the system on the identification and the verification task.

1 Introduction

In recent years, significant progress has been made in recognizing a person based on biometric features [1, 2, 3]. Different biological traits such as face, fingerprint, iris, signature, and voice are being used to identify a person or verify its identity. This chapter addresses the problem of personal identification and verification based on a person's handwriting.

Writer identification is the task of determining the author of a sample handwriting from a set of writers [4]. Related to this task is writer verification, i.e., the task of determining whether or not a handwritten text has been written by a certain person. If any text may be used to establish the identity of the writer, the task is *text independent*. Otherwise, if a writer has to write a particular predefined text to identify himself or herself, or to verify his or her identity, the task is *text dependent*.

If temporal and spatial information about the writing is available, writer identification and verification can be performed *on-line*, otherwise if only a scanned image of the handwriting is available the recognition is performed *off-line*. The system we propose in this chapter performs text independent writer identification and verification using off-line handwritten text lines. Examples

A. Schlapbach and H. Bunke: *Off-line Writer Identification and Verification Using Gaussian Mixture Models*, Studies in Computational Intelligence (SCI) **90**, 409–428 (2008)
www.springerlink.com

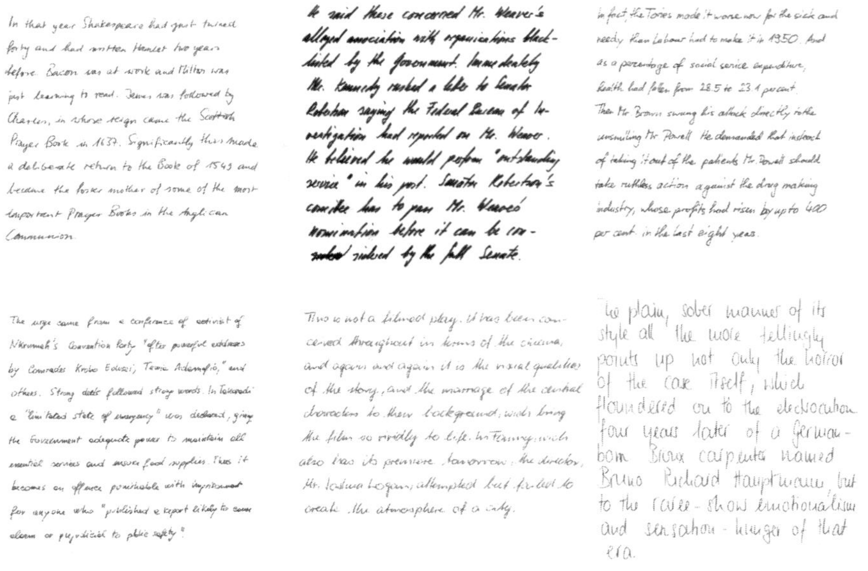

Fig. 1. Examples of text lines

of handwritten text lines from our database, produced by different writers, are given in Fig. 1. Possible applications of our system include forensic writer identification [5], the retrieval of handwritten documents from a database [6] or authorship determination of historical manuscripts [7].

In this chapter we use Gaussian Mixture Models (GMMs) to model a person's handwriting. GMMs provide a powerful yet simple means of representing the distribution of features extracted from text lines written by a person. Formally, a GMM consists of a weighted sum of uni-modal Gaussian densities. While GMMs have been used in speech recognition [8, 9] they have not yet been applied to off-line, text independent writer identification and verification, to the best of our knowledge.

For each writer in the considered population, an individual GMM is trained using data from that writer only. Thus for n different writers we obtain n different GMMs. Intuitively, each GMM can be understood as an expert specialized in recognizing the handwriting of one particular person. Given an arbitrary text line as input, each GMM outputs a recognition score. Assuming that the recognition score of a model is higher on input from the writer the model is trained on than on input from other writers, we can utilize the scores produced by the different GMMs for the task of identifying the writer of a text line or of verifying whether a text line has actually been written by the person who claims to be the writer.

Our approach has several advantages compared to other approaches: GMMs have a mathematically simple, well understood structure and there exist standard algorithms for training and testing [9]. For every writer there is exactly one model which is trained with a set of simple features. We do not need to model characters or words. Therefore we do not need a transcription of the text, but can use any unlabeled text for training and testing. This property makes our system language independent.

The rest of this chapter is structured as follows. In Sect. 2 we present related work in the field of writer identification and verification. The GMMs used by our system are introduced in Sect. 3. An overview of our writer identification and verification system is given in Sect. 4 and in Sect. 5 we present several confidence measures for our system. Results of a number of experiments are presented and discussed in Sect. 6. Finally, Sect. 7 concludes the chapter and proposes future work.

2 Related Work

Surveys covering work in automatic writer identification and signature verification until 1993 are given in [4, 10]. Recently, several additional approaches to writer identification and verification have been proposed.

Said et al. [11] treat the writer identification task as a texture analysis problem. They use global statistical features extracted from the entire image of a text using multi-channel Gabor filtering and grey-scale co-occurrence matrix techniques.

Srihari et al. [12, 13] address the problem of writer verification, i.e., the problem of determining whether two documents are written by the same person or not. In order to identify the writer of a given document, they model the problem as a classification problem with two classes, *authorship* and *non-authorship*. Given two handwriting samples, one of known and the other of unknown identity, the distance between two documents is computed. Then the distance value is used to classify the data as positive or negative.

Zois et al. [14] base their approach on single words by morphologically processing horizontal projection profiles. The projections are partitioned into a number of segments from which feature vectors are extracted. A Bayesian classifier and a neural network are then applied to the feature vectors.

In Hertel et al. [15] a system for writer identification is described. The system first segments a given text into individual text lines and then extracts a set of features from each text line. The features are subsequently used in a k-nearest-neighbor classifier that compares the feature vector extracted from a given input text to a number of prototype vectors coming from writers with known identity.

Bulacu et al. [16] use edge-based directional probability distributions as features for the writer identification task. The authors introduce edge-hinge distribution as a new feature. The key idea behind this feature is to consider two edge fragments in the neighborhood of a pixel and compute the joint probability distribution of the orientations of the two fragments. Additionally, in [17] the histogram of connected-component contours (CO^3) for upper-case handwriting is introduced as a new feature. Combining this feature with the edge-hinge feature achieves better results than each of the features used separately. In [18] this approach is extended to mixed-style handwriting using fragmented connected-component contours.

In a number of papers [19, 20, 21] graphemes are proposed as features for describing the individual properties of handwriting. Furthermore, it is shown that each handwriting can be characterized by a set of invariant features, called the writer's invariants. These invariants are detected using an automatic grapheme clustering procedure. In [22] these graphemes are used to address the writer verification task based on text blocks as well as on handwritten words.

Leedham et al. [23] present a set of eleven features which can be extracted easily and used for the identification and the verification of documents containing handwritten digits. These features are represented as vectors, and by using the Hamming distance measure and determining a threshold value for the intra-author variation a high degree of accuracy in authorship detection is achieved.

Previously, we have proposed to use Hidden Markov Model (HMM) [24] based text recognizers for the purpose of writer identification and verification [25, 26]. For each writer, an individual recognizer is built and trained on text lines of that writer. This results in a number of recognizers, each of which is an expert on the handwriting of exactly one writer. Assuming that correctly recognized words have a higher score than incorrectly recognized words and that the recognition rate of a system is higher on input from the writer the system was trained on than on input from other writers, the scores produced by the different HMMs are used to decide who has written the input text line.

In this chapter, instead of HMM based recognizers, we use GMMs to model a person's handwriting. While GMMs have been used in the speech recognition community [8, 9], they have not been applied, to the best of our knowledge, to off-line writer identification and verification. A GMM can be viewed as a single-state HMM with a Gaussian mixture observation density. The advantages of using GMMs over HMMs are manifold. First, GMMs are conceptually less complex than HMMs consisting of only one state and one output distribution function, which leads to significantly shorter training times. Second, in GMMs only the parameters of the output distribution function have to be estimated during training compared to HMMs where the state transition probabilities have to be estimated as well. Third, neither words nor characters have to be modeled using GMMs, because every writer is represented by exactly one model. Finally, no transcription of the text lines are needed during training.

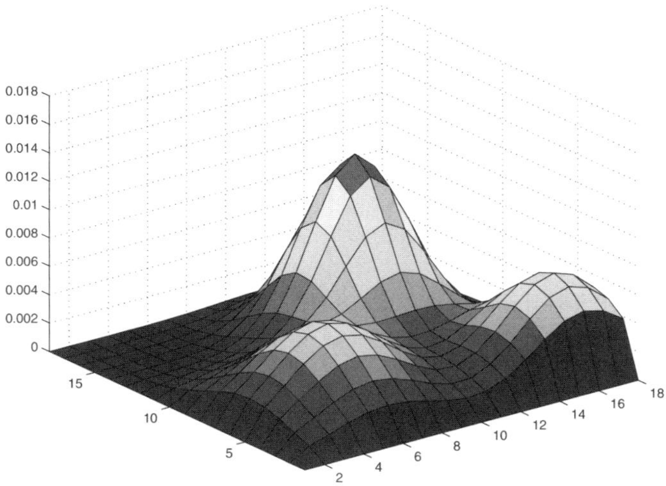

Fig. 2. A two-dimensional GMM consisting of a weighted sum of three uni-modal Gaussian densities

3 Gaussian Mixture Models

We use Gaussian Mixture Models (GMMs) to model the handwriting of each person of the underlying population. The distribution of the feature vectors extracted from a person's handwriting is modeled by a Gaussian mixture density. For a D-dimensional feature vector $\mathbf{x}$ the mixture density for a specific writer is defined as

$$p(\mathbf{x}|\lambda) = \sum_{i=1}^{M} w_i p_i(\mathbf{x}). \tag{1}$$

where the mixture weights w_i sum up to one. The mixture density is a weighted linear combination of M uni-modal Gaussian densities $p_i(\mathbf{x})$, each parametrized by a $D \times 1$ mean vector μ_i and a $D \times D$ covariance matrix C_i:

$$p_i(\mathbf{x}) = \frac{1}{(2\pi)^{D/2}|C_i|^{1/2}} \exp\{-\frac{1}{2}(\mathbf{x} - \mu_i)'(C_i)^{-1}(\mathbf{x} - \mu_i)\}. \tag{2}$$

The parameters of a writer's density model are denoted as $\lambda = \{w_i, \mu_i, C_i\}$ for all $i = 1, \ldots, M$. This set of parameters completely describes the model and enables us to concisely model a person's handwriting.

The GMMs are trained using the Expectation-Maximization (EM) algorithm [27]. The EM algorithm follows the *Maximum Likelihood (ML)* principle by iteratively refining the parameters of the GMM to monotonically increase the likelihood of the estimated model for the observed feature vectors. The algorithm starts with a data set X of N feature vectors $\mathbf{x}_j$, an initial set of M uni-modal Gaussian densities $N_i \hat{=} N(\mu_i, C_i)$, and M mixture weights w_i. Then, in the first step, for each training data point $\mathbf{x}_j$ the responsibility of each component N_i is determined. In the second step, the component densities, i.e., the mean vector μ_i and the variance matrix C_i for each component and the weights w_i are re-estimated based on the training data. These two steps are repeated until the likelihood score of the entire data set does not change substantially or a limit on the number of iterations is reached.

While the general model supports full covariance matrices, often only diagonal covariance matrices are used. This simplification is motivated by the following observations: first, theoretically the density modeling of an M dimensional full covariance matrix can equally well be achieved using a larger order diagonal covariance matrix. Second, diagonal covariance matrices are computationally more efficient than full covariance matrices, and third, diagonal matrix GMMs outperformed full matrix GMMs in various experiments [9]. An example of a two dimensional GMM with a diagonal covariance matrix is shown in Fig. 2.

The Gaussian component densities can either be initialized randomly or by using vector quantization techniques such as k-means clustering [28]. Furthermore, often variance flooring is employed to avoid an overfitting of the variance parameters [29]. The idea of variance flooring is to impose a lower bound on the variance parameters as a variance estimated from only few data points can be very small and might not be representative of the underlying distribution of the data [29]. The minimal variance value is defined by

$$\sigma_{min}^2 = \alpha * \sigma_{global}^2 \tag{3}$$

where α denotes the *variance flooring factor* and the global variance σ_{global}^2 is calculated on the complete data set. The minimal variance, σ_{min}^2, is used to initialize the variance parameters of the model. During the EM update step, if a calculated variance parameter is smaller than σ_{min}^2 then the variance parameter is set to this value.

During decoding, the feature vectors $X = \{\mathbf{x}_1, \dots, \mathbf{x}_T\}$ extracted from a text line are assumed to be independent. The log-likelihood score of a model λ for a sequence of feature vectors X is defined as

$$\log p(X|\lambda) = \sum_{t=1}^{T} \log p(\mathbf{x}_t|\lambda), \tag{4}$$

where $p(\mathbf{x}_t|\lambda)$ is computed according to Eq. 1.

In this work, we use diagonal covariance matrices and the models are initialized using k-means clustering. The GMMs are implemented using the Torch library [30].

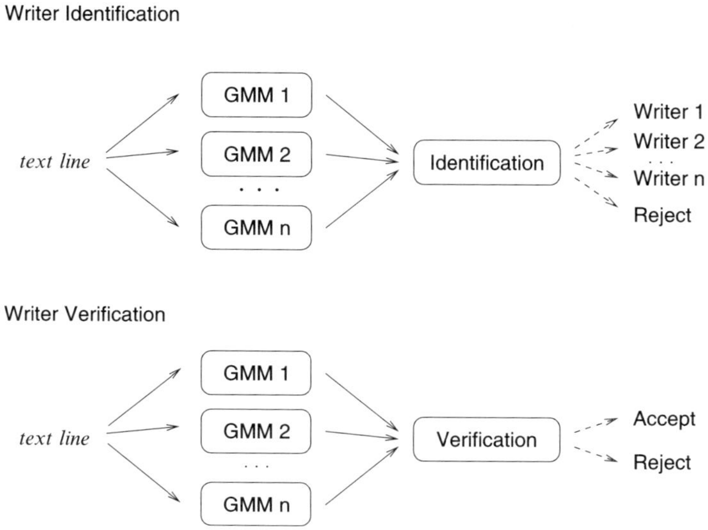

Fig. 3. Schematic overview of the writer identification and verification system

4 System Overview

We use GMMs as the building blocks of our writer identification and verification system. A schematic overview of the system is shown in Fig. 3. For each writer, a GMM as described in the previous section is built and trained with data coming from this writer only. As a result of the training procedure, we get a model for each writer.

A set of features is extracted from each text line to train the GMMs. Before feature extraction, a series of normalization operations are applied to each text line. The operations are designed to improve the quality of the features extracted without removing writer specific information.

For the purpose of normalization, the contrast of the grey-scale images is enhanced first, leading to images with black strokes written on white background. Then vertical scaling and thinning normalization operations are applied, which are described in the following two paragraphs. The aim of vertical scaling is to normalize the height of the text line and thinning assures independence of the writing pen.

To perform vertical scaling a text line is divided into three zones: a zone containing the ascenders, a middle zone, and a zone containing the descenders. These three zones are to be normalized to a predefined height which is important in order to reduce the variability of the features used to train the GMMs. To actually perform this operation, the upper and the lower baseline of the text line have to be determined. To find the two baselines, the histogram of the horizontal projection of the image of the text line is used. The real histogram is matched with an ideal histogram. The location of the

Text line before normalization and thinning

Text line after normalization and thinning

Fig. 4. A text line before and after normalization and thinning

upper and the lower baseline are detected and the three main writing zones are determined. Each of these three zones is then individually positioned and scaled to a predefined height.

Different pens of different width have been used to write the text lines. In order to eliminate the effect of the pen width on the performance of the system, all text lines are thinned using the iterative MB2 thinning algorithm [31]. After thinning, all strokes in a text line image are at most two pixels wide. In Fig. 4 a text line before and after normalization and thinning is shown.

In the next step, features are extracted by a sliding window. The window moves from left to right one pixel per step. For every column of pixels in the sliding window, nine geometrical features are extracted. These features have shown to produce good results on both the text recognition task [32] as well as on the writer identification and verification task [33].

The feature set consists of three global and six local features. The three global features describe the distribution of the pixels in the column, e.g., the fraction of black pixels in the window, the center of gravity and the second order moment. The six local features describe specific points in the column. The features describe the position and the orientation of the upper- and the lower-most pixel, the number of black-to-white transitions in the window, and the fraction of black pixels between the upper- and the lower-most black pixel (see Fig. 5 for an illustration of the six local features). The feature vectors of every column in the sliding window are averaged to produce the final feature vector. At last, the feature vectors which only describe white space are deleted.

The width of the sliding window was optimized in an independent experiment involving 571 text lines from 20 writers. These 20 writers are not part of the data set used to train the GMM models in the subsequent experiments. A fixed number of 100 Gaussian mixture components and a variance flooring factor of 0.001 were used for training. The window width was varied from 2 to 32 by steps of two. The highest writer identification rate of 99.05% was achieved using a window width of 14 pixels. This window width was used in all subsequent experiments to extract the features from a text line.

The sequences of nine-dimensional feature vectors extracted from the text lines are used to train the GMMs. After training, each GMM is especially adapted to the individual handwriting style of a particular writer. During identification, a text line to be classified is presented to the GMM of each

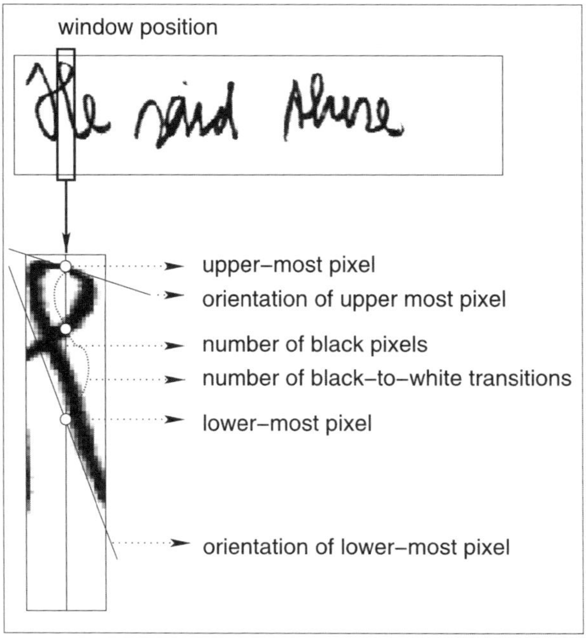

Fig. 5. Six local features extracted from each row in the sliding window

writer. Each GMM outputs a log-likelihood score and a standard deviation for the given text line. These scores are the basis for identification and verification as described below.

5 Confidence Measures

In order to assign a text line to a certain person or to verify the identity of a text line with a claimed identity we need a means of measuring how sure the system is about the given text line. A confidence measure enables us to judge the quality of the recognition and to implement a rejection mechanism based on this measure.

For writer identification we define the following rejection mechanism. If the confidence measure of a text line is above a given threshold, the system returns the identity of the text line with the highest ranked score; otherwise the system rejects the input. Thus if we have n writers, the writer identification problem is a n-class classification problem with a reject option.

The decision criterion for writer verification is similar. If the confidence measure of a text line is above a certain threshold, we assume that the text line was in fact written by the claimed writer; otherwise the input is classified as not being of the claimed identity. In writer verification we deal with a two-class classification problem that is independent of the number of writers under consideration.

Various confidence measures for off-line handwriting recognition have been presented in the literature [34, 35, 36]. In this chapter, three common types of confidence measures are used. The first type of confidence measure is solely based on the score of the model under consideration and therefore is not normalized. The other two types of confidence measures normalize the recognition score based on a *cohort model* and a *world model* approach, respectively. The cohort model approach normalizes the score of the model of the claimed writer with respect to the score of the most competitive writers [37]. The world model approach normalizes the score of the claimed writer by a model which is trained on a large number of samples from many writers [38].

5.1 Confidence Measures for Writer Identification

A text line is presented to each model and the returned log-likelihood scores are sorted. Given a text line t of an unknown author, the simplest confidence measure is to judge the quality of the recognition based on the log-likelihood score of the first ranked model:

$$cmIdent_{\text{LLScore}}(t) = ll_{\text{firstRanked}} \tag{5}$$

The next confidence measure is inspired by the cohort model approach. The confidence measure is calculated from the difference of the log-likelihood score of the first ranked model, $ll_{\text{firstRanked}}$, and the log-likelihood score of the second ranked model, $ll_{\text{secondRanked}}$:

$$cmIdent_{\text{CohortModel}}(t) = ll_{\text{firstRanked}} - ll_{\text{secondRanked}} \tag{6}$$

The third confidence measure uses a world model to normalize the log-likelihood score of the first ranked writer. The world model is trained on a large number of text lines coming from different writers. The confidence measure is calculated on the difference of the log-likelihood score of the first ranked writer, $ll_{\text{firstRanked}}$, and the world model, $ll_{\text{worldModel}}$:

$$cmIdent_{\text{WorldModel}}(t) = ll_{\text{firstRanked}} - ll_{\text{worldModel}} \tag{7}$$

All the confidence measures for writer identification presented in this section need to determine the system which produces the highest log-likelihood score. A text line has to be presented to the model of each writer under consideration. Then the returned scores have to be sorted, which means that the calculation of these confidence measures depends on the number of writers.

5.2 Confidence Measures for Writer Verification

The confidence measures for writer verification are similar to the ones defined for writer identification in the previous section. Compared to the writer identification case where the log-likelihood score of the first ranked system is normalized, the log-likelihood score of the claimed system is normalized instead.

The first simple confidence measure for a text line t is the log-likelihood score of the model of the claimed identity, $ll_{\text{claimedID}}$:

$$cmVerif_{\text{LLScore}}(t) = ll_{\text{claimedID}} \tag{8}$$

The next confidence measure is inspired by the cohort model approach. Based on the ranking of the scores the confidence measure is calculated from the difference of the log-likelihood score of the claimed identity, $ll_{\text{claimedID}}$, and the first best ranked competing writer, $ll_{\text{bestRankedCompeting}}$:

$$cmVerif_{\text{CohortModel}}(t) = ll_{\text{claimedID}} - ll_{\text{bestRankedCompeting}} \tag{9}$$

The third confidence measure implements a world model approach. The difference of the score of the model of the claimed identity and the world model is computed:

$$cmVerif_{\text{WorldModel}}(t) = ll_{\text{claimedID}} - ll_{\text{worldModel}} \tag{10}$$

In comparison to the world model based confidence score in the identification case (Eq. 7), in the verification case we do not need to present the text line in question to all the models, but to the model of the claimed identity and the world model only.

6 Experiments

6.1 Data sets

The text lines used in our experiments are part of the IAM handwriting database [39][1]. The database currently contains over 1,500 pages of handwritten text. For each writer we use five pages of text from which between 27 and 54 text lines are extracted.

Five-fold cross validation is used in our experiments. Cross validation enables us to use all text lines for training without committing the error of training or of optimizing meta parameters on the test set [40]. For each writer, the set of available text lines is split into five sets. The idea is to train the system on three sets, use the fourth set to find an optimal set of meta parameters and then test on the fifth set. This procedure is iterated five times such that each set is used for testing once. The final recognition rate is obtained by averaging the five results from each of the test sets.

[1] The IAM handwriting database is publicly available at: www.iam.unibe.ch/˜fki/iamDB

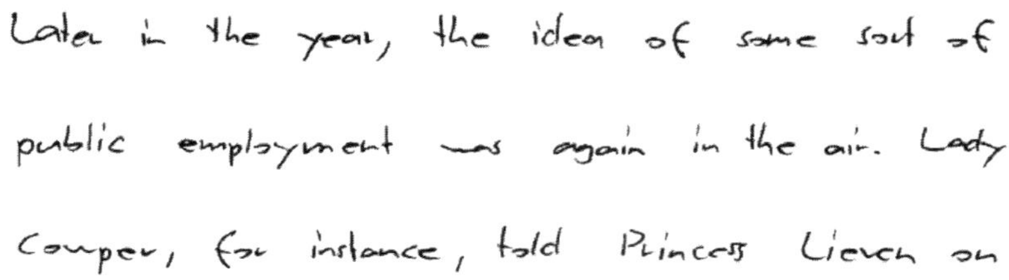

Example of original text lines

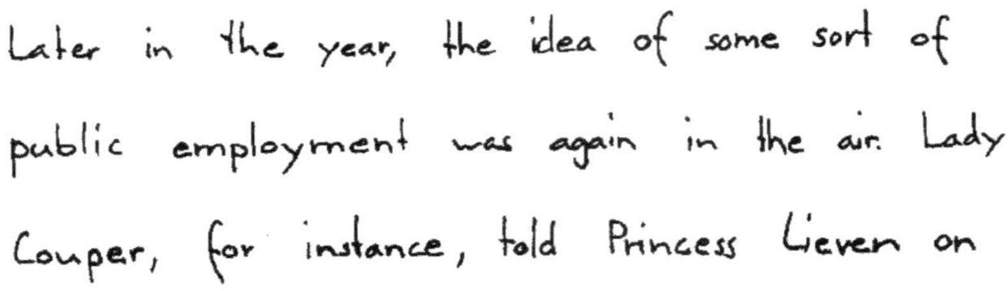

Example of skillfully forged text lines

Fig. 6. Examples of original and skillfully forged text lines

In this experimental setup, the data set consists of text lines from 100 different writers. All in all, 4,103 text lines are available and due to cross validation we can use all the text lines for both training and testing.

A verification system can make two types of errors. First, the system can falsely reject a text written by a client and, second, it can falsely accept a text coming from an impostor [41]. Therefore we need two sets for testing a writer verification system: one set consisting of clients and one set containing impostors. The impostor set can be composed of unskilled forgeries, where the impostor makes no effort to simulate a genuine handwriting, and of skilled forgeries, where the impostor tries to imitate the handwriting of a client as closely as possible [42].

The unskillfully forged test set used in our experiments consists of two disjoint subsets coming from clients and impostors. The unskilled forgeries that form the impostor set are obtained from the database by extracting 571 text lines produced by 20 writers. The writers of these text lines are disjoint from the 100 clients and no model exists that is trained on the handwriting of any of these 20 writers. Based on these text lines the impostor data set is constructed by assigning, to each of these text lines, seven identities of writers known to the system. In total, the impostor data set consists of $7 \times 571 = 3{,}997$ lines and the complete test set of 8,100 text lines. The rationale is that the number of text lines to be accepted should be approximately the same as the number of text lines that have to be rejected, i.e., the two classes under consideration should be balanced.

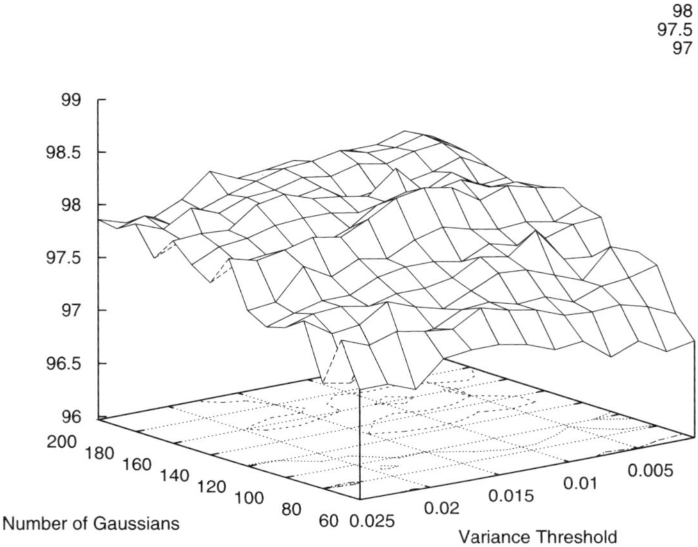

Fig. 7. Writer identification rate as a function of the number of Gaussian mixture components and the variance flooring factor on the validation set

The skillfully forged test set is again composed of two subsets, a client and an impostor subset. The client data set consists of one page of text each from 20 different writers which are part of the 100 clients. A total of 169 text lines are extracted from these 20 pages. The same 20 pages are then skillfully forged. The acquisition protocol is as follows. A person is presented with a page of handwritten text and given 10 minutes to train the writing. Then he or she is asked to forge the text. An example of three original and three skillfully forged text lines are given in Fig. 6. From the forgeries thus created, another 169 text lines are extracted. Hence, in total 338 text lines are used in this test set.

6.2 Writer Identification Experiments

We first conducted an experiment to measure the influence of the number of Gaussian mixture components and the variance flooring factor on the writer identification rate. The number of Gaussian mixture components is varied from 60 to 200 by steps of 10 and the variance flooring factor is varied from 0.001 to 0.025 by steps of 0.002.

The writer identification rate as a function of the number of Gaussian mixture components and the variance flooring factor on the validation set is

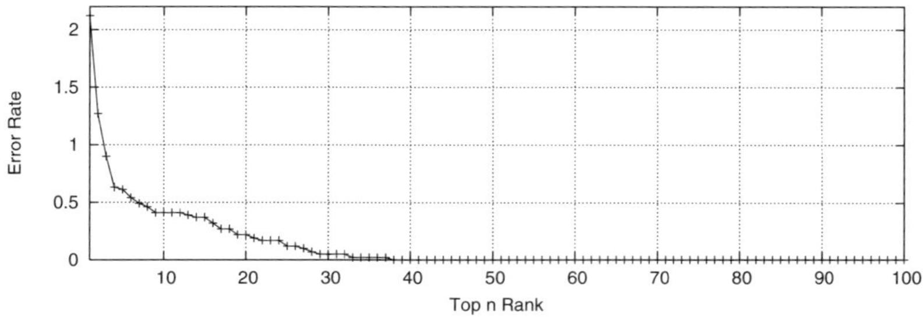

Fig. 8. n-best list for the writer identification experiment

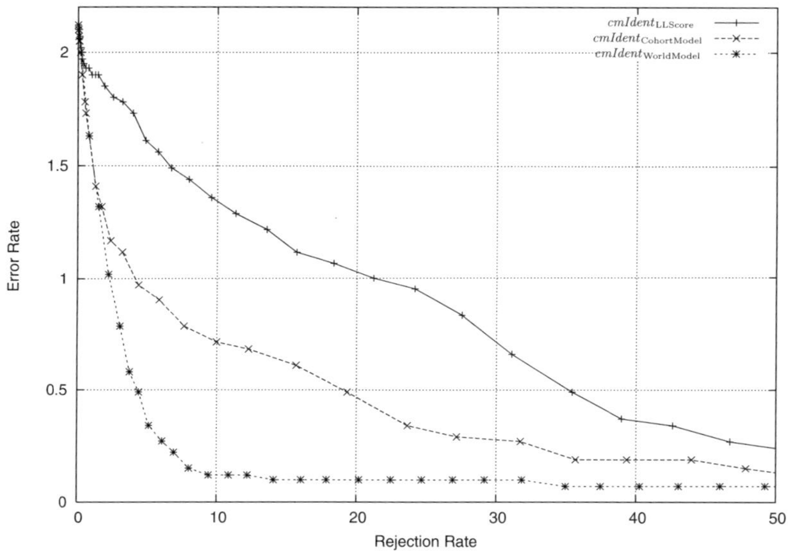

Fig. 9. Error-rejection curves for different confidence measures

shown in Fig. 7. On the validation set, the highest writer identification rate of 98.20% is achieved using 130 Gaussian mixture components and a variance flooring factor of 0.011. An identification rate higher than 97.03% is achieved using 60 Gaussian mixture components or more on the validation set. The two meta parameters optimized on the validation set are then used to calculate the final writer identification rate of 97.88% on the test set. We also use the world model trained with these meta parameters in the subsequent experiments.

In Fig. 8, the n-best list is shown where the writer identification rate based on the first n ranks is plotted. The error rate of the system drops below 1% if the first three ranks, and below 0.5% if the first seven ranks are considered.

Table 1. Equal Error Rates (EERs) for the unskillfully and skillfully forged test set

Equal Error Rate (ERR)	Unskilled Forgeries	Skilled Forgeries
$cmVerif_{\text{LLScore}}$	13.0%	41.0%
$cmVerif_{\text{WorldModel}}$	3.2%	18.6%
$cmVerif_{\text{CohortModel}}$	1.5%	9.3%

The error-rejection curves obtained from the identification test set are shown in Fig. 9. The simple log-likelihood score based confidence score ($cmIdent_{\text{LLScore}}$) produces the lowest performing error-rejection curve. The error rate drops below 1% only if more than 22% of the text lines with the lowest confidence score are rejected. The next best error-rejection curve is produced by the world model based approach ($cmIdent_{\text{WorldModel}}$). The error rate is smaller than 1% if less than 5% of the text lines are rejected. The cohort model based approach ($cmIdent_{\text{CohortModel}}$) yields the best error-rejection curve. Fewer than 5% of the text lines have to be rejected to obtain an error rate smaller than 0.5%.

The observation that the cohort model based approach performs best can be explained by the fact that the normalization is based on the actual text line being presented, i.e., the adequate model to normalize the text line is selected anew for each text line. In comparison, the world model approach normalizes the score of a text line by the score of a general world model which is independent of the text line under consideration.

6.3 Writer Verification Experiments

The results of the writer verification experiments are reported as Receiver Operator Characteristic (ROC) curves in Figs. 10 and 11. An ROC curve describes the performance of a verification system on a test set by plotting the False Acceptance Rate (FAR) against the False Rejection Rate (FRR) [41]. In Table 6.3 the estimated Equal Error Rates (EERs) for the ROC curves are given [41]. The Equal Error Rate estimates the point on an ROC curve where the FAR is identical to the FRR.

In Fig. 10 the ROC curves on the unskillfully forged test set are shown. The ROC curves produced by the simple log-likelihood score ($cmVerif_{\text{LLScore}}$) has the lowest performance with an EER of 13.0%. The world model based confidence measure ($cmVerif_{\text{WorldModel}}$) achieves an EER of 3.2%. The ROC curve based on the cohort model approach ($cmVerif_{\text{CohortModel}}$) performs best and yields an EER of around 1.5%.

The ROC curves on the skillfully forged test set for the GMM based systems are shown in Fig. 11. The ROC curve with the lowest performance results from the simple log-likelihood score confidence measure ($cmVerif_{\text{LLScore}}$) with an EER of around 41.0%. The world model based confidence measure

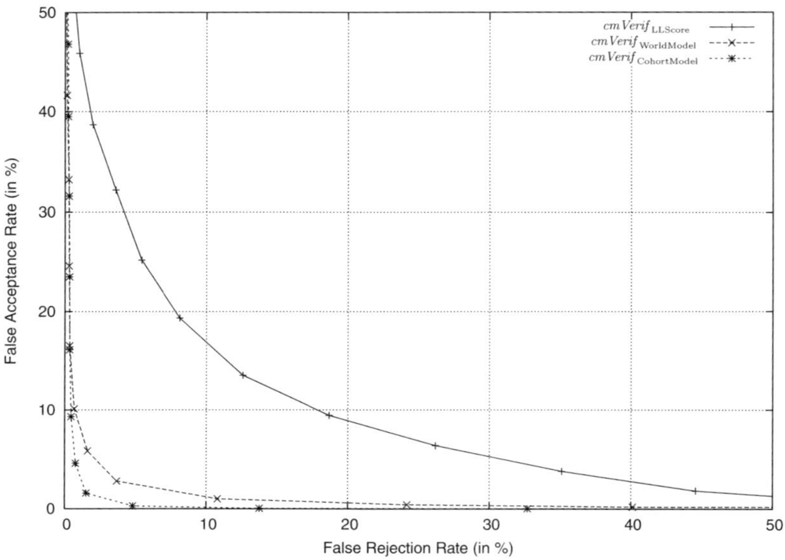

Fig. 10. ROC curves on the unskillfully forged test set

($cmVerif_{\text{WorldModel}}$) yields an EER of around 18.6%. The best ROC curve is produced by the cohort model based confidence measure ($cmVerif_{\text{CohortModel}}$) with an EER of around 9.3%.

In both verification experiments, the ROC curves show the same hierarchy of performance: the simple log-likelihood score based confidence measure yields the lowest performing ROC curve, the next best ROC curve is produced by the world model based confidence measure which itself is outperformed by the cohort model based ROC curve. This behavior is consistent with the writer identification case, where the best error-rejection curve is achieved when the score of a text line is normalized with respect to the score of the most competitive writer.

The calculation of the cohort model based confidence measure however is costly compared to the world model based confidence measure. For every text line, the log-likelihood scores of all writers models have to be computed and sorted to determine the best performing model. In comparison, to compute the world model based confidence measure, only the score of the claimed system and the world model is needed and is independent of the number of client models.

7 Conclusion

We have used Gaussian Mixture Models (GMMs) to address the task of off-line text-independent writer identification and verification. GMMs provide a pow-

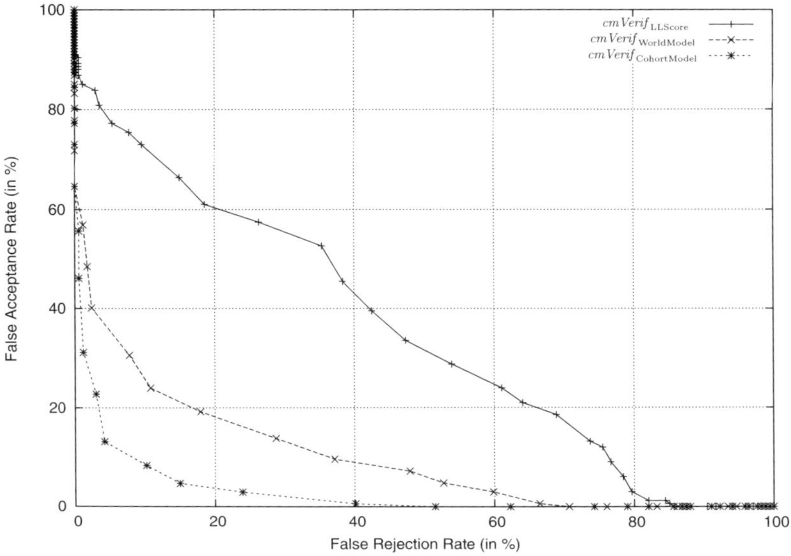

Fig. 11. ROC curves on the skillfully forged test set

erful yet simple means of representing the distribution of features extracted from handwritten text lines. A sliding window extracts a sequence of simple, language independent feature vectors from a text line. The feature sequences are used to train one model for each writer. During recognition, a text line of unknown origin is presented to each of the models. Each model returns a log-likelihood score for the given input. The scores are the basis for writer identification and verification.

On the writer identification task, a text line is assigned to the writer of the first ranked model if the confidence measure is above a given threshold. We achieve a correct writer identification rate of 97.88% in a 100 writers experiment using 4,103 text lines. If we consider not only the first, but the three highest ranked writers, in over 99.0% of all cases the writer of the text line under question is correctly identified. Furthermore, if we reject fewer than 5% of the text lines with the lowest confidence score, the writer identification rate improves to over 99.5% using the best performing confidence measure.

Similarly, on the writer verification task a text line is accepted if its confidence score is above a certain threshold; otherwise it is rejected. Two sets of experiments have been conducted: the unskillfully forged test set contains in total 8,100 text lines from 100 clients and 20 impostors. The skillfully forged test set contains 338 text lines from 20 clients and 20 impostors. An Equal Error Rate (EER) of around 1.5% is achieved on the unskillfully forged test set and an EER of approximately 9.3% is obtained on the skillfully forged test set by the best confidence measure.

Three types of confidence measures have been presented in this chapter: simple score based, cohort model based and world model based confidence measures. For both writer identification and verification, the cohort model based confidence measure performs best. This observation can be explained by the fact that the normalization depends on the actual text line being presented, i.e., the relevant model to normalize the text line is selected anew for each text line. In comparison, the world model confidence measure normalizes the score of a text line by the score of a general world model.

In future work we plan to measure the influence of using less data to train the GMMs. A possible approach would be to use a universal background model [9] and then adapt this model to a specific writer model. Another interesting question is to investigate whether modifications of the world model based confidence measure as presented in [43] would yield performances similar to the ones obtained by the cohort model based confidence measure. Furthermore we plan to compare the performance of this system to the HMM based system developed previously.

Acknowledgement. This research is supported by the Swiss National Science Foundation NCCR program "Interactive Multimodal Information Management (IM2)" in the Individual Project "Visual/Video Processing".

References

1. Jain, A.K., Bolle, R., Pankanti, S., eds.: Biometrics – Personal Identification in Networked Society. Springer (2002)
2. Jain, A.K., Hong, L., Pankanti, S.: Biometric identification. Communications of the ACM **43** (2000) 91–98
3. Kittler, J., Nixon, M.S., eds.: Audio- and Video-Based Biometric Person Authentication. Lecture Notes in Computer Science, Springer (2003)
4. Plamondon, R., Lorette, G.: Automatic signature verification and writer identification – the state of the art. In: Pattern Recognition. Volume 22. (1989) 107–131
5. Srihari, S., Shi, Z.: Forensic handwritten document retrieval system. In: Proc. 1st Int. Workshop on Document Image Analysis for Libraries. (2004) 188–194
6. Baird, H.S.: Digital libraries and document image analysis. In: Proc. 7th Int. Conf. on Document Analysis and Recognition. (2003) 2–14
7. Baird, H.S., Govindaraju, V., eds.: Proc. 1st Int. Workshop on Document Image Analysis for Libraries. In Baird, H.S., Govindaraju, V., eds.: DIAL, IEEE Computer Society (2004)
8. Reynolds, D.A.: Speaker identification and verification using Gaussian mixture speaker models. Speech Communication **17** (1995) 91–108
9. Reynolds, D.A., Quatieri, T.F., Dunn, R.B.: Speaker verification using adapted Gaussian mixture models. Digital Signal Processing **10** (2000) 19–41
10. Leclerc, F., Plamondon, R.: Automatic signature verification: The state of the art 1989–1993. In Plamondon, R., ed.: Progress in Automatic Signature Verification, World Scientific Publ. Co. (1994) 13–19

11. Said, H.E.S., Tan, T.N., Baker, K.D.: Personal identification based on handwriting. Pattern Recognition **33** (2000) 149–160
12. Cha, S.H., Srihari, S.N.: Multiple feature integration for writer verification. In: Proc. 7th Int. Workshop on Frontiers in Handwriting Recognition. (2000) 333–342
13. Zhang, B., Srihari, S.N., Lee, S.: Individuality of handwritten characters. In: Proc. 7th Int. Conf. on Document Analysis and Recognition. Volume 7. (2003) 1086–1090
14. Zois, E.N., Anastassopoulos, V.: Morphological waveform coding for writer identification. Pattern Recognition **33** (2000) 385–398
15. Hertel, C., Bunke, H.: A set of novel features for writer identification. In: Audio- and Video-Based Biometric Person Authentication. (2003) 679–687
16. Bulacu, M., Schomaker, L., Vuurpijl, L.: Writer identification using edge-based directional features. In: Proc. 7th Int. Conf. on Document Analysis and Recognition. (2003) 937–941
17. Schomaker, L., Bulacu, M.: Automatic writer identification using connected-component contours and edge-based features of uppercase western script. IEEE Trans. on Pattern Analysis and Machine Intelligence **26** (2004) 787–798
18. Schomaker, L., Bulacu, M., Franke, K.: Automatic writer identification using fragmented connected-component contours. In: Proc. 9th Int. Workshop on Frontiers in Handwriting Recognition. (2004) 185–190
19. Bensefia, A., Nosary, A., Paquet, T., Heutte, L.: Writer identification by writer's invariants. In: Proc. 8th Int. Workshop on Frontiers in Handwriting Recognition. (2002) 274–279
20. Bensefia, A., Paquet, T., Heutte, L.: Information retrieval based writer identification. In: Proc. 7th Int. Conf. on Document Analysis and Recognition. (2003) 946–950
21. Nosary, A., Heutte, L., Paquet, T., Lecourtier, Y.: Defining writer's invariants to adapt the recognition task. In: Proc. 5th Int. Conf. on Document Analysis and Recognition. (1999) 765–768
22. Bensefia, A., Paquet, T., Heutte, L.: Handwriting analysis for writer verification. In: Proc. 9th Int. Workshop on Frontiers in Handwriting Recognition. (2004) 196–201
23. Leedham, G., Chachra, S.: Writer identification using innovative binarised features of handwritten numerals. In: Proc. 7th Int. Conf. on Document Analysis and Recognition. (2003) 413–417
24. Rabiner, L.R.: A tutorial on hidden Markov models and selected applications in speech recognition. In: Proc. of the IEEE. Volume 77. (1989) 257–285
25. Schlapbach, A., Bunke, H.: Off-line handwriting identification using HMM based recognizers. In: Proc. 17th Int. Conf. on Pattern Recognition. Volume 2. (2004) 654–658
26. Schlapbach, A., Bunke, H.: Using HMM based recognizers for writer identification and verification. In: Proc. 9th Int. Workshop on Frontiers in Handwriting Recognition. (2004) 167–172
27. Dempster, A.P., Laird, N.M., Rubin, D.B.: Maximum likelihood from incomplete data via the EM algorithm. Journal of Royal Statistical Society **39** (1977) 1–38
28. Duda, R.O., Hart, P.E., Stork, D.G.: Pattern Classification. Wiley Interscience (2001)

29. Melin, H., Koolwaaij, J., Lindberg, J., Bimbot, F.: A comparative evaluation of variance flooring techniques in HMM-based speaker verification. In: Proc. 5th Int. Conf. on Spoken Language Processing. (1998) 2379–2382
30. Collobert, R., Bengio, S., Mariéthoz, J.: Torch: a modular machine learning software library. IDIAP-RR 46, IDIAP (2002)
31. Bernard, T.M., Manzanera, A.: Improved low complexity fully parallel thinning algorithm. In: Proc. 10th Int. Conf. on Image Analysis and Processing. (1999) 215–220
32. Marti, U.V., Bunke, H.: Using a statistical language model to improve the performance of an HMM-based cursive handwriting recognition system. Int. Journal of Pattern Recognition and Artificial Intelligence **15** (2001) 65–90
33. Schlapbach, A., Bunke, H.: A writer identification and verification system using HMM based recognizers. Pattern Analysis and Applications **10**(1) (2007) 33–43
34. Marukatat, S., Artières, T., Gallinari, P., Dorizzi, B.: Rejection measures for handwriting sentence recognition. In: Proc. 8th Int. Conf. on Frontiers in Handwriting Recognition. (2002) 25–29
35. Pitrelli, J.F., Perrone, M.P.: Confidence modeling for verification post-processing for handwriting recognition. In: Proc. 8th Int. Workshop on Frontiers in Handwriting Recognition. (2002) 30–35
36. Pitrelli, J.F., Perrone, M.P.: Confidence-scoring post-processing for off-line handwritten-character recognition verification. In: Proc. 7th Int. Conf. on Document Analysis and Recognition. (2003) 278–282
37. Rosenberg, A.E., Deong, J., Lee, C.H., Juang, B.H., Soong, F.K.: The use of cohort normalized scores for speaker verification. In: Proc. Int. Conf. on Spoken Language Processing. (1992) 599–602
38. Matsui, T., Furui, S.: Likelihood normalization for speaker verification using a phoneme- and speaker-independent model. Speech Communications **17** (1995) 109–116
39. Marti, U.V., Bunke, H.: The IAM–database: An English sentence database for off-line handwriting recognition. Int. Journal of Document Analysis and Recognition **5** (2002) 39–46
40. Kuncheva, L.I.: Combining pattern classifiers: methods and algorithms. Wiley-Interscience (2004)
41. Bimbot, F., Chollet, G.: Assessement of speaker verification systems. In Gibbon, D., Moore, R., Winski, R., eds.: Handbook of Standards and Resources for Spoken Language Systems, Mouton de Gruyter (1997) 408–480
42. Plamondon, R., Srihari, S.N.: On-line and off-line handwriting recognition: A comprehensive survey. IEEE Trans. on Pattern Analysis and Machine Intelligence **22** (2000) 63–84
43. Barras, C., Gauvain, J.L.: Feature and score normalization for speaker verification of cellular data. In: Int. Conf. on Acoustics, Speech, and Signal Processing. Volume 2. (2003) 49–52

Index

Printing: Krips bv, Meppel, The Netherlands
Binding: Stürtz, Würzburg, Germany